APPLIED HYDROGEOLOGY
for Scientists and Engineers

Zekâi Şen

LEWIS PUBLISHERS

Boca Raton   New York   London   Tokyo

Library of Congress Cataloging-in-Publication Data

Şen, Zekâi.
  Applied hydrogeology for scientists and engineers / Zekâi Şen.
     p. cm.
  Includes bibliographical references and index.
  ISBN 1-56670-091-4
  1. Hydrogeology.   I. Title.
GB1003.2.S45   1995
551.49—dc20   94-23521
                                                                            CIP

  This book contains information obtained from authentic and highly regarded sources. Reprinted material is quoted with permission, and sources are indicated. A wide variety of references are listed. Reasonable efforts have been made to publish reliable data and information, but the author and the publisher cannot assume responsibility for the validity of all materials or for the consequences of their use.
  Neither this book nor any part may be reproduced or transmitted in any form or by any means, electronic or mechanical, including photocopying, microfilming, and recording, or by any information storage or retrieval system, without prior permission in writing from the publisher.
  CRC Press, Inc.'s consent does not extend to copying for general distribution, for promotion, for creating new works, or for resale. Specific permission must be obtained in writing from CRC Press for such copying.
  Direct all inquiries to CRC Press, Inc., 2000 Corporate Blvd. N.W., Boca Raton, Florida 33431.

© 1995 by CRC Press, Inc.
Lewis Publishers is an imprint of CRC Press

No claim to original U.S. Government works
International Standard Book Number 1-56670-091-4
Library of Congress Card Number 94-23521
Printed in the United States of America 1 2 3 4 5 6 7 8 9 0
Printed on acid-free paper

To my students all over the world . . .

"İnsanların en hayırlısı insanlara faydalı olanıdır,
İlmin en hayırlısı Sürekli ve faydalı olanıdır."

"The best human is who is beneficial to human beings. The best science is which is continuous and beneficial to human beings."

AUTHOR

Zekâi Şen, Ph.D., is Professor and Chairman for the Atmospheric Sciences and Meteorology Department at the Istanbul Technical University, Istanbul, Turkey. Dr. Şen received his Ph.D. degree from the University of London, Imperial College of Science, Medicine, and Technology. He formerly served as Chairman for the Department of Hydrogeology, Faculty of Earth Sciences, at King Abdulaziz University, Kingdom of Saudi Arabia.

PREFACE

Demand on freshwater increases all over the world because of increases in population and human activity, environmental pollution, drought occurrence for long periods, and climatic change. Although over 90% of potable water at any time exists within the subsurface geological formations, water surplus for agriculture, industry, and domestic and trade affairs is obtained from surface reservoirs, natural lakes, rivers, or man-made weirs, diversions, and dams. Groundwater is available almost anywhere in the world at different depths and can sustain human, animal, and plant life especially during the drought periods. Even in areas with significant surface water, the groundwater plays a supplementary role for the surface reservoirs. It is needless to repeat here in detail the vital significance of groundwater reservoirs for domestic, rural, urban industrial, and agricultural uses. At many places in the world the management of groundwater resources without scientific methods has led to continuous depletion of these resources. It is a certain rule all over the world that the exploitation of available groundwater is at the disposal of any individual, factory owner, or society provided that they can invest some money on well drilling. Consequently, in many, especially developing, countries the groundwater resources have reached critical stages and adverse effects have already started to appear. It must be kept in mind that the groundwater resources are, although replenishable, not inexhaustible. Overwhelming demand on groundwater resources has stimulated theoretical researchers and field investigations oriented toward quantification of these resources through basic formulations of plans for its exploitation, management, and conservation.

This book is intended to serve as a guide to researchers, professionals, engineers, earth scientists, and students in the field of groundwater hydrology, hydrogeology, geology, geohydrology, and hydraulics. Without involving detailed mathematics, various methods, procedures, and techniques of assessment of aquifer characteristics and parameter estimations in considering different geological formations will be presented. In the presentation of various methods logical and physical aspects will be included. Useful and practically important interpolations are given the primary concern. Rather than mechanical quantitative evaluations, more qualitative basic reasoning leading to practically effective results are presented. Special emphasis has been given in dealing with geological control of the occurrence and movement of groundwater, especially the hydraulics of hard rocks.

This book, based upon more than two decades of experience in analytical research and field applications, addresses earth scientists and engineers alike. Results of research and interpretative techniques published in various journals, periodicals, and symposia proceedings, some of which are not easily accessible to most earth scientists and engineers, have been consolidated and described in appropriate sections. Numerous interpretative new methods, procedures, and techniques developed and published in scientific journals are being presented for the first time in this book. Among these are the dimensionless-type straight line concept, volumetric approaches, non-Darcian well hydraulics and its application in fractured rocks, hydrogeophysical concepts, etc. In order to increase comprehension a variety of illustrations, examples, and some case studies from the literature have been furnished. Most of the material covered in this book is based on field trips during my stay at the King Abdulaziz University, Faculty of Sciences, Jeddah, Kingdom of Saudia Arabia. There, I have been acting as the chairman of the Hydrogeology Department for about 10 years training students leading to B.Sc., M.Sc., and Ph.D. degrees.

CONTENTS

Chapter 1
Introduction 1

Chapter 2
Hydrogeology of Groundwater Reservoirs 7

Chapter 3
Groundwater Flow Properties 37

Chapter 4
Aquifer Properties 69

Chapter 5
Groundwater Flow Laws 91

Chapter 6
Water Wells 115

Chapter 7
Field Measurements 133

Chapter 8
Steady-State Flow Aquifer Tests 161

Chapter 9
Porous Medium Aquifer Tests 203

Chapter 10
Fractured Medium Aquifer Tests 303

Chapter 11
Well Test 355

Chapter 12
Nonlinear Flow Aquifer Tests 395

Index 433

CHAPTER 1

Introduction

I. GENERAL

Ever since the creation of life man, animals, and plants are dependent continuously on a few precious commodities among which the water constitutes the primary importance. Other vital commodities such as food and oxygen are by-products of water. Man's relationship with water has been vital, complex, and varied from the earliest times. It is for this reason that the Holy Books include divine verses signifying the attachment of human life to water not in the usage sense only but more significantly that living things are created from water. Life in essence is a product of water.

This dependence of man's life on water compelled him to choose his settlement locations at positions where water was easily available. Therefore, early settlements are invariably either along rivers or next to lakes or springs, all of which provide natural water resources. Accordingly, the first activities of man were basically related to water in search of food by means of irrigation, local livestock raising, defense, and communication. In the romantic past of the desert life, the shady oases of local agriculture and the routes of transportation across the arid region were located and controlled by the availability of water. Further expansion of man's own environment gave rise to an increase of the demand on water. Consequently, the interference of human beings on the natural occurrence and movement of water started by dams, dikes, or levee construction so as to augment or translate the available water storage with time. In the arid regions of the world, due to high temperature and rare rainfall occurrences, every single drop of water had to be preserved. For this purpose, human beings excavated surface depressions to save the rain water.

Later, human beings become acquainted with groundwater storage and begin to dig large-diameter wells to reach the available groundwater reservoir. Groundwater is the main theme of this book. Although there are social, economic, political, and juridical impacts of groundwater on man's life, these will not be touched upon. Rather the quantitative description of groundwater reservoir elements, characteristics, flow, and exploitation methods are presented to achieve a sound concept of different analyses as well as their consequences.

The occurrence and exploitation of groundwater depend on the existence of at least three basic components: the source, reservoir, and abstraction of water. The conceptual model for these components is shown in Figure 1. In fact, such a model is valid, in general, for any water resources. However, depending on the resource each component will have different properties and interpretations. For instance, in the case of groundwater, the source component is completely dependent on the hydrological cycle that provides water through a catchment and subsequent infiltration process. Hence, investigation of the source requires hydrological knowledge only. Because the reservoir is under ground, the subsurface geology becomes the dominant media for its existence. Finally, the abstraction phase is an indicator of man's direct or indirect interaction with the groundwater through the man-made structures such as wells,

Figure 1 Groundwater activities.

dikes, fountains, channels, kanats, etc. It is, therefore, inevitable that this last phase is concerned with hydraulics.

In short, the aforementioned descriptions bring out the three significant disciplines, i.e., hydrology, geology, and basic sciences such as physics, chemistry, and mathematics, to manage the groundwater resources in an optimum manner. In the following sections the mutual relationships among these disciplines with the groundwater are discussed in detail. The historic developments of these disciplines are given in a concise manner by Tolman (1937).

II. HYDROLOGY AND GROUNDWATER

Hydrology is the study of water in the earth. Broadly, it deals with the occurrence, distribution, and movement of water in addition to physical as well as chemical relationships with the surrounding environments. In general, there are three distinctive environments—atmosphere, biosphere, and lithosphere—as schematized in Figure 2. An essential and comple-

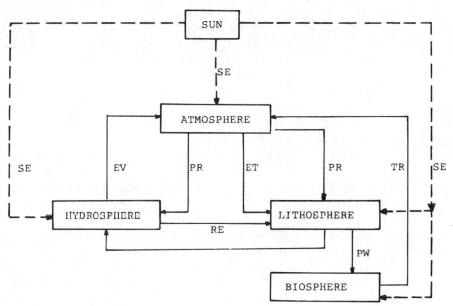

EV : Evaporation
ET : Evapotranspiration
GF : Groundwater Flow
PR : Precipitation
PW : Plant Water Use
RE : Recharge
SE : Solar Energy
TR : Transpiration

Figure 2 Water environments.

mentary part in this figure is the hydrosphere which refers to environments of sole water such as lakes, rivers, and oceans. The water of the earth circulates in these environments from the hydrosphere (oceans) to atmosphere then to the lithosphere. The circulation including complex and independent processes such as evaporation, precipitation, runoff, infiltration, and groundwater flow is called "the hydrological cycle".

In the atmosphere water is in the vapor form and mixed with various gases. Biosphere is the plant world that is entirely dependent on water. However, the lithosphere is the general name for the rock domain and it is the main environment that will be dealt with in this book from the water storage and movement points of view. The groundwater reservoirs are a significant part of the lithosphere as will be explained in Chapter 2.

The transitions of water from one environment to another and its transfer among these environments do occur naturally in the universe continuously with time. These transitions as well as transfers constitute the basis of water movement in nature. The driving forces of such a movement are sun's radiation and earth's gravity. The collection of these endless movement routes is referred to as the hydrologic cycle. The natural components of this cycle that connect different environments are shown in Figure 2. This cycle is the only source that brings water as precipitation from the atmosphere to the lithosphere that includes the groundwater reservoirs. It is a gift to the human being that all of the interenvironmental connections occur with no cost at all but at places in a very unpredictable manner as well as in rare amounts. However, the contrast in the practice of groundwater hydrology between arid and humid zones of the world is in general less spectacular than in surface hydrology.

The part of hydrological cycle within the lithosphere that transforms the precipitation into the groundwater is shown in Figure 3. Depending on the rock properties some part of the precipitation crosses the earth's surface, hence indicating the infiltration process whereby the water moves rather vertically due to the gravitational forces. One part of the infiltration water remains near the earth surface layers of lithosphere as soil moisture whereas the complementary part advances deeper into the earth which is referred to as the percolation process. This process ends up within the groundwater storage where the water movement is negligible; if not, the direction of flow is almost horizontal.

The present-day contact of hydrological cycle with the groundwater reservoirs gives rise to replenishable, i.e., fresh, groundwater storage. On the other hand, groundwater out of contact with the hydrologic cycle for some geological time period is referred to as fossil water which is highly mineralized. For instance, deep-seated groundwaters in the eastern part of the Arabian Peninsula have been deposited during the pluvial times of Pleistocene.

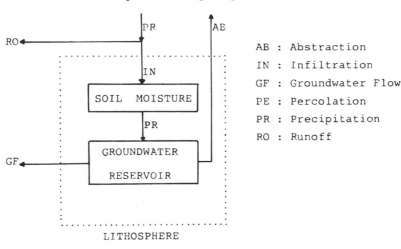

Figure 3 Hydrologic cycle components in the lithosphere.

The hydrological study of groundwater includes depiction of recharge and discharge quantities, storage fluctuations, and groundwater divide line delineations. Detailed information about groundwater hydrology is given by Todd (1980) and Bouwer (1978).

III. GEOLOGY AND GROUNDWATER

Geology is the study of rocks and one tends to think of rocks in terms of mineral composition. This is natural because all of the rocks on the earth's surface are outcrops and they are dry. However, at depths below only a few meters from the surface, they are partially or fully saturated with water. If it is not very salty, some of this water is important as a source of fresh water for domestic, industrial, or agricultural purposes. Identifications of such source areas are possible only after comprehensive surface and subsurface geological studies.

The reservoir component in Figure 3 is dependent on the geological composition of subsurface more than any other agent. The geological setup of the groundwater reservoirs in forms of igneous, sedimentary, and metamorphic rock proportions; voids between minerals and grains; and joints, fractures, and faults as structures all play different roles in developing groundwater quantity as well as quality variations. Hence, the formation of these various rock types and their prior changes are vital issues for the groundwater exploration and exploitation. The distribution and composition of rocks affect the porosity, groundwater velocity, and ionic constituents of water.

It is, therefore, inevitable that any groundwater study should include in the early stages a detailed geological prospecting of the area. In general, a geological study should include lithological phase concerning the mineral composition, grain size, sorting and packing; stratigraphic phase describing the age, unconformities, and geometrical relationships between different lithologies; and structural features such as fissures, joints, fractures, folds, and faults. The collection of this information gives a rather clear picture of the subsurface geology leading to an understanding of the distribution of the various water bearing formations.

The infiltrating water from the catchment seeks its way with difficulty toward the groundwater storage and then horizontally toward outlet points such as wells or springs at small velocities. During such a journey it touches different types of minerals and rocks that give the groundwater its characteristic quality. In general, the zones close to the outcrops have comparatively good-quality water that becomes rather brackish and saline as it gets closer to the abstraction zone. It is possible to label the geological formations as the transformers of groundwater from source to abstraction zone with a change in the quality. On the other hand, the quantity and especially the continuity of groundwater are more dependent on the hydrology than the geology.

In fact, as it reaches the ground surface, water is extremely homogeneous but its penetration into the rocks gives a heterogeneous appearance due to the haphazard distribution of minerals in the geological formations.

All the branches of geology are of significance in the study of groundwater. Structural and photogeology help to identify the rock types, whether porous or nonporous, massive, bedded, jointed, folded, faulted, fractured, etc. In addition, weathering and rock texture, cleavages, and fault breccia provide potential zones of groundwater in the igneous and metamorphic rocks.

Sedimentology is another branch of geology that concentrates mainly on the sedimentary rock formations, their regional settings, stratigraphic sequences of different lithologies, grain sizes, shapes, and distributions in relation to the granular porosity, solution cavities in limestones and dolomite, joint sets in the shales, and especially synclines from the groundwater occurrence point of view. The most important groundwater storage in the sedimentary rocks is within the sandstones, limestones, and unconsolidated and glacial deposits.

Prior to any study on groundwater the geological setup and various features in a region are the vital steps for better understanding of groundwater occurrence, movement, recharge, and discharge areas. In general, these studies are initiated by rather cheap field reconnaissance trips of the surface geology inspections that are then succeeded with subsurface investigations to various degrees of detail depending on the significance of the study, financial resources availability, and the stipulated time. To have a clear idea of the subsurface geological setting and its relation to the groundwater, either cross sections in some preferred directions or three-dimensional pictures of the subsurface geology are necessary and they assist to make meaningful correlative interpretations.

Any groundwater study with incomplete geological information leads to erroneous conclusions. It is, therefore, strongly advised that hydrological studies should be succeeded with the geological ones to achieve good judgment of the ground and water occurrences and their interrelationships.

Unfortunately, it is a common tendency that engineers do not consider basic geological facts in their groundwater resources evaluations and accordingly their estimations are in error; likewise, geologists neglect to check the accuracy of their conclusions with the engineering measurements. However, hydrogeologists and geohydrologists are trained on an interdisciplinary manner and they are usually equipped with fundamentals of geology, hydrology, physics, and chemistry as well as fluid mechanics, mathematics, and computer sciences. The importance of combined geologic and hydrologic training together with engineering concepts has been realized long ago by Meinzer (1932). On the other hand, Chapman (1981) addresses the basics of understanding the mechanical role of water in relation to geology.

IV. SCIENTIST AND GROUNDWATER

Quantification and practicable uses of the groundwater system in Figure 1 require scientific observations, investigations, and evaluations of various components. Therefore, a multitude of disciplines and specialists are involved in any large-scale groundwater study.

There are specific scientists who are concerned with each of the components. For instance, hydrologists pay great attention in studying the physical occurrences of the source component of the groundwater system; geologists are mainly interested in the rock composition of the groundwater reservoir domain; and finally, hydraulic engineers concentrate on the abstraction part of the system in Figure 1. Although such specialists are essential ingredients in any groundwater study, unfortunately, they cannot collectively come out with a scientific presentation of the groundwater system unless there is a common terminology and mutual understanding among them.

In order to deal with the overall groundwater system, interdisciplinary specialists such as groundwater hydrologists, geohydrologists, hydrogeologists, and water engineers have recently emerged. Unfortunately, there is confusion regarding the tasks of each of these scientists. The groundwater hydrologist and geohydrologist are used synonymously and they are scientists who deal with the three-component groundwater system but are more interested in the source component of such a system and less worried about the geological composition of the reservoir. Their main research methods are experimental and laboratory modeling works.

On the other hand, hydrogeologists are more concerned with the geological makeup of the groundwater reservoir with less emphasis on the source and abstraction (except springs). Their research methods are mainly the field works in the forms of data as direct measurements for quantitative evaluations (pumping tests, piezometric level, joint and fracture measurements, etc.) and water sample collection for the qualitative studies. At the end, with all of the information available, hydrogeologists prepare a detailed report for the groundwater reservoir concerning the existence of exploitable groundwater storage, flow rates, and its quality variations with their suitability to different domestic, agricultural, and industrial usages.

Groundwater problems are mainly of practicable nature. In order to supply groundwater in meeting man's needs, it is necessary to construct man-made structures such as wells, ditches, fountains, channels, pipes, subsurface dams, etc. These are concerned with the abstraction and supply of groundwater to the centers of demand. The scientist who deals with the movement of water toward or within the aforementioned structures is called a water engineer. He is very much involved with the abstraction component of the groundwater system. The main subject is hydraulics for the design of structures but computer modeling and optimization are the best economical solutions of available alternatives. Also, the groundwater management falls very much within the interest of the water engineer. He is least concerned with the geology and for him the porous medium idealizations together with the applicability of Darcy's law are enough to solve various groundwater problems.

It seems from the above discussions that the greatest uncertainty related to the groundwater studies lies within the geologist's or hydrogeologist's work due to the unknown underground composition of the rocks and the geologic structures. However, they use geophysical techniques of different versions (electrical resistivity, seismic reflection, and refraction) or direct bore hole logs in reducing the uncertainty.

The main purpose of this book is not to address any one of the aforementioned specific professions but to provide rather simply an understandable basis of groundwater occurrences and abstractions. It is hoped that such an approach is helpful to anybody concerned with the groundwater and especially to those who are not interested in the phases of mathematics but rather physical implications in addition to straightforward mechanical applications leading to practically quantitative answers.

REFERENCES

Bouwer, H., 1978. *Groundwater Hydrology,* McGraw-Hill, New York.
Chapman, R. E., 1981. *Geology and Water. An Introduction to Fluid Mechanics for Geologists. Development in Applied Earth Sciences,* Martinus Nijhoff/Dr. W. Junk, London.
Meinzer, O. E., 1932. Outline of Methods for Estimating Ground-Water Supplies, U.S. Geologic Survey, Water Supply Paper, 638-C, 104.
Todd, D. K., 1980. *Groundwater Hydrology,* John Wiley & Sons, New York.
Tolman, C. F., 1937. *Ground Water,* McGraw-Hill, New York.

CHAPTER 2

Hydrogeology of Groundwater Reservoirs

I. GENERAL

The lithosphere is a continuous composition of different rock masses, of which some include void spaces as well as solids whereas others consist of solids only. The void space provides a place where groundwater can be stored and kept under suitable conditions for use at times of demand. Any rock mass in the lithosphere consisting of mutually exclusive but complementary volumes of voids and solids will be referred to as a reservoir. It is, therefore, possible to consider the lithosphere under two very broad and mutually exclusive categories: reservoirs and nonreservoirs. A reservoir contains voids, allowing liquid into its main body. Of course, many sedimentary formations, such as sandstones, limestones, clay, etc., have the reservoir property whereas most igneous rocks are of the nonreservoir type.

The term "reservoir" does not necessarily mean that a rock can yield water easily in practical terms. This brings one to the question, "does the reservoir allow water to move with ease at time of need?" If it does not yield water easily it is called "impermeable"; otherwise, it is "permeable". In other words the mere existence of a reservoir implies water availability but not accordingly an abstraction route that is the property of permeable reservoirs only. Clay, for instance, qualifies as a reservoir but is not permeable. The aforementioned division of the lithosphere from the water interaction points of view is summarized in Figure 1. Furthermore, some permeable reservoirs do not accommodate water. A good example of this type is the shale layers. Depending on the geological evolution of the void spaces, permeable reservoirs may be subdivided into three major groups: porous, fractured, and karstic media.

II. POROUS MEDIUM

The most important and commonly known reservoirs are composed of granular materials such as unconsolidated and consolidated sands. Such a medium includes countless irregular voids of random sizes and shapes comprising pore spaces, which are also referred to as the interstices between the individual solid particles of sand or pebble. Each pore is connected with adjacent ones by constricted channels of different sizes. Collectively, pores and channels may form a completely interconnected network of voids through which the water can move in various directions. Such a setup in rock mass will be referred to as the porous medium.

The smaller the grain sizes of solid particles the more regular the flow path and hence the resistance of solid to water passage at almost every point, so that any water particle is under continuous resistance from the solid matrix. Herein, "resistance" means the difficulty met by water in moving through the medium. On the other hand, in a coarse-grained medium

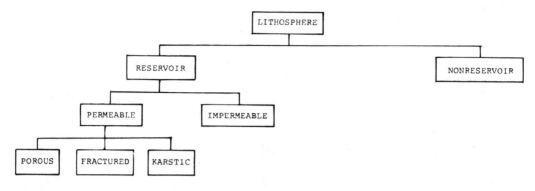

Figure 1 Different parts of the lithosphere.

the water will meet less resistance from the solids but the flow path will be more irregular and the flow rates will have greater amplitudes of fluctuation (see Figure 2). The above is tantamount to saying that water will flow more easily in a coarse-grained than in a fine-grained medium. Furthermore, it follows that very fine-grained reservoirs will resist the water movement more than fine- or coarse-grained reservoirs. Most study and research have been devoted to the porous medium than to any other type, partly because saturated sand deposits are the most common aquifers and partly because this medium is amenable to sophisticated studies.

The porous media dealt with in this book will be the relatively fine- or coarse- grained rocks providing good reservoirs; very fine-grained media will be considered as impervious layers similar to clays.

In general, only the sedimentary rocks of different origins provide porous media for groundwater occurrence and movement. In what follows, each of the potential porous media for groundwater storage will be explained from the geological point of view. Figure 3 represents the main geological formations that provide a porous medium for groundwater storage and transmission.

A. ALLUVIAL FANS

These formations are of fluvial origin and occur where a stream leaves a steep valley and slows down as it enters a plain. The resulting cone of gravel and sand point upstream are shown in Figure 4. Alluvial fans are very noticeable and abound especially in arid and semiarid regions. The growth of alluvial fans was initiated during the Pleistocene epoch when the climate was more humid and rainfall more abundant. A good account of alluvial fans has been presented by Rachocki (1981). The extent of an alluvial fan depends on drainage basin slope, size, climate, and the character of the rocks in the source area. Individual forms may have radii up to several kilometers. The gravels that make up the porous medium tend to be

Figure 2 Water movement in the porous medium.

HYDROGEOLOGY OF GROUNDWATER RESERVOIRS

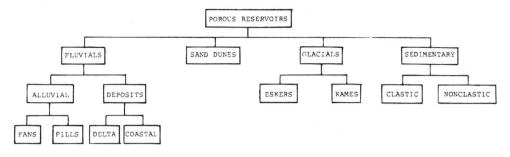

Figure 3 Major porous medium formations.

poorly sorted and angular to poorly rounded. Fine-grained debris is deposited further downstream and may be cross bedded, massive, or thick bedded. Groundwater flow in alluvial fans is replenished by percolation of river water. Most often the water appears in the form of springs; otherwise, it may continue its journey further downstream where it emerges as surface flow. The volume of voids in the porous medium of an alluvial fan becomes smaller by compaction, especially in zones composed of fine material.

Alluvial fans provide groundwater in coastal desert areas. In arid regions, they reach dimensions larger than in humid regions and play a more important role as potential aquifers. Groundwater in their higher parts is phreatic, whereas in the lower parts it is confined. In between, the phreatic surface intersects the ground surface as seepages and springs. Some seepage may be found near the fringes of a fan where subsurface drainage encounters the surface. Alluvial fans may provide an ideal groundwater source for irrigation.

B. ALLUVIAL FILLS

These are formed as a result of weathering and the flow of water in fills or where the relief is favorable and the rainfall sufficient to provide the driving force for movement. Gravels, which are the coarsest product of erosion, are moved shorter distances from their sources and are deposited in more restricted areas than are sand, clay, or mud (Figure 5). Fluvial gravels are widespread, especially in arid regions, where they fill the valleys of rivers, surface depressions, or fault zones. The thickness of an alluvial fill may reach several hundred meters. Such fills are initiated by the formation of alluvial fans. They are commonly coarse, interbedded with coarse and fine sands, and show a progressive decrease in size downstream, accompanied by a notable increase in roundness. Generally, bedding and sorting are comparatively better

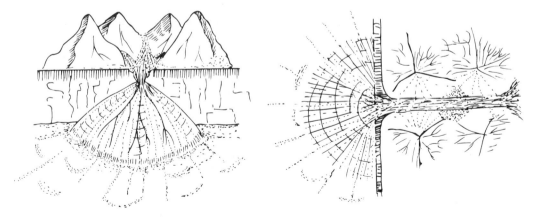

Figure 4 Alluvial fan deposits.

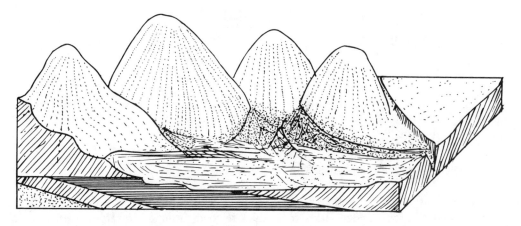

Figure 5 Alluvial fill formation.

than in alluvial fans. The alluvial fills in arid regions are known as "wadis". Wadis are natural watercourses but are dry most of the time. At times, however, they become conveyors of flash floods that carry away large amounts of sediment, leaving a marked imprint on the desert landscape. Usually, the medium is a coarse-grained type, as the fine grains are either washed away by flash floods or blown away during the long dry spells.

In general, the groundwater is found in the voids of gravels. The alluvial fills make up potential groundwater reservoirs for local uses. In fact, in arid regions they are the primary locations for water-well excavation to supply the nearby villages. The convenient features of a wadi are the floodplain, terraces, meander scrolls, swamps, and natural drainage toward the river, with the result that silt-loaded flood water remains on the flood plain swamps develop. Evaporation of the swamps causes salt deposition, which creates salinity problems when irrigation is later practiced on the swampy soil. The drainage system on the wadi surface may have been altered by various factors such as climate changes, tectonic activity, and fluctuations in floodwater discharge. Because the porous medium is so heterogeneous, the rainfall has a direct relationship with the amount of groundwater stored in alluvial fills.

When the natural pressures are insufficient to rise to the surface the groundwater must be pumped. For best results, well locations should be chosen in rather a coarse gravel zone, with a large area and deep alluvial fills. The withdrawal of water creates a depression cone where volume is directly proportional to the porosity and permeability of the porous medium.

Permanently flowing water often exists in the gravels below the surface of a large wadi. In areas of considerable annual fluctuations in the level of the groundwater table, wells may have only seasonal value and continuous irrigation is not possible. Such is the case in the western part of the Kingdom of Saudi Arabia, where alluvial filled wadis run more or less perpendicularly to the Red Sea (Al-Sayari and Zoti, 1978). In general, groundwater in the upstream parts of wadis is less saline than that in the downstream. A significant portion of the surface water that collects in the wadi at times of substantial rainfall eventually infiltrates the ground. With modern drilling and pumping methods, much of this water can be utilized for agriculture. The alluvial fill of plains and valleys, consisting of gravel, sand, and clay of volcanic origin, may present important groundwater possibilities, but these deposits are as diversified and variable as those in any region. Some coarse- to medium-grained pyroclastic deposits have a very high porosity.

Although significant groundwater reservoirs may be found within these deposits, they occur far below the surface. Surface depression in the alluvial fills may have ponded water, which indicates a close proximity to the water table. Especially in arid zones, excess of evaporation over rainfall causes the capillary zone to move downward, leaving behind its dissolved salts or alkaline deposits.

C. SAND DUNES

Because of the long dry spells in the desert regimes, the deposits of water are worked over by the wind (Figure 6). Dunes are accumulated wind deposits, consisting of sand-size particles that are rather well sorted and silt. Unequal side slopes reflect the dominant wind direction. The various morphologies of dunes, such as dome shaped, transverse, barchan, parabolic, star, and longitudinal, are explained by Stearns and Friedman (1972). The fundamental principles of sand transport due to wind have been reviewed by Bagnold (1941) and Cooke and Warren (1973).

Sand dunes can be regarded as one of the most isotropic and homogeneous deposits in nature. Sand-dune materials are of uniform size and allow rapid infiltration and percolation of rainfall. The geological layers underlying sand dunes may offer suitable groundwater supplies. Eolian dunes include only sand and rarely gravels. From this point of view they can be considered as a fine grained porous medium. Thick sand dune sequences of porous, permeable, cross-bedded sandstones with few impermeable barriers such as shales provide important potential reservoirs for water, gas, oil, and hydrothermal metalliferous deposits. Because of the poorly graded nature of sand dunes and their association with rapid interval drainage, no seasonal high-water table is found near the earth's surface. It was stated by Al-Sayari and Zotl (1978) that the Rub-Al-Khali desert in the Kingdom of Saudi Arabia covers groundwater at an average depth of 500 m. The Rub-Al-Khali is the largest single reservoir of sand dune cover with more or less continuous eolian accumulation rarely interrupted by eroded remnants of other relief. It provides a huge volume of porous reservoir with favorable conditions for percolation and preservation of groundwater. Unfortunately, only during intense rainfalls does the water infiltrate the thick cover to reach the groundwater storage. A geological account of sand-dune deposit evolution is presented by Anton (1983).

D. GLACIAL DEPOSITS

Any deposit that owes its origin more or less directly to the grinding action of glaciers is referred to as a glacial deposit (Geikie, 1894). Glacial deposits are generally referred to in the literature as till. They provide a poorly sorted porous medium, which has clasts of many sizes including bolders and therefore may provide a potential groundwater reservoir. For the northern parts of the U.S., Canada, Europe, and the former Soviet Union, deposits formed by the continental glaciers furnish significant water resources.

Particle-size distributions provide instructive data on the mechanics of till formation. Detailed investigations by Dreimanis and Vagners (1971) have stated that rock components in Pleistocene tills show characteristics of grain. The grain-size curves depend on the number

Figure 6 Sand dune formations.

of rock types making up the till and the number of minerals making up each rock type. The curves for most tills are bimodal, suggesting that there is some inherent asymmetry in the process of rock breakdown during glacier transport. Figure 7 shows three successive stages in glacial transport. Near the source of a till the coarse fraction is large and the mineral fraction small, as in Figure 7a. At a later stage, both fractions become more or less equal, (Figure 7b), whereas after long-distance transport the curve possesses the shape in Figure 7c, where the mineral fraction is greater than the rock fragment fraction. This is a good indication that the void ratio of a glacial deposit is high at the till source but decreases with the distance traveled away from it. Much of the debris transported by glaciers is either deposited near the down-glacier margins or laid out as outwash along the downstream course (see Figure 8).

Tills in contact with ice differ from outwash sediments in their generally greater range of particle size often showing abrupt changes between coarser and finer materials. On the other hand, outwash sediments differ from till in being stratified and relatively well sorted.

Two most important forms of glacial deposit from the groundwater point of view are askers and kames. Terraces along ridges of glaciofluvial material lying roughly parallel to the direction of former ice flow are usually termed askers which include much sand and gravel. The ridge may be sinuous or straight. Askers have very granular, coarse-textured soils that are useful for construction. There are two distinctive stages in the formation of askers: (1) deposition in a subglacial drainage tunnel (Figure 9a) and (2), collapse of the glacial-fluvial sediments with disappearance of the glacier (Figure 9b). On the other hand, kames are steep-sided mounds. Like askers, they too are composed of sand and gravel, but they generally contain more fines than askers. Because of their coarse composition, elevated position, and steep slope, kames do not commonly hold water. Usually, they are formed on both banks of a valley. Their formation involves two major stages. In the first during the existence of the valley glacier, melt-water streams run along the sides of the valley as in Figure 10a, building up lateral terraces. In the second, with disappearance of the glacier, existing glaciofluvial deposits on both valley sides collapse to form kames (Figure 10b).

More information concerning glacial deposits is given by Sugden and John (1976), Eyles (1983) and Goldthwait (1975).

E. DELTA DEPOSITS

A modern definition describes a delta as "the subarea and submerged contiguous sediment mass deposited in a body of water (ocean or lake) primarily by the action of a river" (Moore and Asquith, 1971, page 2563) (see Figure 11). Fundamentally, deltas are of terrestrial deposition, not marine. However, marine sediments may be incorporated in delta fronts intercalating with alluvial deposits, if phases of subsidence alternated with phases of delta make up. Deltas are formed when a stream channel reaches a large body of water, either an ocean or a lake. Because deltas are at the downstream ends of a drainage basin, both the gradient and the flow velocity decrease and suspended sediments and bed loads consequently settle down. Unconsolidated deposits in deltas grade seaward, starting with fine sand followed by silt, open-marine clay, and mud.

Deltas are always associated with water and, because of the flat topography, the water table occurs within a few meters of the ground surface. Generally, deltas contain fairly well-sorted granular deposits and are well drained along the surface, since any rainfall infiltrates to the water table close below. The groundwater table elevations in deltas are fairly constant, reflecting the elevation of the nearby body of water. Therefore, deltas formed near freshwater lakes are excellent locations for groundwater exploitation. However, there is always saltwater intrusion into the fresh groundwater body from the oceans. The extent of intrusion depends on the difference in elevation between the groundwater table in the delta and the ocean surface as well as the nature of the delta deposits.

HYDROGEOLOGY OF GROUNDWATER RESERVOIRS

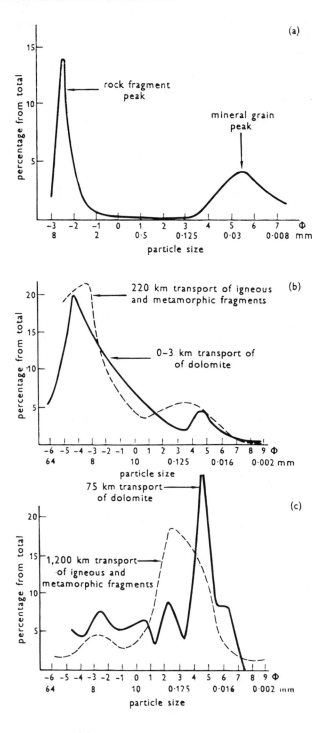

Figure 7 Different stages in glacial transport.

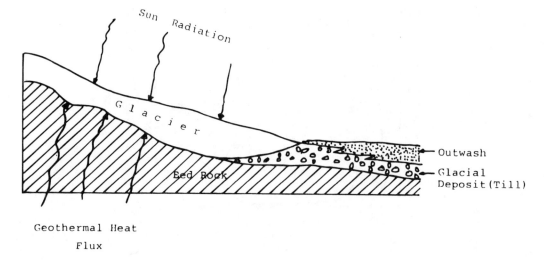

Figure 8 Glacial transport stages.

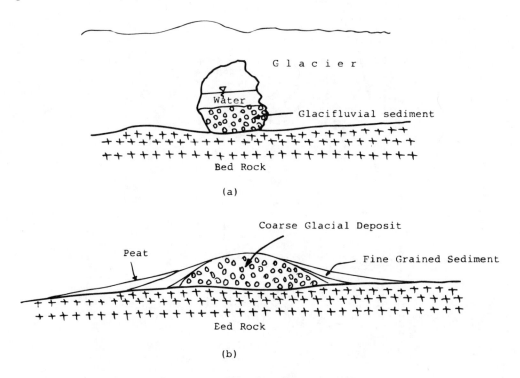

Figure 9 Asker formation. (a) Tunnel deposition; (b) collapse deposition.

F. COASTAL PLAIN DEPOSITS

Coastal plains are found all over the world as unconsolidated sediments, bounded on the continental side by a highland such as a cliff, reef, hill, or escarpment, and separated on the marine side by a shoreline from a surface water body, either a lake or an ocean, as shown in Figure 12. Coastal plains include deposits of both continental and marine origin. Close to the foothills of the highlands continental deposits predominate, gradually giving place to marine

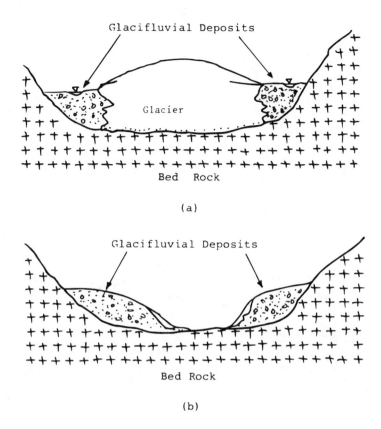

Figure 10 Kame formation. (a) Lateral deposition; (b) glacier deposition.

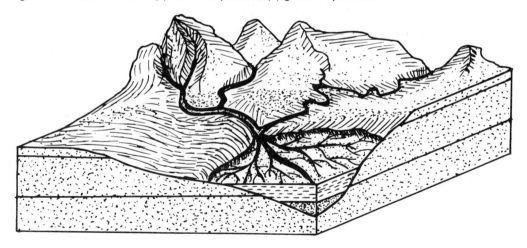

Figure 11 Schematic representation of a delta.

deposits seaward. With the regular tidal fluctuations these two types of deposit become intercalated. Their source material is supplied by river, ice, wind, and coastal erosion.

Rather fresh groundwater occurs in some areas of the coastal plain where there are no valley deposits in the hinterlands. This water is provided directly by infiltration from the rainfall and indirectly by inflow of water from the adjacent hills. If there is no adequate replenishment, the fresh groundwater is replaced by saline waters. The fresh groundwater

Figure 12 Coastal deposit formation.

table is usually higher than the adjacent surface water bodies and follows the plain surface topography smoothly (Figure 13). Coastal-plain formations can act as a groundwater reservoir by holding the freshwater supplies slightly above sea level and the saltwater table. In such situations, overabstraction of the freshwater may allow the saltwater to contaminate the groundwater supply. The greater the distance from the sea, the higher the groundwater table elevation and the greater the freshwater depth.

G. CLASTIC SEDIMENTARY ROCKS

Sediments derived from preexisting rock material by weathering and erosional processes are referred to as clastic rocks. They are recognized by their clastic textures where neither chemical nor biological precipitation nor accumulation of organic material has been involved in their formation. Clastic rocks are subdivided according to particle size. There are three main divisions: conglomerate, in which gravels predominate, sandstone, and siltstone both compacted forms of clay layers.

Gravels are unconsolidated accumulations of rounded rock particles larger than sand (>2 mm). In practice, however, any accumulation where fragments made up less than half the constituents are referred to as gravel. Conglomerate is an indurated version of gravel accumulations that have undergone modifications after deposition. On land, gravels accumulate as a result of weathering, provided the runoff is sufficient to carry the rock fragments. They are

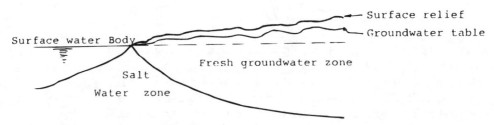

Figure 13 Groundwater level in coastal plain.

moved shorter distances from the source area and are not as widespread as sands or silts. Just as the coastal gravels resultings from wave action accumulate along the coastline, land gravels accumulate in large lakes.

Sandstones are the most important category of the clastic sedimentary formations, extending over vast areas and in places reaching thicknesses of about 10 km. They are formed from the products of every kind of rock erosion by wind or water. Because their matrix material is sand grains, they usually contain abundant void spaces where fluids such as water, oil, and gas can accumulate and migrate. This capability gives them economic importance.

The transmission of fluids in sandstone, however, depends on a number of factors. First, there is the void ratio and degree of interconnectedness of the voids. Second, there is the sorting, texture, and fabric of the sedimentary rocks concerned, which result in different bedding facies that can influence fluid transmission in different directions. Third, there is the degree of compaction at various points within a clastic sedimentary rock which also effects transmission and storage. Hence, on a regional scale fluid flow in sandstones can behave differently at different points.

The presence of local or extensive beds of siltstones or shales within the sandstone, however, can affect the fluid flow even more than above-mentioned differences in composition. For instance, extensive layers of siltstone may make the whole sandstone reservoir behave as a stratified reservoir where there may be no fluid exchange between adjacent layers or vice versa. The most favorable formations for water storage and transmission are found mainly in clastic sedimentary rocks, which furnish about 80% of the groundwater supply from sedimentary formations. During the process of accumulation, either the fine particles may have been washed away from the coarser grains (where the transporting flow or wind blow was rather weak) or heavier and coarser particles may have been transported and deposited over the earlier fine-particle deposits. As a result of successive depositions, a stratified reservoir builds up over geological time, with each bedding plane separating two different lithological units. The bedding planes clearly distinguish these intermittent depositions. Such a bedding plane is called an unconformity.

Tectonic movements give various forms to the sedimentary rocks, the most significant of which are the syncline and the anticline. Synclines are domes in which groundwater can be found in abundance, anticlines do not provide potential areas for groundwater storage.

III. FRACTURE MEDIUM

In general, fractures are defined as discontinuous planar features within any rock mass that may have developed as a result of pressure and temperature differences during and/or after the formation of the rock.

Fractures occur chiefly in dense crystalline or cemented rocks. Major fractures are of supercapillary size and/or fed by tributary fractures that are commonly capillary size. A fracture can be defined as a secondary structure in the form of a planar or nonplanar surface within a rock mass along which there is no cohesion. For the creation of a fracture, therefore, some force must be exerted on the rock mass after its evolution. However, fractures are referred to in terms of relative strength of the force involved; if appreciable displacement has occurred it is called a fault, and if there is no noticeable displacement it is called a joint. A fissure is a fracture whose faces have moved apart. A detailed account of fractures is presented by Davis (1984). From a geological point of view, the generation of fractures can be attributed to the following different events:

1. Tectonic movements that cause the earth crust to deform.
2. Significant erosion of the overburden so that fracturing takes place along different planes or weak points.

3. Change in rock volume due to the loss or gain of water, especially in shale or shaly sands.
4. A change in rock volume due to temperature differences, especially in the igneous rocks.

As LeGrand (1979) states, some problems that may arise in fractured rock systems are

1. Mapping the hydrologic fractures, except on large-scale maps.
2. Interpolating and extrapolating hydrologic conditions.
3. Determining the unevenly distributed properties of the rocks.
4. Determining the relationship of the surfaces and clay media with the underlying rock-fracture media.
5. Predicting the groundwater supply.
6. Predicting waste management potential.
7. Determining water-level characteristics.

The most significant fracture features are orientation, aperture, density, and roughness. These characteristics depend on the resistance offered by the rock to the force involved. For instance, in hard rocks fractures are extensive, large, and dense compared with those in softer rocks, which are of limited extent, relatively small, and less dense. In either type of rock, however, fractures (unless filled with cementing material) facilitate the flow of water through the rock. Furthermore, compared with a porous medium, a fractured medium offers relatively less resistance to water movement because the water molecules are not subjected to barriers within the fracture. It is logical to say that, the longer the fractures and the smoother their faces, the more easily the water is transmitted. Since fractures dominate the movement of groundwater, a knowledge of their geological settings is indispensable in hydrology and hydrogeology, as well as in water engineering.

Fractures within a porous medium can thus increase its void space as well as its capacity for water transmission. For instance, sandstones and limestones can be among the fractured-porous media that are referred to as "double-porosity" media (Barenblan et al., 1960). In such a dual system, the fractures are assumed to function mainly in water transmission without water storage capacities, whereas a simple porous medium is regarded mainly as groundwater reservoir storage with comparatively less transmission capability.

Igneous and metamorphic rocks and consolidated deposits that occur within several hundred meters of the land surface tend to be fractured to some degree. In these rocks the chief avenues for water movement are the planar openings that are referred to interchangeably as joints or fractures.

Igneous rocks that have fractures still lack any significant porosity, so that their storage capacities are negligible even though their water transmission abilities may be high. Another extreme case can occur when the fractures are numerous and randomly distributed such as grains in a porous medium; where this kind of fracturing has occurred as at large scale, the groundwater movement will be very similar to that in the porous medium.

In some cases, there may be an extensive major fracture within either a porous or a fissured medium that calls for a special treatment from a water movement point of view.

Fractures can occur in any type of rock, in different forms and patterns depending on the rock type. Significant numbers of fractured reservoirs occur in the igneous, metamorphic, and indurated clastic rocks. The porosities and hydraulic conductivities of the fractures and, therefore, the yields of wells penetrating fractured rock reservoirs differ according to the rock type in which the fracture network is formed. Fractures are generally more numerous at rock surfaces but become rarer with depth. Therefore, the yield of a well in a fractured rock reservoir decreases with increasing depth, as shown in Figure 14. Deep-seated fractures are produced by earth stresses originating at great depths. Among such fractures are faults, shear zones, breccia, regular and irregular fractures, and fissures. The extent of open fractures

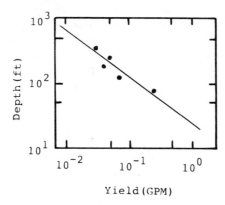

Figure 14 Yield variation with depth (Turk, 1963).

depends on the ability of the rock to hold back the overburden. Igneous and metamorphic rocks have voids for water storage in the forms of either individual extensive fractures, fracture sets, or sets of clusters. In many parts of the world fractured rocks are the main source of domestic water supply.

Fractured (jointed or faulted) regions of impermeable igneous and metamorphic rocks may store water within the joints, yet any well in such a region may penetrate impermeable rock to a depth below the groundwater table and find a joint that yields water under pressure.

Crystalline rocks are heterogeneous and anisotropic water-bearing systems whose permeability varies markedly within short distances. Therefore, application of the porous-medium approach in quantifying their hydraulic properties can lead to erroneous conclusions. Hydrologists and geologists have long known that water wells in crystalline rocks in general and those of pre-Cambrian age in particular have low yields (Meinzer, 1923). In mountainous regions, because of highly dissected relief groundwater flow is relatively faster than in porous medium and numerous springs occur at the surface, usually with low yields.

The direction of large joints and fracture zones can be observed in areas where normal and reverse faults exist.

The water transmission characteristics of a fractured reservoir depend on the width, roughness, continuity, spacing, and filling of the fractures and on the kinematic viscosity of the water.

A. INTRUSIVE IGNEOUS ROCKS

Among the most significant groundwater reservoirs in an igneous rock environment are granite, gabbro, syenite, and diorite, which may have joints parallel to the rock surface delimiting different sheets and concentric joints forming conical surfaces. In their original form, these rocks do not have any significant property of water storage and transmission. Metamorphism and structural changes, however, can enhance their usefulness to the water supply. Aplites are fine-grained intrusive rocks but in places they may have a dense pattern of fractures that enables them to provide local pockets of groundwater reservoir. Pegmatites have weak cohesion among their individual crystals and therefore may fracture easily with any tectonic force. Depending on the intensity of local tectonic phenomena, granite rocks may be more highly fractured, whereas in other regions diorite and gabbro will be less fractured. Shear jointing is very common in igneous rocks and is virtually ubiquitous in intrusive rocks such as granites and granodiorites (Larrson, 1972).

In the arid regions of the world, the weathered surface of the fractured granite basement and the overlain product of decomposition as well as deposition provide the best aquifer as shown in the cross section in Figure 15. These rocks were formed under great pressure deep in the earth. They have been relieved of a great overburden of rock, so that decomposition cracks, joints, and planes have developed as potential water-bearing openings. Weathering action leads to cracks, swelling, and solution of minerals or ions, all of which can improve the porosity and transmissivity properties of intrusive rocks.

B. EXTRUSIVE IGNEOUS ROCKS

These rocks include less granular porosity than the intrusive rocks because they crystallize at the surface rather rapidly, which results in fine grains. Hence, the main petrographical properties depend on the secondary structures within the rock such as cooling joints, shrinkage cracks, vesicular porosity caused by gas bubbles, fractures produced by buckling, and the voids that appear between successive extrusion periods. Basalt and andesite are the most important extrusive rock reservoirs in which groundwater may flow freely. Large tabular openings occur in lavas subject to flowage during solidification. Shrinkage cracks provide more or less vertical passages that help the surface water penetrate the subsurface. In this way either the underlying porous layers are replenished or if the underlying stratum is impervious springs are formed at the bottom of hills, as shown in Figure 16. Basalt flow may cover extensive areas to an incomparatively uniform thickness. The most aquiferous basalts are of relatively recent age such as Quaternary or late Tertiary. Older basalts tend to be kindly weathered and to have their secondary openings filled by clay minerals that have destroyed their water-transmitting capacities.

In some areas of the world, such as western and southern Saudi Arabia and India, recent basalts cover the underlying drainage pattern, forming buried channels where the groundwater is trapped under confining conditions. On the other hand, basalt flows block the main stream channel, resulting in a surface depression in which sedimentation of silt, clay, or volcanic ash provides an impervious layer. In addition, silica-rich rhyolite of high viscosity has erupted also to form impervious layers. Another interesting point about extrusive rocks is that they can intrude as dikes or salt dams to form vertical impervious barriers.

The weathering of basalt with time leads to the formation of secondary minerals that subsequently fill up the primary pores within the rock mass. It is, therefore, to be expected that the water-transmitting capacity of extrusive rocks decreases with time. For instance, the Deccan basalts in India were subjected to weathering for a very long period and hence have low permeabilities.

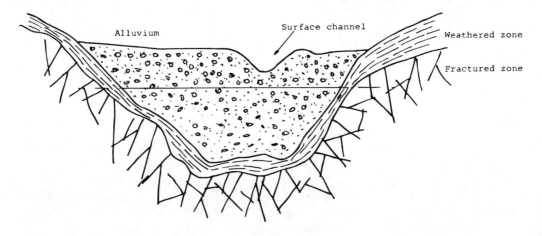

Figure 15 Wadi cross section.

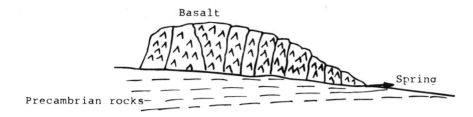

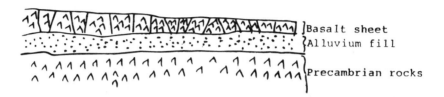

Figure 16 Basalt covers.

C. METAMORPHIC ROCKS

Weathering process or any type of water activity may have considerable influence on the storage and water-transmission properties of a rock mass. Volcanic activity, deposition, vegetation, and the weathering effects of climate transform the primary rock and its fractures into a more void-rich reservoir type. In magmatized gneiss and other metamorphic rocks, sheet jointing is fairly common. The metamorphic rocks are mainly gneiss, schist, quartzite, marble, phyllite, and slate. Phyllite, slate, and schist have very poor water storage and transmission properties. In general, the most favorable coordinates for groundwater accumulation occur in fractured marble reservoirs. These rocks are formed from earlier rocks by pressure and/or heat action, both of which alter and compress them so that their original void spaces are destroyed. However, when these altered rocks appear on the earth's surface there is a good chance of fracturing and, hence, an increase in water storing and transmitting capabilities, first as for igneous rocks (see Figure 17). In general, three fracture patterns occur in the metamorphic rocks; these are exemplified schematically in Figure 17 for gneiss, phyllite, and quartzite.

The water-bearing capacity of these rocks depends almost entirely on the secondary porosity that develops as a result of fracturing and also weathering. Because of the rock mass overburden, the effect of weathering and, consequently, the water-transmitting property of these rocks decrease with depth.

D. CONSOLIDATED SEDIMENTARY ROCKS

Once sediments have been consolidated they are subjected during later time periods to fracturing and hence the development of secondary porosity in the same way as the igneous and metamorphic rocks. In fact, most of the sedimentary formations owe their significant water storage and transmission capacities to tectonic faulting and folding.

IV. KARSTIC MEDIUM

Unlike the previously discussed media, a karstic domain is a product of the chemical reactions between the rock and water, that is, for a karstic medium to develop such interaction

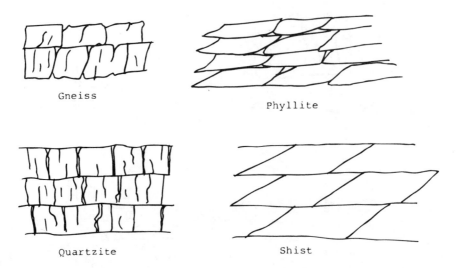

Figure 17 Fracture pattern in metamorphic rocks.

is essential, whereas for other types of medium it may not be involved at all. For instance, a porous medium may develop as a result of deposition from the wind effect, and fractures may be caused by tectonic forces only. The formation of a karstic medium requires very special sediments—limestones, dolomite, gypsum, halite, anhydride, and other soluble rocks—to constitute the reservoir. Through time, because of the rock solubility and the effects of the various geological processes that come into play, the terrain develops a special, unique topography that is defined as karst (Milanovitch, 1981). For instance, limestone, no matter how hard, is dissolved by water, so that caves or even river channels develop underground while drainage sinks, rifts, and shallow holes appear on the surface, all of which leave the land dry and relatively barren. The fractured limestone, dissolution of carbonates by carbonic acid present in the atmosphere, and the soil give rise to the enlarged fractures, conduits, caverns, or caves referred to as a karstic terrain. Very often the surface water is in direct contact with the groundwater through numerous sinkholes and outlets. Larger rock fragments may create blockages in the karstic network that divert water through the granular body or points or fine fractures, which bring a delayed flux to the main system where the subsurface flow is very rapid. Beneath the base level, which is the lowest level of groundwater during dry seasons, the fractures are saturated with carbonate so that this zone has no open fractures. However, the base level changes over geological time, so that below the present base level a paleokarst region may occur as in Figure 18. It is obvious from this figure that solution cavities terminate either as freely flowing springs or as concealed outlets under the superficial Quaternary and Tertiary sediments or under surface water bodies such as lakes, rivers, and seas. In a karstic terrain the surface runoff may be carried totally or partially as subterranean flow. Springs in limestone terrain can be interconnected to topographic depressions caused by collapsed sinkholes at higher elevations. The water-level fluctuation in the sinkhole indicates recharge and discharge to the karstic formations.

Karstic formations are fully developed in the humid and semi-arid regions where the lakes are usually interconnected with the underlying solution-cavity network. During wet periods of the year, this network transports water from the surrounding terrain into the lake, that is, recharge to the lake occurs as shown in Figure 19a. In the dry seasons, however, the water level in the lake is higher than the surrounding subsurface flow level and therefore the same network feeds the karstic aquifer (see Figure 19b).

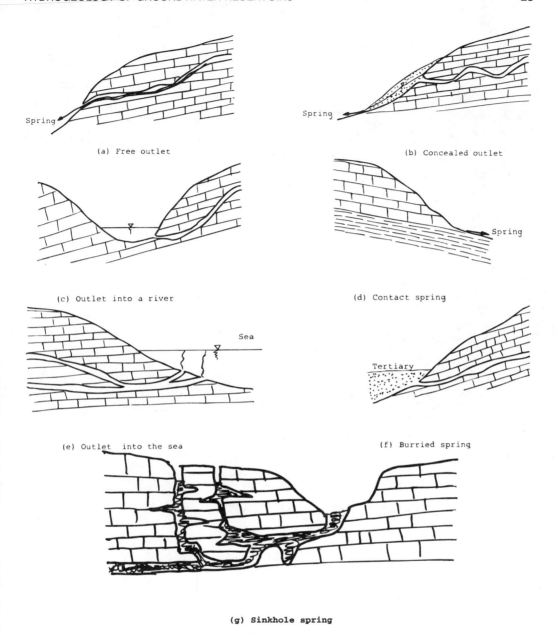

Figure 18 Karstic formations.

Sometimes the sinkholes in humid regions take all the surface water, which disappears underground as in Figure 20.

On the other hand, in arid zones oases are formed within the karstic reservoirs, as represented in Figure 21. Herein, the solution cavity network transports the groundwater flow from the deep-lying water-bearing formations toward these water bodies.

Contact of the limestone with water and carbon dioxide at the same time leads to its solution and the development of various complicated subsurface caverns as well as surface features. In scientific terminology, the collection of phenomena characterizing regions of soluble limestone rocks is known as a "karst". It is for this reason that nonclastic formations

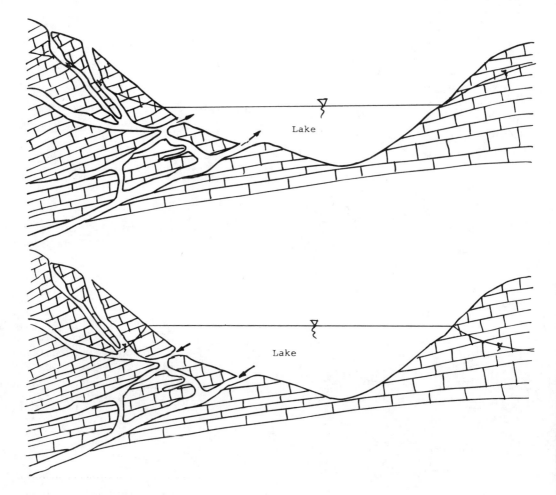

Figure 19 Recharge to and discharge from a lake.

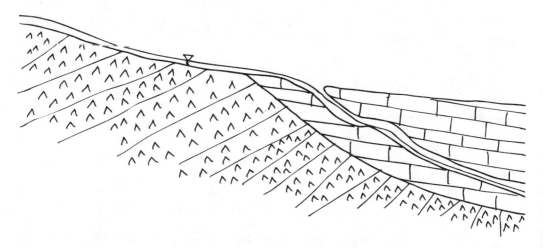

Figure 20 Sinkhole.

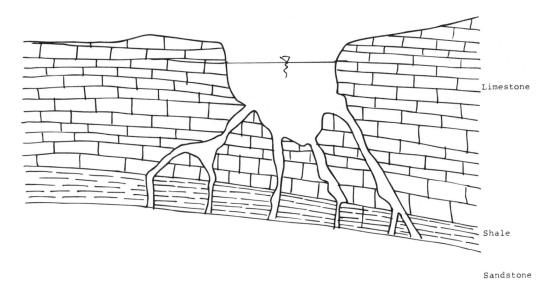

Figure 21 Oases within a karstic formation.

are called karst reservoirs. The characteristics of such reservoirs differ from the previous ones in many respects:

1. The process involved is chemical rather than mechanical or tectonic.
2. Wide channels develop within the karstic reservoirs.
3. Overall permeability decreases rapidly with increasing depth, especially below the water table. However, the zone above the water table is a very transmissive and cavernous medium.

From the hydrogeological point of view, solution cavities provide more storage capability than fractures. A detailed account of the karstic medium is presented by Moreaux et al. (1984).

The general mechanism in the formation of karstic reservoirs includes four successive stages: sedimentation, consolidation, fracturing, and chemical solution, as schematically shown in Figure 22. Depending on the spread of these events, the region is either fully or partially karstified. Areas of extensive karstification contain widely spaced interconnecting solution channels. Fully karstified areas generally have the following reservoir characteristics:

1. An interconnected network of highly permeable channels near the water table.
2. Rapidly decreasing karstification within increasing depth below the water table.
3. Exceptionally high permeability zones in and around the valleys.
4. A very cavernous unsaturated zone.
5. Rather saline water in the lower and less-permeable parts of the reservoir.
6. Low storage of freshwater after long periods of dry weather.

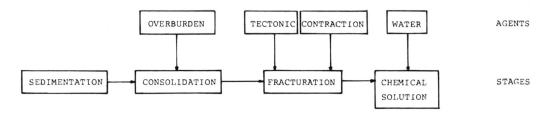

Figure 22 Karstification processes.

In the early stages, the groundwater percolates downward, through either fractures or the primary porosity voids in the rock, to reach the water table, after which it moves in all lateral directions until it discharges on the surface (see Figure 23).

Lastly, in any engineering work on or within the karstic reservoirs, special attention must be given to individual solution cavities to avoid possible loss of capital and net benefit. For instance, Figure 24 shows a schematic longitudinal cross section of a dam location where the water of a filled surface reservoir is emptied simply through a vertical sinkhole beneath the impervious layer.

Typical karstic features can be found in evaporite contact with water. On the other hand, dolomites are natural reservoirs without fractures and are much less soluble. Therefore, true karstic formations are not observed in the dolomitic rocks. Solution fractures occur along the rare fractures, and increase the water transmission capacity of these rocks only locally because they lack network of conduits. The same applies to chalk or marly limestones.

V. SUBSURFACE WATER

Groundwater storage results from the various hydrological and geological events that occur throughout geological time. Some accumulations are renewable whereas others are not. Some

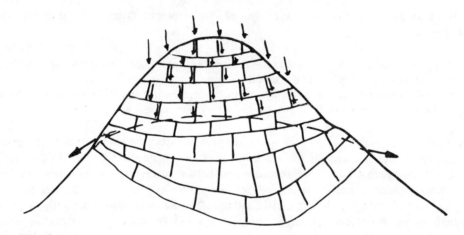

Figure 23 Early karstic development.

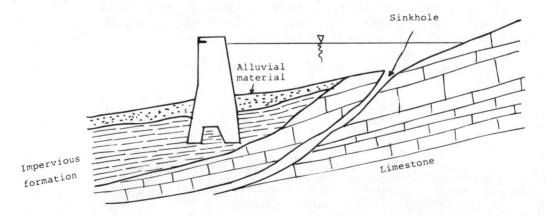

Figure 24 Damage on dams.

may be extracted economically whereas others may not. According to location, depth, and formation, groundwater may have different qualities, pressures, and temperatures. To explore, evaluate, and exploit a region's groundwater resources, it is necessary to understand each type individually as well as the ways the various types interact. Therefore, subsurface waters and their existence in the geological formations must be classified from different points of view. The classification presented here will be based on origin, saturation, and the pressure characteristics of mutual water and geological interactions.

VI. ORIGIN OF WATER

In general, the groundwater of the world is derived from three sources as shown schematically in Figure 25. First, most groundwater is derived from the atmosphere in the form of rainfall, snow, hail, humidity, etc. Water of this type is referred to as meteoric water. The hydrological cycle is the mechanism that transports this water through the atmosphere, hydrosphere, and lithosphere. Meteoric water reaching the earth's surface either infiltrates directly through porous, fractured, and karstic media or accumulates as rivers, lakes, and ponds from which, with time and/or suitable locations, it reaches the groundwater storage. Return of groundwater to the surface occurs either naturally by means of springs or subsurface flow toward surface water bodies or artificially through wells.

Water for domestic, agricultural, and industrial use is extracted mainly from such meteoric water, derived from rainfall and infiltration within the present-day hydrological cycle. The chemical composition of meteoric groundwater changes during its passage through various rocks. The changes depend mainly on such factors as the temperature and pressure variations, the minerals it encounters, and the time duration for water and mineral contact.

A second source, called connate water, is water that was entrapped in the interstices of a sedimentary rock at the time the rock was deposited. It may have been either land water or ocean water. Its main characteristics are that it occurs at great depths, has not undergone the present-day hydrological cycle (but only the one operating at the time the sediments were deposited), and tends to be rather salty.

Connate water is thus fossil water that has been cut off the hydrological cycle for at least an appreciable part of a geologic period. White (1967) has stated that it consists of the fossil interstitial water of unmetamorphosed sediments and extrusive volcanic rocks, and water that has been driven from the rocks. According to Meinzer (1923), connate water has remained since burial with the rock in which it occurs.

Third, groundwater may have come from the earth's interior with no previous existence as atmospheric or surface water. This is a juvenile type of water that is believed to be derived from igneous processes within the depths of the earth. Having never taken part in any previous hydrological cycle, it can contribute unusually to the meteoric groundwater it joins.

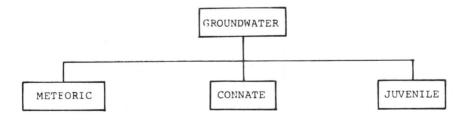

Figure 25 Groundwater sources.

VII. ZONES OF WATER

Subsurface waters can be classified into various groups according to their physical occurrence within the medium. The part of the lithosphere where each void space is filled with water is referred to as the saturation zone. Above this zone is the aeration or unsaturated zone, where the voids include a mixture of water, moisture, and air (see Figure 26). There is a free exchange of air and moisture in this zone. Any water collected within the unsaturated zone is called vadose water. Nevertheless, there are three different zones for vadose water. The uppermost layer, which is necessary for plant life, is the soil moisture zone. The thickness depends on the type of soil as well as the prevailing climate of the area. The movement of water in this zone is upward or downward, depending on the surrounding conditions. The soil suction and the gravity are the two dominant forces here. Beneath the soil moisture zone comes the intermediate zone that is bounded at the bottom by the capillary fringe. The water in this last zone is held by the attraction force between the molecules of two different materials, in this case water and soil. The capillary fringe overlies the zone of saturation and contains capillary interstices, some or all of which are filled with water that is part of the water in the zone of saturation but is held above it by capillary forces acting against gravity.

From the pressure point of view, the aeration and saturation zones have different characteristics, in that the pressure gradient in aeration zone is less than atmospheric pressure, whereas in the saturation zone the voids are filled with water at a pressure greater than atmospheric. As a result, the subsurface depth where atmospheric pressure prevails will be the separating boundary between the saturated and unsaturated zones, and is referred to as the groundwater table. Of course, with any increase in depth below the groundwater table there is an increase in pressure.

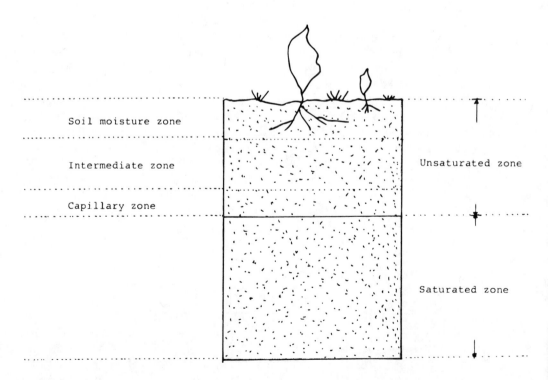

Figure 26 Water zones in the lithosphere.

VIII. WATER-BEARING FORMATIONS

Any lithospheric layer that includes water is called a water-bearing formation and, as mentioned earlier, all reservoir rocks are of this type. To facilitate the solution of groundwater problems, therefore, further distinction and classification of the reservoirs is needed. According to water storage and transmission properties, geological formations can be classified into four hydrogeological units as aquifers, aquitards, aquicludes, and aquifuges.

A. AQUIFER

Meinzer (1923) defined an aquifer as a geological formation, group of formations, or part of a formation that contains sufficient saturated permeable material to yield significant quantities of water to wells and springs. There are at least three separate definitions that are useful for practicing hydrogeologists. First, an aquifer is a saturated geological unit that can transmit water easily in significant amounts under normal conditions. In this statement, "saturation", "easiness", and "significance" are among the basic water-related description of aquifers. A second definition is that an aquifer is a water-bearing geological formation or stratum that yields a significant quantity of water for economic extraction from wells. This definition appeals to groundwater engineers, who are not concerned in detail with geology but rather with hydraulics and economics.

Still a third definition is that an aquifer is a zone of rock through which groundwater moves. This last statement attracts nonspecialists who are concerned with the movement of groundwater. Rock formations that serve as aquifers are gravel, sand and sandstone, alluvium, cavernous limestone, fissured marble, fractured granite, weathered gneiss and schists, heavily shattered quartzite, vesicular basalt, and jointed slate. Two properties shared by the definitions of aquifers are the storage and transmission of water. Where either or both of these properties are lacking in a geological formation, three different types of hydrogeological unit can emerge—aquiclude, aquifuge, and aquitard.

B. AQUICLUDE

This is saturated geological formation which, although capable of absorbing water slowly, will not transmit it fast enough to yield a significant supply for a well. In other words, it can store water but cannot transmit it easily. Clay lenses and shale layers are examples. Some aquicludes cannot transmit water at all.

C. AQUIFUGE

A geological formation with noninterconnected openings or interstices is called the aquifuge because it neither absorbs nor transmits water. Basalt and granites are good examples.

D. AQUITARD

Any geological formation of a rather impervious nature that transmits water at a slower rate than an aquifer but for water supply is called an aquitard. Aquitards represent a transition between aquicludes and aquifuges. The aquitards in nature are clays, shales, and intact crystalline rocks. Often they are called semipervious layers.

From a hydrogeological point of view, these four individual units of water-bearing geological formations are not in themselves meaningful. In nature, their combinations occur at different levels, resulting in a complex groundwater domain. It is, therefore, essential to interpret a geological log, i.e., a stratigraphic sequence of lithologies from water storage and transmission points, and then produce the closest counterpart for each lithology in terms of aquifer, aquitard, aquiclude, and aquifuge. The end product is a succession of hydrogeological units that will be referred to as hydrostratigraphic units. The transition from a geologic log to the hydrostratigraphic sequence is represented in Figure 27. The main points to be considered in such a task are

1. An aquifer is a lithological unit with appreciably greater transmissivity than adjacent units, which commonly stores and transmits water that is recoverable in economically usable quantities. In the light of this explanation, because sandstone and alluvium lithologies have relatively greater transmissivity than shale and limestone, they are identified as aquifers. Hence, there exist three types of aquifer in the geologic log.
2. The lithological units of low permeability that limit the aquifer are commonly called aquicludes or confining beds.
3. Any appreciable quantities of water that move through the confining beds are called aquitards or semipervious layers or leaky layers. Hence, in Figure 27 shale lithologies correspond to aquitards in the hydrostratigraphic sequences.

On the other hand, hydrostratigraphy helps to classify aquifers into unconfined, confined, leaky, and perched forms according to the present-day water pressure.

IX. AQUIFER TYPES

Aquifers can be classified according to prevailing subsurface geological composition, or to hydrological conditions, or to groundwater pressure. The areal extent of a geological layer,

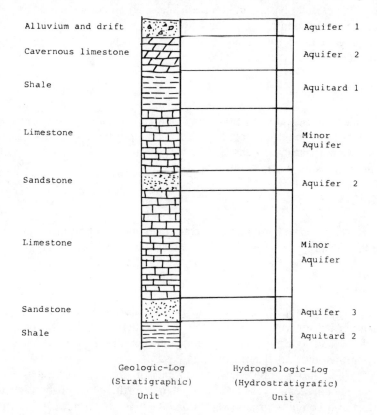

Figure 27 Geologic log vs. hydrogeologic log.

HYDROGEOLOGY OF GROUNDWATER RESERVOIRS

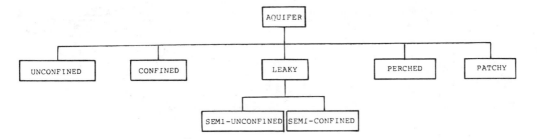

Figure 28 Aquifer classification.

the types of adjacent layers, and the number of layers involved are among the important geological factors that should be considered in any groundwater study. The important hydrological factor is the recharge mechanism which may be direct or indirect. Finally, the groundwater pressure may be either equal to or greater than the atmospheric pressure. Any proper aquifer classification must include all three types of data; no one of them is sufficiently meaningful by itself. Unfortunately, geologists tend to consider only the geologic factors, which can lead to erroneous conclusions, whereas engineers most often stick to the hydrological and pressure conditions, so that their evaluations can also be in error. In this book, all the above-mentioned factors are considered to provide an interdisciplinary understanding. The conventional classification of aquifers is given in Figure 28.

A. UNCONFINED AQUIFER

Geologically speaking this type of aquifer is two layered and aerially extensive. The lower layer has aquifuge characteristics, whereas the upper layer is practically a water-bearing formation as shown in Figure 29. Furthermore, in an unconfined aquifer the groundwater table forms the upper boundary of the saturation zone. In addition, the water table is defined as a connection of points where the absolute pressure equals the atmospheric pressure and the relative pressure, i.e., the pressure due to the water only, equals zero. From the hydrological point of view, unconfined aquifers are subject to direct recharge from infiltration, i.e., they are directly connected with the hydrological cycle. Among these three factors the atmospheric pressure is the most significant in deciding whether the aquifer is unconfined. For instance, if the groundwater table is below the upper aquifuge level, as shown in Figure 30, the aforementioned geological and hydrological conditions do not exist directly, but the reservoir pressure is still atmospheric and the aquifer therefore is unconfined. This example shows how

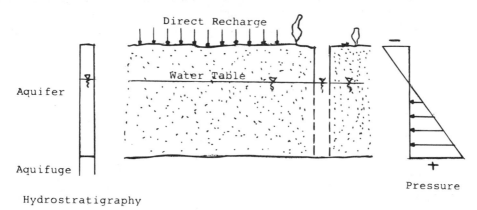

Figure 29 Idealized unconfined aquifer.

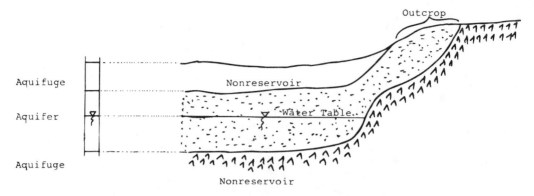

Figure 30 Unconfined aquifer.

geological and hydrological considerations above can be misleading in identifying an aquifer type. It is necessary also to know the position of the groundwater level in relation to the geological layers. As a practical rule, it can be concluded that, if the groundwater level falls within the same lithological unit, the reservoir is unconfined.

In an unconfined aquifer the groundwater is directly affected by only atmospheric change. For instance, its surface level will vary in accordance with the amount of rainfall and especially the rate of infiltration. Its chemical composition and temperature also vary temporally as well as spatially. Groundwater in unconfined aquifers occurs at shallow depths and is therefore readily susceptible to contamination. The surface at the water table usually assumes a wavy form and is horizontal only in plains. In fact, any area of horizontal groundwater surface is referred to as the groundwater basin. Depending on the surface slope and the ease of movement through the rocks, a certain amount of groundwater flow takes place. On the other hand, sand dunes may provide fresh groundwater aquifers in an unconfined manner as in Figure 31. The groundwater is replenished both by direct infiltration into the sand and by inflow from adjacent heights. Its surface follows the dune topography in a smooth manner. In general, groundwater in the voids of alluvial fill gravels forms an unconfined aquifer.

B. CONFINED AQUIFERS

Such an aquifer differs geologically from the unconfined type in having at least three layers, of which two are aquifuges with an aquifer between. As a result of this layering, direct recharge is not possible. If the aquifer is completely saturated, the water is confined between the two aquifuges, and the extra pressure thus exerted on the groundwater elevates its pressure above the atmospheric pressure at any point within the aquifer. The two aquifuges are referred

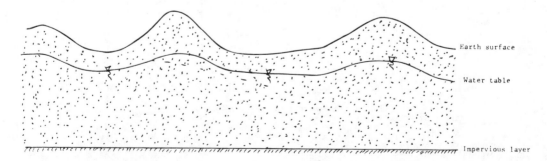

Figure 31 Unconfined aquifer in sand dunes.

HYDROGEOLOGY OF GROUNDWATER RESERVOIRS

to as the upper and lower confining layers. An idealized form of a confined aquifer is shown in Figure 32.

Any well drilled in a confined aquifer will have a water level elevation above the aquifer itself. In other words, the points of atmospheric pressure are within or above the overlying aquifuge. Without wells this elevation remains merely an imaginary line which may even be above the earth's surface provided that the appropriate structural and hydrogeological conditions are met. The main geological features needed to form a confined aquifer are various synclines, monoclines, depressions, troughs, grabens, and tectonic fracture zones. Such aquifers are observed also in Quaternary sediments. They always have three basic components: a recharge, a transition, and a discharge area. The recharge area corresponds to the outcrop area which is located at high elevations and consequently receives direct infiltration from rainfall and surface runoff. In the discharge area, the aquifer outcrops at lower elevations generally comprises a number of springs. The transitional zone constitutes the main portion of the aquifer, where the groundwater is found between the two confining layers.

Furthermore, part of a geological formation may behave as a confined aquifer while adjacent parts may be unconfined. Such a situation occurs in the synclinal type as schematized in Figure 33. Even the extent of an aquifer, whatever its type, can change with time. For example, an increase in the recharge at an outcrop area will increase the extent of a confined aquifer, whereas abstraction of water from well A will reduce the confined aquifer domain and increase the extent of an unconfined aquifer. Figure 34 represents confined and unconfined aquifers within the same hydrolithological unit. Any drop in the water table below point A will transform a confined aquifer into an unconfined type. Because the groundwater in the confined aquifer is under extra pressure, any well penetrating the finite confining layer will cause the water to rise to the BB' level.

C. PERCHED AQUIFERS

Especially in Quaternary deposits, a layer of fine clayey sand or silt overlies coarse sandy materials. If this sand or silt layer is flat or basin shaped, it may support a body of water that is called a perched aquifer. This is, in fact, a local unconfined aquifer that can provide small amounts of groundwater for short periods. It is generally located above the water table of an

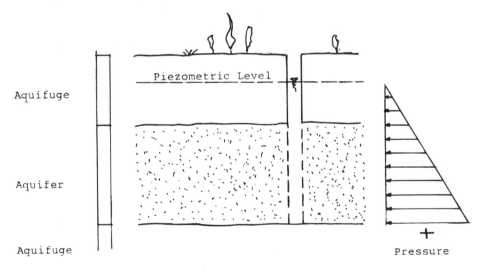

Figure 32 Idealized confined aquifer.

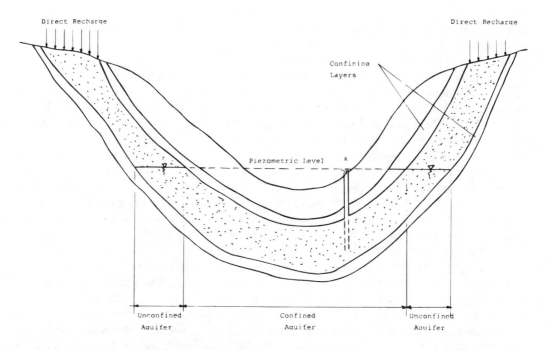

Figure 33 Adjacent aquifers in a syncline.

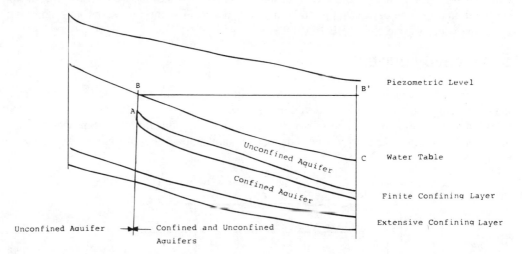

Figure 34 Confined and unconfined aquifers.

extensive unconfined aquifer (see Figure 35). If the water table rises above the impervious lens, the extra sedimentary unit acts as one unit-unconfined aquifer.

Perched aquifers are more common than is often supposed. Sometimes they are only a few centimeters thick or occur only at a major infiltration event. They do not provide a reliable supply of water as sometimes a well can be deepened right through the impermeable layer, so that most of the perched water drains away.

D. LEAKY AQUIFER

In all the aforementioned examples, the confining layers are assumed to be aquifuges. However, should any one of them be an aquitard, the water exchange may start between

HYDROGEOLOGY OF GROUNDWATER RESERVOIRS

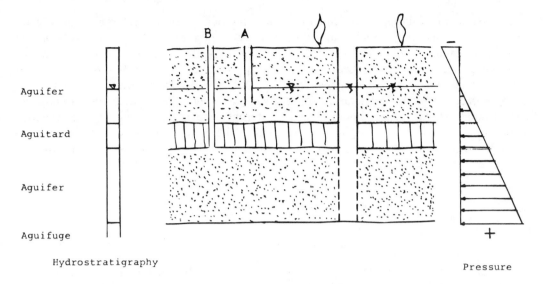

Figure 35 Perched aquifer.

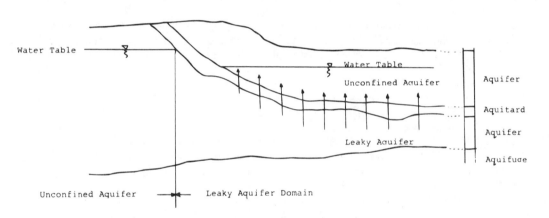

Figure 36 Idealized leaky aquifer.

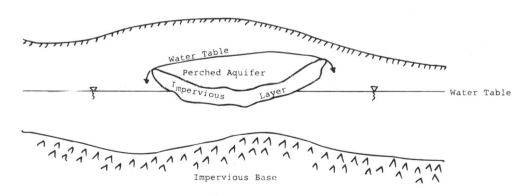

Figure 37 Multitude of aquifers.

successive aquifers through the aquitard. Such a geological setup is referred to as a leaky aquifer. It is, therefore, possible to define a leaky aquifer as one where, in addition the characteristics already cited for confined and unconfined aquifers, there are also groundwater exchanges as a result of leakage between different aquifers through aquitards. From a geological point of view, the hydrostratigraphic sequence in Figure 36 exemplifies a leaky aquifer. Leakage occurs through the aquitard, provided these two water-table elevations do not coincide. Downward (upward) leakage requires that the water-table elevation on the left (right) be above (below) the elevation on the right (left). Of course, the direction of leakage offers clues to the relative positions of the two piezometric surfaces. Figure 37 shows different types of aquifer existing in the same area.

REFERENCES

Al-Sayari, S. S. and Zotl, J. H., 1978. *Quaternary Period in Saudi Arabia,* Springer-Verlag, Vienna.

Anton, D., 1983. *Modern Eolian Deposition of the Eastern Province of Saudi Arabia. Eoline Sediments and Processes. Developments in Sedimentology,* Vol. 38, Elsevier, Amsterdam, 365.

Bagnold, R. A., 1941. *The Physics of Blown Sand and Desert Dunes,* Methuen, London.

Barenblatt, G. E., Zheltov, I. P., and Kochina, I. N., 1960. Basic concepts in the theory of seepage of homogeneous liquids in fissured rocks, *J. Appl. Math. Mech.,* 24(5), 1286.

Cooke, R. V. and Warren, A., 1973. *Geomorphology in Deserts. Batsf, London,* University of California Press, Berkeley.

Davis, G. H., 1984. *Structural Geology of Rocks and Regions,* John Wiley & Sons, New York.

Dreimanis, A. and Vagners, U. J., 1971. Bimodal distribution of rock mineral fragmentations in basalt tills, in *Till, A Symposium,* Goldthwait, R. R., (Ed.,) Ohio State University Press, Columbus, 237.

Eyles, N., 1983. *Glacial Geology,* Pergamon Press, Elmsford, NY.

Geikie, J., 1894. *The Great Ice Age,* 3rd ed., London.

Goldthwait, R. P., 1975. *Glacial Deposits,* Benchmark Papers in Geology, Vol. 21, Dowden Hutchinson and Ross.

Larsson, I., 1972. Ground water in granite rocks and techtonic models, *NordicHydrology,* 3, 111.

LeGrand, H., 1979. Evaluation techniques of fractured rock hydrology, *J. Hydrol.,* 43, 333.

Meinzer, O. E., 1923. The Occurrence of Ground Water in the United States, U.S. Geologic Survey, Water Supply Paper 489.

Milanovich, P. T., 1981. *Karst Hydrogeology,* Water Resources, Fort Collins, CO.

Moore, G. T. and Asquith, D. O., 1971. Deltas: term and concept, *Bull. Geol. Soc. Am.,* 82, 2563.

Moreaux, E. L., Wilson, M., and Memon, B, 1984. Guide to the Hydrology of Carbonate Rocks, UNESCO.

Rachocki, A., 1981. *Alluvial Fans,* John Wiley & Sons, New York.

Stearns, D. W. and Friedman, M., 1972. Reservoir in Fractured Rock, American Association of Petroleum Geologists, Reprint Series, No. 21.

Sugden, D. E. and John, B. S., 1976. *Glaciers and Landscape. A Geomorphologic Approach,* Butler and Tanner.

Turk, L. J., 1963. The Occurrence of Groundwater in Crystalline Rocks, M. Sc. thesis, Stanford University, Palo Alto, CA.

White, D. E., 1967. Magmatic, connate and metamorphic waters, *Bull. Geol. Soc. Am.,* 68, 1659.

CHAPTER 3

Groundwater Flow Properties

I. GENERAL

Groundwater is in continuous movement from a recharge to a discharge area in accordance with laws governing water flow in lithosphere. The laws give the rate of energy loss against resistance from the flow medium. The conservation of energy principle helps calculation of losses. On the other hand, because of water recharge and discharge from the aquifer, the volume of groundwater in motion changes temporally and spatially according to the principle of mass conservation.

Groundwater movement is a dynamic process depending on the medium and the fluid properties as well as their mutual interactions, giving rise to various flow types under different geological and hydrological conditions. Dynamism of the process originates naturally in large scale from the hydrological cycle and in relatively small scales from human interference by wells, surface and subsurface dams, galleries, and agricultural and similar activities. Groundwater in the lithosphere is overwhelmingly under a continuous dynamic state even though the velocities may be very slow. Further, there are regional static groundwater states with no movement either for short or long time spans. The static groundwater locations within the lithosphere are a result of suitable subsurface geological composition. These regions were referred to as the groundwater basins in the previous chapter. Existence of an impervious dike or fault on the way of movement is a convenient condition for a static region as shown in Figure 1.

Two complementary branches in water sciences concerning the dynamic and static states are hydrodynamics and hydrostatics, respectively. In practice, in the regions in which the piezometric level is horizontal, principles of hydrostatics are applicable. Such hydrostatic situations are rarely considered in the groundwater studies. Their wide applications are within the domains of engineering geology, civil engineering, and to a minor extent of hydrogeology in the case of deep-seated groundwater reservoir. On the other hand, practical implication of a hydrodynamic state occurs where the piezometric levels are inclined, which implies essentially groundwater flow. Only in this case are the flow properties and laws valid; otherwise, they all cease to function in any static state.

Complex geological structure coupled with the dynamic groundwater processes present challenging barriers to the groundwater scientists. Their decompositions into some basic classifications with realistic simplifications and physical implications enable the researchers to overcome these barriers within the practically acceptable error percentages. Hence, it is the main purpose of this chapter to identify, define, and physically explain and combine flow and aquifer material properties in order to quantify the groundwater level fluctuations.

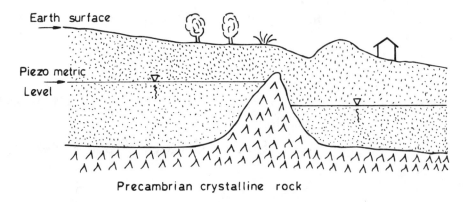

Figure 1 Static groundwater.

II. ENERGY OF WATER

As has been already mentioned in Chapter 1 water circulates in the universe without an end on different paths that constitute the hydrological cycle. Keeping this point in mind, it is interesting to consider the following several practical questions:

1. Where does the moving water get its energy?
2. Which types of energy are involved in different paths of hydrological cycle?
3. What is the dominating energy type along each path?
4. What are the effects of water during its journey in the cycle?
5. Which types of energy play a distinctive role in the groundwater movement?
6. What are the roles of geologic formations and topographic and geomorphological factors in the energy in groundwater movement?
7. How does the energy relate to flow properties and what are the flow laws?

Practical descriptive and physical answers to these questions will be sought in this and the following two chapters without involved mathematics. The purpose is to search for a basic understanding of the groundwater movement physics to the level where one can benefit from such an understanding effectively in studies concerning earth sciences generally, and groundwater engineering particularly.

In order to perceive easily the energy of circulating water and its transformations into different types, we will, for the time being, concentrate our discussion on a single water particle. Generalization of the derived ideas on all particles is a straightforward conclusion for any body of water which is, indeed, a collection of water particles.

Water in the universe has three distinctive energies: the potential, kinetic, and the heat energies. Potential energy is related to the position (elevation) of the particle relative to a reference level. In practice, the mean sea level is adopted for comparisons on the world-wide basis. However, in the groundwater studies any arbitrary fixed point within or at the vicinity of the study area can be selected. We can, therefore, take an arbitrary horizontal plane as datum. The potential energy, E_p, of unit mass of water is expressed as

$$E_p = gz \qquad (1)$$

where z is the elevation of water mass position from the datum as shown in Figure 2 and g is the acceleration due to gravity. Physically, Equation 1 gives the work required to move unit mass of water vertically along a distance equal to z. On the other hand, any moving

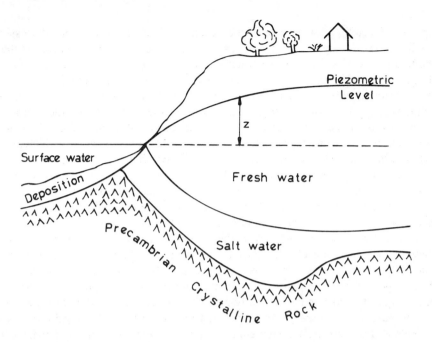

Figure 2 Potential energy.

object has kinetic energy depending on its mass as well as the velocity. Basically, the kinetic energy, E_k, per unit of mass is defined as proportional with the square velocity, v, as

$$E_k = \frac{1}{2} v^2 \tag{2}$$

Any water particle can have simultaneously both potential and kinetic energies.

It is important to notice at this stage that a driving force is needed for the water particle to acquire either a position and/or velocity to have its energy. What might be this driving force? Before answering this question let us think about the forces to which any water particle is subjected during its movement. Of course, there are friction forces among water molecules as well as between the water and wetted surfaces of river beds, channels, solid particle or fracture planes, and cavity walls. It is a common experience that the friction causes loss of energy in terms of heat. Flowing water loses energy continuously. So far as the friction between the water molecules is concerned, the energy loss is negligible because of very low water viscosity. Hence, the significant energy loss is a result of friction between the water and the surface areas of reservoir medium in the case of groundwater.

The basic physical principle of energy conservation at any time and equivalence of total energies at different positions bring a restriction on the total energy. As a consequence of these principles it is obvious that energies should be transformable continuously from one type to others during movement.

III. ENERGY CYCLE

Since our concern is water in nature we know already that any water particle in the hydrological cycle has velocity and/or position simultaneously along various paths. It is, therefore, illuminating for our further discussions to identify different types of energy in the hydrological cycle.

Let us consider any water particle in the ocean. In light of the previous discussions this particle will not have energy since it is at rest. By adopting the mean sea level as the datum for all the hydrological cycle the same water particle will not have potential energy. In other words, large water bodies such as oceans and seas are water reservoirs with zero energy. In order for the water particle to acquire any type of energy there should be some external driving force. What is the source of this force? Indeed, it is the sun that provides through solar radiation an endless source of heat energy to the water particles at the surface of water bodies.

Solar radiation causes the water particles to vibrate near the free surface. Increase of heat means increase in vibration velocity, hence increase in the kinetic energy of the water particle that may escape from the free surface into the air in the form of gaseous state. In classical hydrology, this escape is referred to as the evaporation process. In our current discussion this process is the transformation of heat energy into the kinetic energy. Hence, the ocean surfaces are the main locations for transformation of heat to kinetic energy whereas the potential energy is still zero. Escaping water particles with their kinetic energy rise up in the atmosphere to certain heights and form the clouds. During this rising, continuously the kinetic energy is transformed into the potential energy. It is, therefore, possible to conclude those water particles in clouds have potential energy only, that is, clouds are reservoirs of potential energy whereas the kinetic energy of a particle is practically zero. Clouds as accumulators of water molecules mostly in the gaseous state are given very small kinetic energies by wind currents so that they move inland. During such a movement the potential energy remains the same. Occurrence of raindrops or snowflakes in the cloud due to cooling and condensation transforms the potential energy into the kinetic energy by which the raindrops fall on the earth surface due to gravitational force. From the time that the water particle reaches the earth surface the topography, geomorphology, and surface geology play effective roles in energy transformations. However, the subsurface geology and structures rule the energy transformations for the groundwater movement. In general, after reaching the lithosphere there is a continuous increase in the kinetic energy at the cost of potential energy loss until the water reaches the oceans. During this time irreversible heat losses also occur. Figure 3 presents the energy cycle for water.

IV. GROUNDWATER ENERGY

The slowest rate of energy transformation zone along the energy cycle occurs in the subsurface and especially in the saturated zone. Groundwater encounters excessive resistance from the medium as already explained in Chapter 2. More detailed quantitative information on this point is referred to in the next chapter. The question is how these resistances effect the groundwater energy. A physical answer is that excessive friction causes retardation in groundwater velocity and consequently the kinetic energy is very small and practically negligible. As a consequence, logically, the potential energy of groundwater is always more than the kinetic energy. Similar to clouds the groundwater reservoirs are also potential energy storages. However, on surface water the resistance comes from the periphery of water channels only and therefore the total resistance to surface water is smaller than the groundwater. This is the physical reason why surface waters move quickly and have large kinetic energies. So far as the water is concerned the earth's surface is a faster energy transformer than subsurface.

Any water particle is under atmospheric pressure. We have seen in Chapter 2 that in saturated zone the water particle is under pressure more than the atmospheric pressure. The pressure is force per unit area and the question now is whether this force means any additional energy, and of which type. An illuminating answer to this question is that a stone freely left to fall has less energy than a stone left to fall with extra weight attached to it. This simple example indicates that extra pressure, p, on water particle means extra energy. The type of this extra energy is potential energy.

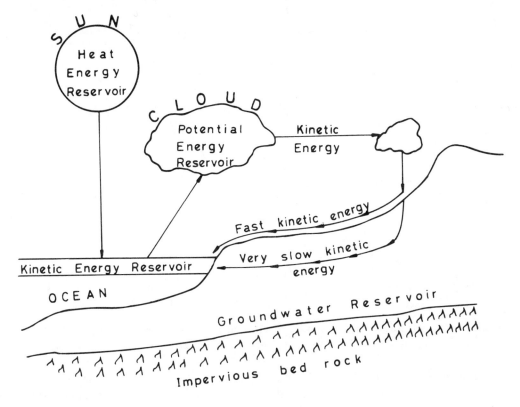

Figure 3 The energy cycle.

The total energy, E_T, is the summation of potential and kinetic energies. It can be written for per unit volume of fluid as

$$E_T = \frac{1}{2}\rho v^2 + \rho g z + p \qquad (3)$$

where ρ is the water density and p pressure. The density of water is at 0°C is practically equal to one. The first term on the right hand side (r.h.s.) is the kinetic energy and others are potential energies due to the position and pressure. To render this expression into a readily applicable form for the field applications both sides are divided by the specific weight of the fluid, $\gamma = \rho g$. Hence,

$$E_S = \frac{v^2}{2g} + \frac{p}{\gamma} + z \qquad (4)$$

where E_s will be referred to as the specific energy which has the dimension of length. This equation is known as the Bernouilli's theorem and is valid for incompressible liquids only. In a practical sense, it is the total energy height on the vertical of concerned point from the datum as shown in Figure 4. In groundwater hydraulics the specific energy is called total hydraulic head. It has three contributing parts as shown on the r.h.s. of Equation 4. The first term is the velocity head, representative of kinetic energy and dependent on the flow velocity only. The second term is the pressure head which is the height of a column of liquid of

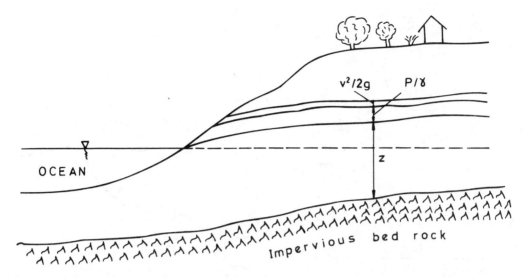

Figure 4 Specific energy height.

specific weight supported by a pressure, p. The third term is called the elevation head with reference to a certain datum. Some of the practical interpretations of total head are

1. That if a hole is drilled in the groundwater reservoir the maximum elevation of groundwater surface coincides with the total head level. This head is referred to generally as the piezometric level in any type of aquifer. However, in unconfined aquifers the water table level terminology is also used interchangeably.
2. That under normal conditions of groundwater flow the velocity is so small that the velocity head is not appreciable. Groundwater velocity is practically low, even if it reaches 50 m/sec the velocity head is only $v^2/2g = 1.3$ cm.
3. That the pressure at any point in the groundwater domain can be evaluated with the hydrostatic principles without taking into consideration hydrodynamics since the groundwater velocity is negligibly small. The hydrostatic pressure at any point below the water surface is equal to the density of water times depth of water above this point (see Figure 5).

$$p = \gamma h \tag{5}$$

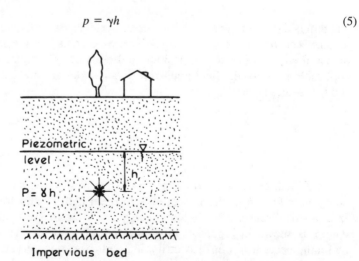

Figure 5 Hydrostatic pressure.

The total head, H, or potentials above a datum is expressed after discarding velocity heads from Equation 4 as

$$H = \frac{p}{\gamma} + z \quad (6)$$

It is obvious from this expression that for a given point the hydraulic head increases with the increase of pressure on that point.

V. PIEZOMETRIC SURFACE

So far we have restricted our attention to hydraulic head on a single vertical within the flow domain. In order to expand the discussion into a multitude of points it is necessary and sufficient first to explore the relative positions of hydraulic heads at two distinct verticals. Any specific line or curve that connects two heads is referred to as the piezometric line or level. The first question is whether the two heads should be equal to each other. Equivalence means that the energies at these two points are equal and hence there is no energy loss. This situation may occur on two occasions only:

1. If the fluid is ideal there is friction neither among the fluid particles nor between the fluid and its container even though the fluid is in motion. This is an ideal theoretical simplification that never occurs in nature.
2. The groundwater is a real liquid and therefore friction occurs within the water as well as between the water and medium; only in motion does it lose energy in the form of heat. The heads at two points are equal provided that the water is at rest. Further, equivalence of heads implies that the piezometric line is horizontal.

If one of the above points does not occur then there is always a difference in heads and the groundwater moves in the direction of decreasing head or potential.

If three or more points are considered then their hydraulic heads define a surface referred to as the piezometric surface. This surface represents the areal variation of the hydraulic head of an aquifer and is defined by the levels to which water will rise in wells. The geometric shape of this surface defines the regional groundwater movement directions completely. In unconfined aquifers the hydraulic heads coincide with the actual groundwater surface and therefore it is called groundwater table.

Similar to the earth's surface, the piezometric surface has its own topographic features but they vary with time, depending on recharge and discharge situations. In confined aquifers it is a conceptual surface that does not exist physically. However, it is important to remember that it is not the level to which the water would rise if it were free to do so (Chapman, 1981), but, as mentioned above, it is the locus of all points to which water would rise in a manometer (hole or pipe) inserted into the aquifer at any point. The practical implications of the piezometric surface will be discussed further in Section VIII.

VI. HYDRAULIC GRADIENT

The relationship between any two hydraulic heads at cross sections A, B, and C can be written for uniform and homogeneous medium (see Figure 6a):

$$H_A = H_B + (\Delta H)_{AB} \quad (7)$$

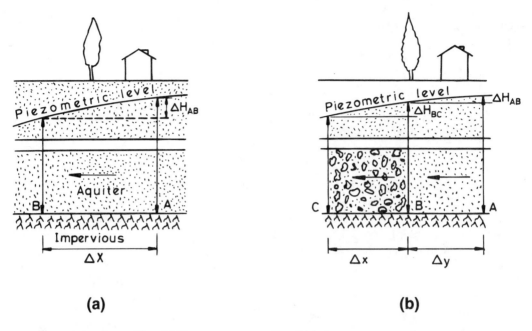

Figure 6 Hydraulic gradient. (a) Homogeneous aquifer; (b) heterogeneous aquifer.

or for an anisotropic medium shown in Figure 6b as

$$H_A = H_B + (\Delta H)_{AB}$$
$$H_B = H_C + (\Delta H)_{BC} \quad (8)$$

or

$$H_A = H_C + (\Delta H)_{AC}$$

in which the total head loss, $(\Delta H)_{AC}$, is the summations of head losses in different facies, i.e.,

$$(\Delta H)_{AB} + (\Delta H)_{BC}.$$

Geological, topographical, and hydrological features determine the regional variations in the piezometric level and hydraulic head changes along different directions. The change in the head per unit horizontal distance is the hydraulic gradient. In general, it indicates the slope between hydraulic heads at two verticals. By definition hydraulic gradient, i, is expressed as

$$i = \frac{(\Delta H)_{AB}}{\Delta x} = \frac{H_A - H_B}{\Delta x} \quad (9)$$

where Δx is the horizontal distance between points A and B. Indeed, the hydraulic gradient is the ratio of hydraulic head rise (or fall) to the horizontal distance and it is a dimensionless value. However, its physical implication is the rate of potential energy dissipation per unit weight of fluid. Hence, the greater the hydraulic gradient the greater is the flow velocity and consequently the greater is the expected energy loss. The difference in the slopes of hydraulic head straight lines in Figure 6b is due to geological facies change. If there is no facies change between two points, then the hydraulic head changes linearly between these points. A practical question is whether the piezometric line is dependent on the surface or subsurface topography.

GROUNDWATER FLOW PROPERTIES

The answer is that it is a function of neither of these topographies. Figure 7a shows the cross section of an alluvial fill in a subsurface depression filled with Quaternary deposits. There, the groundwater is at rest and therefore the piezometric line is horizontal which indicates that there is no relationship whatsoever to the aquifer bed or earth surface topography. Anticline and syncline are continuous parts of any geological formation such as the one depicted in Figure 7b.

The hydraulic gradients are generally comparatively high in recharge areas close to influent streams over deep water tables, along the banks of streams, and in extensively pumped areas. The gradients are low in groundwater discharge areas, valley bottoms, and extensive plains. Groundwater movement in the sandstone formation is possible when the piezometric level along the formation decreases (or increases) continuously. The groundwater recharge in high lands of surface topography causes the water table to rise with respect to the wadi and hence

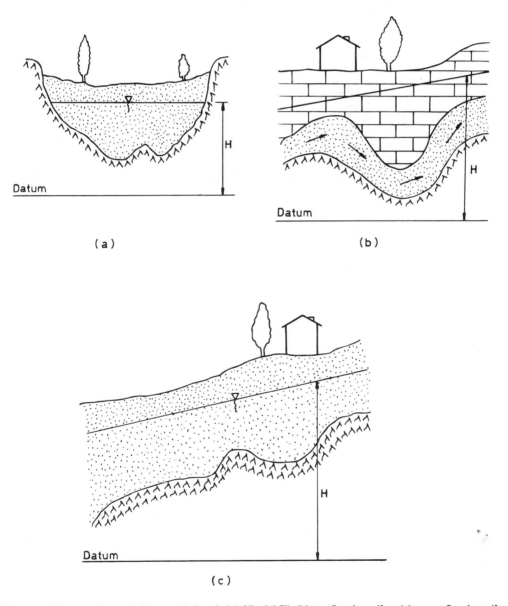

Figure 7 Topography and piezometric level. (a) Alluvial fill; (b) confined aquifer; (c) unconfined aquifer.

the flow takes place from high lands toward low lands. Figure 7c shows that the bed rock topography is immaterial so far as the flow direction is concerned. On the contrary, surface flow is completely dependent on the geomorphology and topography.

Another question regarding the position of piezometric level is whether the geological structures affect the piezometric surface. If the geological structure is discontinuous, including faults, fractures, joints, dikes, solution cavities, etc. (but not anticlines and synclines), their impact is appreciated directly on the piezometric line. Figure 8 represents some of the common situations encountered in groundwater studies. In the discontinuity points the value of hydraulic gradient is theoretically very large, i.e., $i \to \infty$.

The springs are located at the end points of piezometric lines or they are at the crossroads between the earth surface and piezometric level (Figure 9).

Finally, it is worth remembering that the hydraulic gradient definition is not dependent on the spatial variations only. Simultaneously, time changes in the piezometric level may take

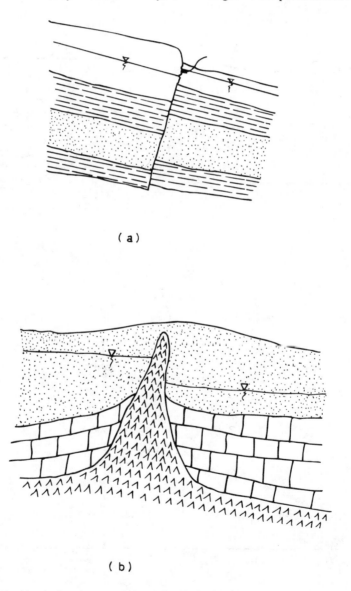

Figure 8 Discontinuities in the piezometric level. (a) Fault; (b) dome.

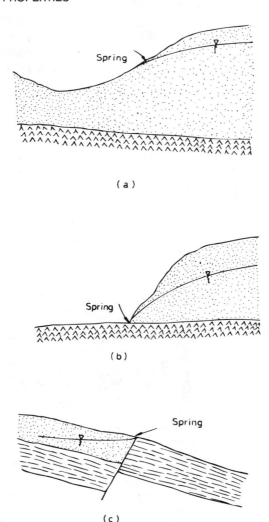

Figure 9 Types of springs. (a) Depression; (b) contact; (c) fault.

place that are not accounted by the hydraulic gradient at all. Similarly, it is possible to define timewise hydraulic gradient provided that at least two measurements, $(H_A)_t$ and $(H_A)_{t+\Delta t}$ of the hydraulic head are made Δt time apart. Hence, the timewise hydraulic gradient, $i_{\Delta t}$ is

$$i_{\Delta t} = \frac{(H_A)_t - (H_A)_{t+\Delta t}}{\Delta t} \qquad (10)$$

This concept is useful in classifying the groundwater movement according to time as will be discussed in Section X.

VII. DISCHARGE AND SPECIFIC DISCHARGE

Above-mentioned discussions give intitutive impression that either increase in hydraulic gradient and/or decrease in the medium resistance leads to an increase in the groundwater velocity. The relationship between the velocity, hydraulic gradient, and the resistance is discussed later in this chapter.

Physically, the velocity is equal to the distance over time required to cover this distance. In groundwater studies there are three different velocity types. First, the real velocity, v_r, is defined as the ratio of flow line path length over time as shown in Figure 10. By definition,

$$v_r = \frac{L}{t} \tag{11}$$

in which L is the real length of flow line and t is the time required for a water particle to travel from section A to B. The tortuosity of medium gives a random behavior to streamline. In addition, at any cross section the distribution of void through which water actually flows is also random. This double effect of randomness makes the real velocity measurements rather impossible. Accordingly, the real velocity is very difficult to calculate in the practical applications and therefore it has an academic significance only.

Second, the actual velocity, which is the ratio of horizontal distance, D, between two sections over the time required by any water particle to travel this distance, is illustrated in

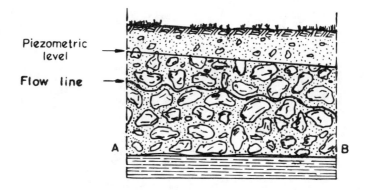

Figure 10 Real groundwater velocity.

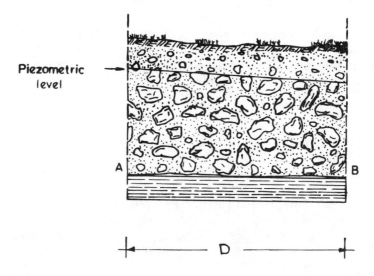

Figure 11 Actual velocity.

GROUNDWATER FLOW PROPERTIES

Figure 11. Hence the expression of the actual velocity, v_a, is

$$v_a = \frac{D}{t} \tag{12}$$

The straight line between these two sections over the measured time gives an approximation of the actual groundwater velocity. However, repeated applications of such methods are not possible in practice.

In order to overcome difficulties in groundwater velocity calculations, and to have practically applicable measurements, one should think of expressing the velocity in terms of easily realizable dimensions of the medium. To this end, let us multiply the nominator and denominator of Equation 12 by the cross-section area, A, of the medium perpendicular to the flow direction.

$$v_a = \frac{AD}{At} \tag{13}$$

where the nominator is equal to the volume, $V = AD$, of the water past and its ratio to time is referred to as the discharge, Q,

$$Q = \frac{V}{t} \tag{14}$$

Substitution of Equation 14 into Equation 13 gives the hydrologic definition of velocity which will be called from now on "specific discharge", "filter velocity," or "Darcy velocity" and denoted by q (see Figure 12)

$$q = \frac{Q}{A} \tag{15}$$

In specific discharge definition water is assumed to cross the whole aquifer section as if there were no grains. Therefore, q is not real velocity but a fictitious one that has practical value rather than physical merit. It can be related to the real groundwater velocity, v_r, by considering velocity definition in terms of discharge as

$$v_r = \frac{Q}{A_v} \tag{16}$$

where A_v is the area of voids in the same cross section. Elimination of Q between Equations

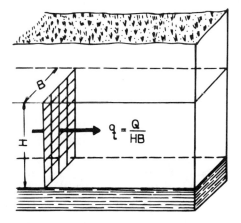

Figure 12 Specific discharge.

15 and 16 leads to

$$q = v_r \frac{A_v}{A} \tag{17}$$

The ratio term on the r.h.s. varies between zero and one and is equal to the porosity, n, of the medium. Hence, finally,

$$q = nv_r \tag{18}$$

The significant interpretations that result from this equation are that (1) the specific yield (filter velocity) is always smaller than the real velocity and (2) the greater the porosity the closer is the specific discharge to real velocity.

VIII. FLOW NETS

Knowledge of piezometric surface gives full-scale information about the energy losses in the flow domain. This surface provides information about the direction of flow and stagnant zones. In practical applications it is important to find ways of describing the complete picture of the piezometric surface. Because it is an irregular surface similar to topographic maps, the easiest way to describe it is to map this surface. Then comes the question of lines to describe the surface. What type of lines should be adopted in the mapping? From the groundwater point of view, there are two general lines and/or zones, namely, those along which there is no flow (i.e., constant energy) and otherwise. A simple analogy is that, for instance, if somebody tries to descend from a mountain, there are two paths for him; either he moves through the path of the same elevation or moves from a high elevation to lower one. In the first alternative his potential energy remains the same since he remains at the same elevation whether he moves or not. However, in the second way he loses energy but descends continuously. In descending, he may have many paths steep or smooth but the best one among them is the path along which he moves minimally while spending the same energy. Hence, such a path is perpendicular to the constant elevation line. The water molecule movement under the potential energy is exactly the same as the mountaineer's walk to descend the mountain. By remaining at the same elevation the mountaineer transverses the contour line of a certain elevation (locus of equal elevation points). In the same way a water particle having the same potential energy gives rise to equal hydraulic head lines which are referred to as the equipotential lines. Like the contour lines the equipotential lines cannot intersect each other. Contour lines drawn at convenient intervals are enough to describe the topographic surfaces and similarly equipotential lines are the necessary and sufficient lines to represent piezometric surface completely. According to a mountaineer, movement means essentially crossings from one contour line to another. If the mountain surface gave the opportunity at equal levels he should prefer the path that leads to the next contour in the shortest way. This is approximately tantamount to saying that walking paths are more or less perpendicular to the contours; in fact, ideally they should have this property in order to have minimum work. In light of this discussion in a similar manner the water molecule prefers the minimum work path that is perpendicular to the equipotential lines in a homogeneous and isotropic medium. It is obvious that another set of lines for water movement can be drawn on the same map showing the flow lines which are also referred to as the streamlines. The existence of two sets of lines, namely, equipotential and stream lines, brings out a net which is referred to as the flow net. Hydraulic head field measurements for flow net construction will be discussed in Chapter 7. None of the lines can have corners but they are smooth traces only.

In a flow net, streamlines can be regarded as imaginary impervious barriers because there is no cross-flow through them. In practice, once the flow net is constructed the following

GROUNDWATER FLOW PROPERTIES

geological and/or hydrological information can be obtained. Interpretation of water table or piezometric contour maps provides useful information to identify areal variations in transmissivity and thickness of aquifers, and to quantify inflow and outflow through aquifers.

1. In the case of unconfined aquifers, the water table separates the saturated zone from the capillary water zone. Its fluctuations help to calculate additions (recharges) and subtractions (discharges) from the aquifer storage.
2. The slope of piezometric surface is related to the rate of percolation and/or hydraulic conductivity of the aquifer. In fact, the slope varies directly with the velocity, and inversely with the ease of movement which will be called permeability, as will be explained in the next chapter. This statement is very fundamental in the groundwater map interpretations.
3. The piezometric level intersection with the earth's surface delimits the surfaces subject to influent and effluent seepage as well as springs as shown in Figure 13. Also, permanently moistured areas such as swamps indicate the intersection of the piezometric level with the earth surface.
4. Geologically in confined aquifers water table separates the belt of weathering, oxidation, rock decomposition, and the section from the underlying belt of mineral precipitation and rock cementation.
5. Hydraulic heads can be known for any desired point within the flow net by interpolation.
6. Hydraulic gradients can be known between the two successive contour lines and change of hydraulic gradient between any two points can be seen by taking cross sections (see Figure 14).

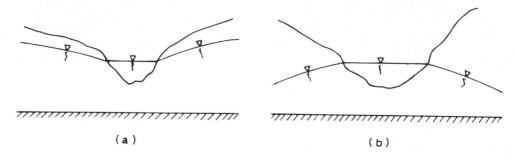

Figure 13 Seepage. (a) Effluent; (b) influent.

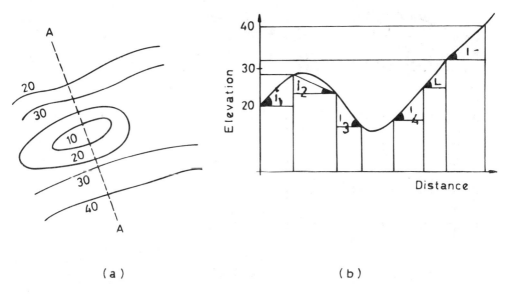

Figure 14 Hydraulic gradient from flow net. (a) Plan view; (b) cross section.

7. Closed equipotential lines indicate either recharge or discharge zones (see Figure 15).
8. Convergence of equipotential lines indicates difficulty in the water movement within the media whereas divergence means easier groundwater movement (see Figure 16). When an aquifer is known to be uniform in thickness, areas in which the equipotential lines are widely spread will show that the aquifer is more permeable. On the other hand, if the aquifer is known to be uniformly permeable, wider spacing may be due to thickening of the aquifer.
9. Discontinuity of equipotential lines corresponds to geological discontinuities such as faults, dikes, fractures, etc. (see Figure 17).
10. Existence of two or more equal value equipotential lines indicates stagnant water zones (see Figure 18).
11. Divergence and then reconvergence of flow lines show the zones of impervious lenses (see Figure 19).
12. Parallel equipotential lines imply parallel streamflow lines and hence parallel (one-dimensional) flow (see Figure 20).
13. Highest points imply a groundwater divide line as in Figure 21.
14. Refraction of streamlines implies the change in the geological lithology (see Figure 22).
15. The groundwater direction at any point is tangental (perpendicular) to the streamline (equipotential line) at the same point as in Figure 23.

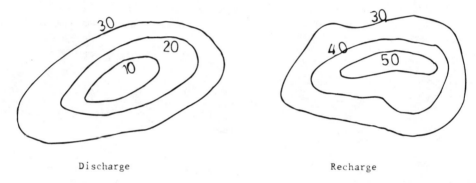

Figure 15 (a) Recharge zones and (b) discharge zones.

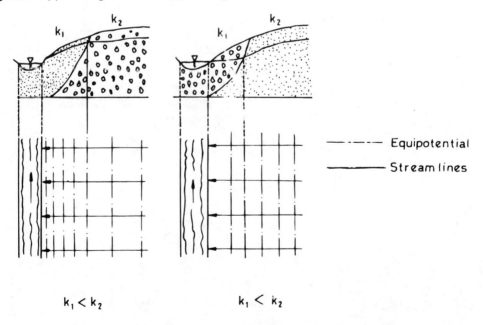

Figure 16 (a) Flow with ease and (b) flow with difficulty.

GROUNDWATER FLOW PROPERTIES

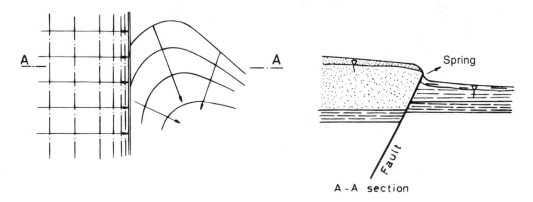

Figure 17 Fault zone.

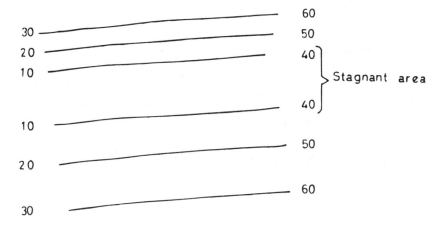

Figure 18 Stagnant water zone.

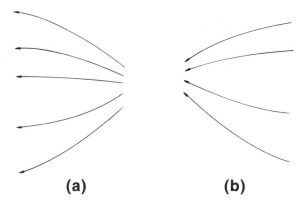

Figure 19 (a) Divergent and (b) convergent flow lines.

16. The aquifer area between the two streamlines is referred to as the streamtube and the tube walls can be considered as impervious boundaries since there is no cross flow through these boundaries.

In impervious fractured rocks the water table is the surface at contact with the water body in the fractures and the aeration zone. In fact compared to the porous medium, there is no

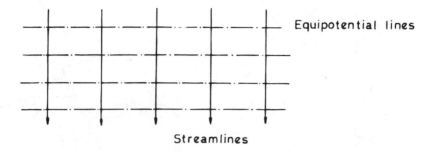

Figure 20 Parallel flow net.

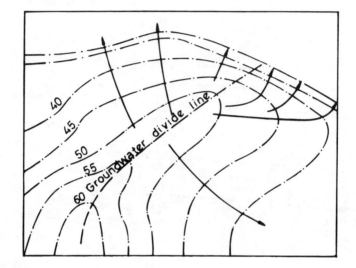

Figure 21 Groundwater divide line.

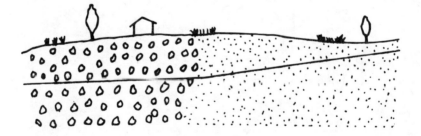

Figure 22 Change in the material.

continuous water table but it is interrupted by the impervious rock blocks among fractures and constitutes only a small portion of the surface projected through the water table-air contact in the network of fractures as shown in Figure 24. The groundwater fills the fractures in an irregular network consisting of intersecting tabular members with tabular enlargements at the intersections. The hydraulic gradients exist along the fractures but gradients must not be calculated normal to the fractures with impermeable walls.

On the other hand, pervious fractured rocks such as sandstones show continuity of the water table at the equilibrium state. The movement of water is easier in fractures than the remaining rock body and therefore water withdrawal receives immediate response from the

GROUNDWATER FLOW PROPERTIES

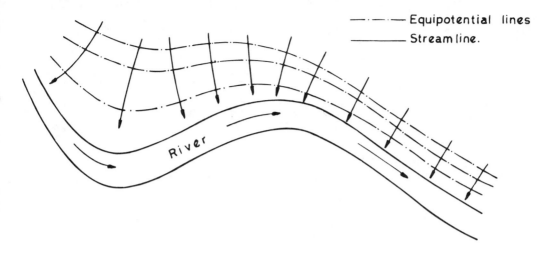

Figure 23 Groundwater flow direction.

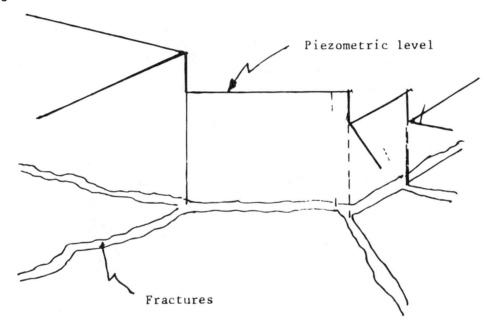

Figure 24 Water table in impervious fractured rocks.

fractures. The response from blocks is comparatively slow. Within the rock mass there are two coexisting water table patterns that are mutually exclusive but complementary as in Figure 25. In fact, the irregular fracture water table is surrounded by water table patches within the rock fragments.

The piezometric surface in karstic media is composed of mutual and complementary water and piezometric surface patches as shown in Figure 26.

IX. SPACE CLASSIFICATION OF GROUNDWATER

In general, any groundwater particle moves in three dimensions within the lithosphere at any point just as an airplane in the sky. Accordingly, it will have quantities such as the specific

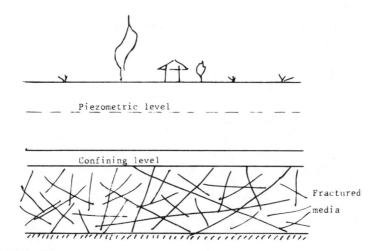

Figure 25 Water table in pervious fractured rocks.

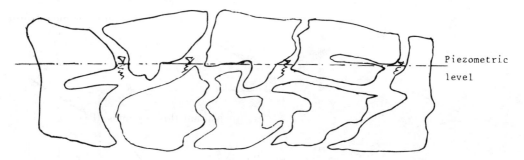

Figure 26 Piezometric level in karstic media.

discharge which can have three components: longitudinal, lateral, and vertical. Any physical quantity that can be measured and mapped in three-dimensional space is said to have a field of that quantity. So far as the groundwater is concerned, the most important quantity is its velocity; therefore, one can mention the groundwater velocity field. Other important fields in the groundwater are the gravity, the potential, and the pressure fields. These fields are not independent from each other but all are interrelated and cause groundwater movement.

Our main interest herein is in the groundwater velocity field. Depending on the number of velocity components according to the problem at hand, the groundwater flow may be three, two, or one dimensional. Similarly, if the streamlines are straight lines, planar curves, or space curves, the groundwater flow is said to be one-, two-, or three-dimensional, respectively. Prior to attempting to solve any groundwater problem its flow dimension should be known or assumed. Of course, as expected, the smaller the dimension, the easier will be the solution of the problem at hand. Space classification is concerned with the geometry of flow lines. The complete picture about the groundwater movement can be obtained quantitatively by means of the flow nets in a plane in addition to flow nets in cross sections along various directions. This leads to three-dimensional visualization of the groundwater movement in nature.

A. THREE-DIMENSIONAL FLOW

By definition it is the groundwater flow that evolves in such a way that the specific discharge at any point has three components perpendicular to each other as shown in

GROUNDWATER FLOW PROPERTIES

Figure 27. Three-dimensional flow appears in nature either statically from the geological setup or dynamically from the hydrological effects or artificially from man-made structures. In a multi-aquifer system differences in water transmission properties of each layer lead to water exchange between the layers and hence three-dimensional flow occurs as shown in Figure 28.

Among the hydrological reasons of three-dimensional flow are the recharge and discharge as in Figure 29. Human activities also give rise to three-dimensional flow, as shown in Figure 30.

The simplest form of three-dimensional flow is the spherical flow of nonpenetrating well into a semiinfinite aquifer. This type of flow will be referred to as the spherical flow (see Figure 31). Existence of natural depressions such as lakes, rivers, wadis, etc., in an unconfined aquifer give rise to regional flow patterns. These depressions are subject to recharge from rainfall so the conditions change with the season during the year. In wet seasons the free water surface on both sides of the stream has higher elevations than the stream as shown in Figure 32. In this figure the water in the river has the lowest hydraulic head and therefore there is an equipotential surface close to the river bed and the flow is almost radial to the river. The highest hydraulic heads are at points on the free water surface beneath the hills. These are the groundwater divide points where the piezometric surfaces are horizontal with vertical flow direction. In addition, at these points the direction of flow changes from flow to the river to flow to the adjacent catchments. Around the water divide points the piezometric surfaces are concave upward similar to the flow under the river.

B. TWO-DIMENSIONAL FLOW

Most of the practically interesting groundwater flows are either two dimensional or may be fairly well approximated by one of the two-dimensional flow systems. Two-dimensional flows are expressible in terms of piezometric maps. Maps do not show the vertical component of the flow that is negligible in most of the cases. Flows toward fully penetrating wells in uniform thickness porous medium aquifers are two dimensional. In the confined aquifers the flow is perfectly two dimensional for fully penetrating wells. With vertical component of flow being equal to zero, Figure 33 represents in the vertical cross sections and plan view of the flow lines for these types of flow. Flow toward point sink or finite diameter circular cross-sectional well is referred to as the radial flow which constitutes the simplest form of two-dimensional well problems.

In radial flow, equipotential lines are circles whereas the flow lines are straight lines. Furthermore, flow toward a fully penetrating finite length vertical fracture has elliptical equipotential lines and the flow is elliptical (see Figure 34). In an unconfined aquifer not only

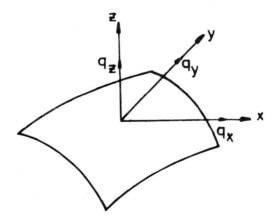

Figure 27 Rises and falls in piezometric level.

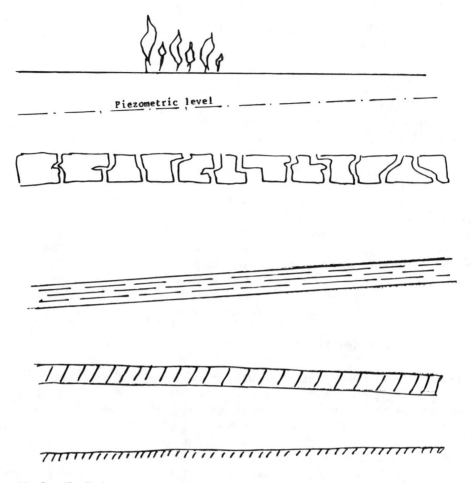

Figure 28 Specific discharge components.

is the flow three dimensional near the well, but also the area of concentric cylinders across which the water flows decreases toward the well, that is, the saturation thickness is not constant. However, if the drawdown in the well is negligibly smaller than the saturation thickness, then the flow may be assumed as two dimensional.

It is clear from these discussions that the flow net depends not only on the geology but at times completely on the man-made structures such as wells. This is one of the main reasons why in designing a new well one should try to have shapes that give rise to simple flow nets.

The second type of two-dimensional flow problems are related to the fractures where apertures are comparatively very small compared with two other dimensions.

C. ONE-DIMENSIONAL FLOW

The simplest form of flow geometry is concerned with one-dimensional flow situations, which do not very often arise in practice. The groundwater flow in the alluvium fills of valleys has only the longitudinal velocity component and therefore it is treated as a one-dimensional

GROUNDWATER FLOW PROPERTIES

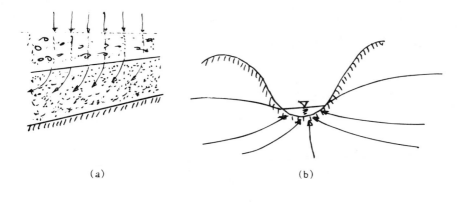

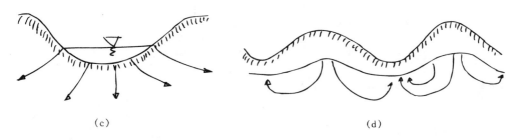

Figure 29 Multiaquifer flow.

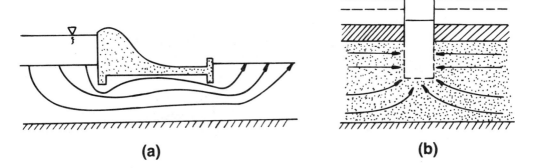

Figure 30 Groundwater seepage due to (a) dam; (b) well.

flow. Extensively long major fractures or man-made slots cause the groundwater in the host-permeable rock to flow perpendicularly to it.

X. TIME CLASSIFICATION OF FLOW

The basic time-dependent property of flow is the groundwater velocity (specific discharge) and it is a vector quantity that has direction as well as magnitude. The flow nets provide information about the direction and magnitude of specific discharge. The tangent of streamline at any point shows the specific discharge direction at this point. On the other hand, the relative position of two successive equipotential lines indicates the direction and it is proportional to the magnitude of specific discharge.

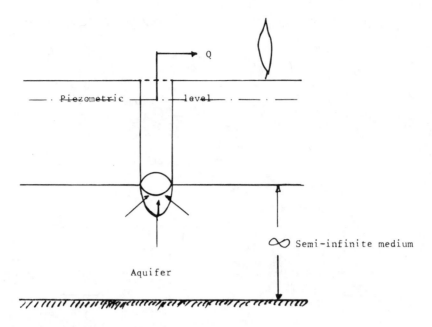

Figure 31 Human interference.

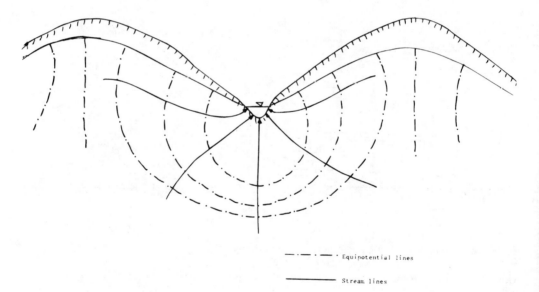

Figure 32 Spheric groundwater flow.

The flow net changes with time due to the effects of hydrologic cycle or human activities and as a result the flow velocity changes its direction and the magnitude with time. Change of streamlines only means change in the specific discharge direction but not the magnitude. However, the magnitude will be altered by changes in equipotential lines without any change in the direction (see Figure 35). Most often in nature streamlines as well as equipotential lines change with time to a certain extent. Due to the involvement of time the flow nets are valid only for the time instances for which they are drawn.

GROUNDWATER FLOW PROPERTIES

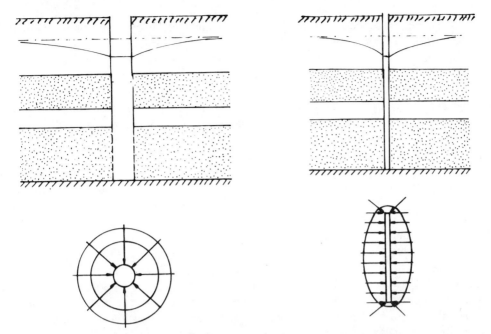

Figure 33 Regional groundwater flow.

Figure 34 Two-dimensional flow in confined aquifer.

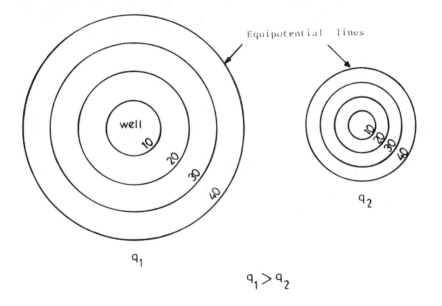

$q_1 > q_2$

Figure 35 Elliptical flow lines.

A. STEADY-STATE FLOW

If at any point neither the direction nor the magnitude of specific discharge changes with time, the flow is said to have a steady state and the flow net is constant throughout a certain time period. The groundwater level does not change with time. This time period may be long or short, depending on the circumstances. Constancy of flow net implies fixed hydraulic gradients and thickness of the saturation zone. Subsequently, at any cross section the hydraulic gradient, cross-sectional area, and the velocities do not change with time, which suggests that the discharge is constant. If we expand the discussion to two different cross sections with their respective constant discharges it is not difficult to conclude that these discharges are equal to each other. For any difference, there should be either an increase or a decrease in the storage and likewise in the equipotential lines. This is a very important conclusion, which proves that, in the steady-state case, input discharge to a control section is equal to output discharge. On the other hand, such a situation is possible only when there is a continuous supply to the aquifer equal to the output discharge from the same aquifer. Furthermore, in the steady-state case the groundwater level does not change with time. It is rather difficult to observe the aforementioned conditions naturally so as to have a steady-state flow situation. However, examples in Figure 36a, b, and c are illuminative for steady state. Figure 36a shows a circular island that is surrounded by an inexhaustible water source. If a constant amount of discharge, Q, is pumped out after some time, the whole of this amount comes from the immense water source and the groundwater level does not change. Therefore, the flow attains a steady-state case. In Figure 36b a leaky aquifer gets its water from the overlying aquifer through aquitard. If water is pumped out from the lower aquifer, after some time all of it comes from the above aquifer through leakage and then the piezometric level of the pumped aquifer remains the same with time. In fact, this is the unique realistic case where the flow toward wells is in steady state. Figure 36c shows infiltration to an unconfined aquifer and abstraction of water from it. When the abstracted water amount becomes equal to the infiltration amount over the aquifer then the flow will be in steady state. Finally, Figure 36d indicates a contribution of surface flow body to the groundwater provided that the water abstracted from the aquifer is equal to the recharge from the surface body of water; hence, the groundwater level in the well does not change with time and therefore a steady-state flow prevails in the aquifer.

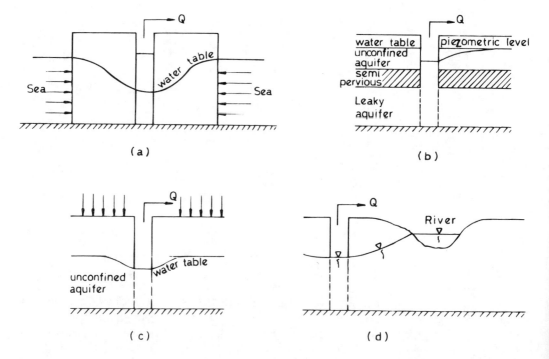

Figure 36 Steady-state flow. (a) Circular island; (b) leaky aquifer; (c) surface infiltration; (d) surface water.

GROUNDWATER FLOW PROPERTIES

As will be noted from the above-mentioned examples, the steady state predominates as long as the equality of input and output waters into the aquifer continues. However, such a condition is difficult to be met in nature. Chapter 8 is concerned with the analytical solutions of groundwater problems in steady-state flow that are easier to tackle than the unsteady-state flow cases.

B. UNSTEADY-STATE FLOW

Contrary to what is discussed in the previous section the unsteady flow has the following properties:

1. In a flow net either the streamline or equipotential line or both change their position with the passage of time.
2. The discharges at different sections have different values.
3. Both of the previous points imply that the specific discharge magnitude and/or direction change with time.
4. The groundwater level fluctuates with time.
5. There is not always continuous external recharge but even if there is it may not be enough to balance the amount of water abstracted from the aquifer.
6. In the unsteady flow case changes occur in the storage within the aquifer. The water movement alone is not enough to represent the groundwater flow; in addition to this, the storage capacity should be known. There are numerous unsteady-state flows that occur in nature every day. Few of the practical examples are shown in Figure 37.

C. QUASI-STEADY-STATE FLOW

As stated before, it is almost impossible to encounter steady-state flows in nature. In groundwater hydraulics transient steady-state flow is also used synonomously with the quasi-

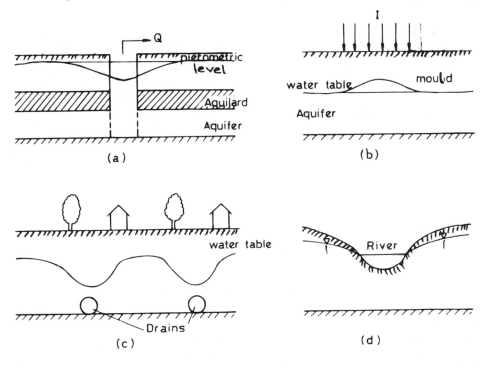

Figure 37 Unsteady-state flow. (a) Confined aquifer; (b) groundwater mould; (c) groundwater drainage; (d) recharge to a river.

steady state flow. However, some of the unsteady-state flows can be considered as quasi-steady state which is an approximation of the steady-state flow. Such terminology is very useful for practical purposes. The groundwater problems that fall within the quasi-steady-state domain can be solved similar to the steady-state case, which are mathematically easier to tackle as will be shown in Chapter 8. In this book, quasi-steady state has been defined as the situation where variations of groundwater level with time are negligible. Field conditions may require considerably long periods to reach the steady-state flow. Such long periods are not always required because quasi-steady-state flow may be reached earlier.

D. PSEUDO-STEADY STATE FLOW

This terminology is used for the groundwater flow from matrix blocks to fractures in hard and soft rocks. Its elaboration will be presented in Chapter 10.

XI. FORCE CLASSIFICATION OF FLOW

So far in our discussions we have not concentrated on the forces that cause the water to move. In fact, for any object to move, there must be a force exerted on it and accordingly during the movement there appears to be retarding forces. The flow of water in an aquifer is governed mainly by two principle forces: the component of the force of gravity along the piezometric surface tending to accelerate the water and the friction forces between the water and the material of the aquifer medium. In addition to friction there are inertial and adhesion forces that try to retard the water movement. The magnitudes of retarding forces are functions of the specific discharge, hydraulic gradient, and medium characteristics such as the grain size, fracture aperture and frequency, and solution cavity sizes and shapes. Dynamic classification of flow depends on the ratio of retarding forces to the accelerating forces which is equivalent to the gravitational forces. In groundwater studies only the friction and the inertial forces are important as resistances to the motion. Invariably, at the beginning of any groundwater movement, friction forces play the major role as retarding forces and with the acceleration of water body they leave the role to the inertial forces. However, the transition from the frictional to the inertial forces is not an abrupt phenomenon but there is a transitional zone where they both contribute as a retarding force at different proportions.

On the basis of force ratios all water flow falls into one of the two major categories as laminar flow or turbulent flow. Such a distinction is very important for groundwater resource evaluations because of the fact that it reflects to a significant extent the behavior of the aquifer medium. We will discuss in the following two sections their individual definitions, physical appearances, and implications in groundwater movement.

A. LAMINAR FLOW

Generally, in the initial phases of any groundwater flow phenomenon the type of flow is said to be laminar. The word "laminar" implies that the flow takes place as a multilayer flow with no cross flow between the neighboring layers, (Figure 38). This is in fact the general definition of the laminar flow. Physically, for flow to occur, the water must overcome the retarding force which is the friction between the water molecules due to its viscosity as well as between the water and the aquifer medium. Because the viscosity of water is small, the major retardation comes from the latter type of friction. Hence, in laminar flow the gravity and friction forces are dominant. Their ratio as the friction force to the gravity force indicates the intensity of laminar flow (Bear, 1972). The smaller this ratio the weaker is the laminar flow.

GROUNDWATER FLOW PROPERTIES

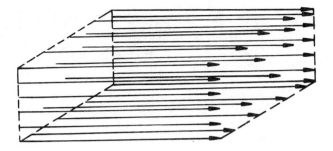

Figure 38 Laminar flow.

Practically, if the groundwater velocity and the average grain size are small, then the groundwater flow is expected to be of laminar type. In addition, small hydraulic gradients are also among indications of laminar flow. In a laminar flow groundwater moves steadily and slowly through the interconnected voids of rocks so that each water particle moves along a regular path without crossing other paths. On a macroscopic scale velocity and direction of flow are constant but on a microscopic scale both may differ from particle to particle within the pores.

B. TURBULENT FLOW

Generally, this type of flow is observed after the completion of laminar flow. The word "turbulent" means violent, disorderly, or uncontrollable, which implies that the flow takes place not in the form of layers between parallel streamlines but rather in a random pattern form with local eddies (Figure 39). In the turbulent flow the water molecules begin to rush past each other in a random fashion. In turbulent flow water particles move irregularly crossing and recrossing at random accompanied by eddy formation.

On the other hand, from a physical point of view, in the turbulent flow the friction forces become less influential and instead inertial forces eventually take their place. Hence, the ratio of inertial forces to the gravity force is an indicator for the turbulent flow. This ratio is referred to as the Froude number. The larger this number the greater is the scale of turbulence in the flow domain. Practically, turbulent flow occurs at high velocities in coarse-grained porous medium or in the fractured or karstic media. Also, areas of steep hydraulic gradient give rise to turbulent flow such as in the vicinity of a pumping well.

XII. REYNOLDS NUMBER

In the above discussions of laminar and turbulent flow we understood that the dynamic type of flow is a function of the velocity (specific discharge), ratios of accelerating to retarding

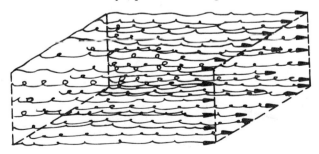

Figure 39 Turbulent flow network.

forces, medium characteristics, and the water properties. The Reynolds number relates all of these variables in one decision variable so as to determine whether the flow will be laminar or turbulent. The Reynolds number, Re, is a dimensionless number that expresses the ratio of inertial to friction forces during flow. It is defined as

$$Re = \frac{q\gamma d}{\mu} \qquad (19)$$

where γ and μ are the density and dynamic viscosity of water, respectively, q and d are the specific discharge and a representative length of the porous medium which is taken as a mean pore dimension or in the practical studies as a mean grain diameter. For groundwater $\gamma = 0.999 \times 10^3$ kg/m³ and $\mu = 1.15 \times 10^{-3}$ pa-sec and the substitution of these values into Equation 19 gives

$$Re = 8.7 \times 10^5 qd \qquad (20)$$

The graphic representation of this expression is presented in Figure 40 for different specific discharge valves.

XIII. DEPRESSION CONE

When groundwater is at rest the hydraulic head levels within the well and the aquifer are equal and piezometric level is horizontal. Abstraction of water from the well lowers the piezometric surface around it and a hydraulic gradient is established, resulting in a convergent radial flow toward the well. The piezometric surface around the well assumes the shape of an inverted cone as shown in Figure 41. The well is at the apex of the cone and the base, conforming to the original piezometric surface at some radial distance. The cone of depression is defined by Theis (1938) as "the geometric solid included between the water table or the piezometric surface after a well has been discharging and the hypothetical position of water table of piezometric surface would have had if there had been no discharge by the well." The height of the depression cone at any radial point is called drawdown at this point. The base area of the depression cone defines the "area of influence" and the radius of the base is

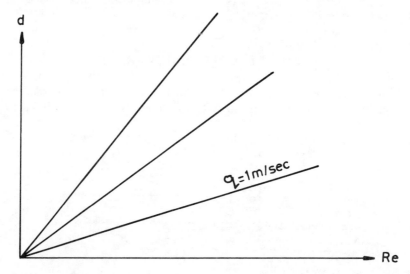

Figure 40 Re-d relationship.

GROUNDWATER FLOW PROPERTIES

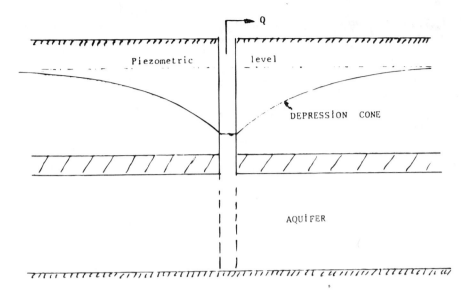

Figure 41 Depression cone.

referred to as the "radius of influence". In homogeneous, horizontal, isotropic, and uniform aquifers the base has a circular shape.

As the water abstraction continues the depression cone keeps expanding indefinitely if the withdrawal is not compensated fully by recharge to the aquifer. Recharge is possible by one of the following situations.

1. Recharge from the rainfall, irrigation, or other sources within the area of influence.
2. Intercepting surface water sources.
3. Adequate natural discharge from the aquifer itself.
4. Vertical leakage of water from adjacent aquifer or aquiclude. Existence of these situations leads to steady-state groundwater flow after prolonged pumping; otherwise, the flow is said to be in quasi-steady state at long times.

In confined aquifers the depression cone does not have any physical meaning but imaginary geometric shape only. However, in unconfined aquifers it is called equivalently "dewatered zone" which gives rise to delayed yield within the aquifer as will be explained in Chapter 9.

The depression cone extension depends on the type of aquifer, its transmissivity, the pumping duration, the discharge rate, the length of the well screen, and whether the aquifer is stratified or fractured. In confined aquifers with abstraction of water through the well the hydraulic head loss (drawdown) propagates rapidly due to the compressibility of the aquifer material and water. The drawdowns are observable at several hundred meters. On the contrary, in unconfined aquifers, the drawdown increments are relatively smaller and the drawdowns are appreciable only at the well vicinity within about 100 m. In leaky aquifers the rate of drawdown expansion will be in between the confined and unconfined aquifers.

The longer the continuous water abstraction duration, the more the cone of depression will continue to expand. If the discharge rate is low, the depression cone will be shallower and smaller than if the discharge is high. The change of hydraulic properties along short distances in heterogeneous aquifers requires more piezometers than porous medium.

REFERENCES

Bear, J., 1972. *Dynamics of Fluids in Porous Media*. Elsevier, New York.
Chapman, R. E., 1981. *Geology and Water. An Introduction to Fluid Mechanics for Geologists*, Dr. W. Junk, The Hague.

Dennis, J. G., 1972. *Structural Geology.* Ronald Press, New York.

Lomize, G. M., 1951. *Flow in Fractured Rocks,* Gesenergoizdat, Moscow (in Russian).

Louis, C., 1969. A Study of Groundwater Flow in Jointed Rock and Its Influence on the Stability of Rock Masses, Rock Mech. Res. Rep., 10, Imperial College, London.

Regan, L. J. and Hughes, A. W., 1949. Fractured reservoirs of Santa Maria District, California, *Am. Assoc. Petroleum Geol. Bull.,* 33, 31.

Snow, D. T., 1965. A Parallel Plate Model of Fractured Permeable Media, Ph.D. thesis, University of California, Berkeley.

Snow, D. T., 1968. Rock fracture spacing openings and porosities, *Am. Soc. Civil Eng. Proc., J. Soil Mech. Found. Div.,* 94(1), 73.

Stearns, D. W. and Friedman, M., 1972. Reservoirs in fractured rock, *Am. Assoc. Petroleum Geol.,* Reprint Series No. 21, 1972.

Theis, C. V., 1938. The significance and nature of the cone of depression in ground water bodies, *Econ. Geol.,* 33, 889.

Witherspoon, P. A., Wang, J. S. Y., Iwai, K., and Gale, J. E., 1980. Validity of cubic law for fluid flow in a deformable rock fracture, *Water Resour. Res.,* 16(16), 1016.

CHAPTER 4

Aquifer Properties

I. GENERAL

Quantitative evaluation of groundwater potential is possible only after a set of basic parameters is properly defined and physically related to some aquifer properties. Reliable interpretations and conclusions about the whole aquifer performance in an area are possible, provided that the accurate determination of basic parameter values is achieved. In addition to temporal variability of groundwater occurrence, distribution, and movement, there is distinctive spatial variability due to the geological composition. Consequently, the basic parameters will also be functions of space and time. Fortunately, the parameters relating to the aquifer medium are independent of time compared to human life but they are certainly changing with geological time scales which is not significant from groundwater abstraction point of view. Hence, medium-dependent parameters do change with space and give rise to the questions of anisotropy and heterogeneity which will be described later.

On the other hand, a subset of parameters does change with time on hourly or even minutely basis. These parameters are under the continuous effect of either the hydrological cycle or human activities such as pumping, irrigation, dam construction, excavation, etc. To have a good control on these parameters, basic variables for their calculations should be monitored continuously.

It is, therefore, not difficult to imagine that the reliability in groundwater resources evaluation is a function of the reliabilities of basic parameters which constitute the fundamental data for any scientific study. Furthermore, the confidence in a groundwater expert on his conclusions and interpretations is directly proportional not only to the reliability of basic parameters but equally important to his understanding of these parameters from the geological, hydrological, and especially physical points of view.

This chapter will emphasize the definitions, physical foundations, and feasible validity ranges of basic parameters in light of geological, hydrological, and practical field discussions.

II. GRAIN SIZE DISTRIBUTION

The grain size composition of any porous medium has a major effect on its regional properties. Intuitively, the greater the grain of the same size the more will be the voids between them. Thus the water movement in coarse-grained medium is always easier than fine grained under normal conditions. Mixtures of different grain sizes at various proportions result in different void-solid ratios in the medium. From the groundwater point of view, increase in interconnected void space is very convenient but nondesirable for foundation engineering purposes. Especially in construction, engineers mix different sizes of materials to attain prespecified properties of the medium, but the groundwater specialist has to determine the properties of the medium that has already been mixed in nature due to various past geological phenomena. However, the method used for grain size analysis is the same in both cases. Gradation of grain sizes is achieved by means of mechanical analysis. The analysis is conducted through a nest of standard sieves with the coarsest on the top and the finest at the bottom covered with a lid on the top and a pan at the bottom as receiver. In order to perform the test a representative sample of porous medium, about 150 to 500 g is taken by quartering oven-dried and exact weight poured into the top sieve. The whole nest is shaken for about 5 min and the total percentage passing each sieve is then calculated and the data are plotted on a semi-logarithmic paper (Figure 1). The logarithmic horizontal axis is for the grain sizes and along the same axis three categories of grains are shown as gravel consisting of the particles that individually may be boulders, cobbles, or pebbles; sand which may be coarse, medium, or fine; and, finally, mud consisting of clay and various silts. The main benefits from a grain size distribution curve are

1. To know the proportions of different sizes of particles in the aquifer material.
2. To understand the dominant porous media whether coarse, medium, or fine grained.
3. To deduce something of the geological history of the material. In this, the very fine material is probably due to wind-deposited sand dunes; the very graded mixture of boulders, sand, and mud would be identified as a glacial till, etc.
4. To determine some design quantities such as average grain size, effective diameter, and uniformity coefficient (Driscoll, 1986).
5. To learn much about the properties of the medium and to estimate some of the parameters as will be discussed later in Chapter 9.

III. POROSITY

The simple property of a medium is the percentage of voids that may accommodate fluid. The quantitative measure of this percentage is the porosity which is defined as the ratio of volume of void space to the total volume. Mathematically, porosity, n, is defined as

$$n = \frac{V_t - V_v}{V_t} \times 100 \tag{1}$$

where V_t and V_v are the total sample and void volumes, respectively. It is obvious from

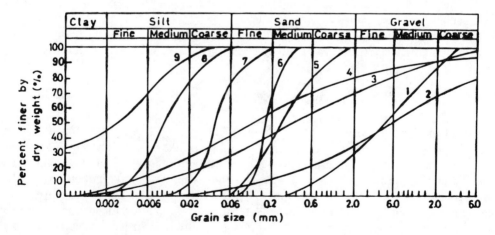

Figure 1 Grain size distribution.

Equation 1 that porosity has a value that varies between 0 and 100% physically depending on the geometrical formation of the grains. The porosity definition statement above includes "void space" which can be of two types. First, the voids among the grains and second any fracture or solution cavities within the same media. The former type of voids known as a primary opening and formed through a process known as digenesis. The latter type of voids are secondary openings and are formed either due to physical forces or tectonic movement or otherwise or as a result of chemical reactions between the rock and mining water.

Sedimentary rocks, which are formed from sediments by processes involving physical and chemical changes such as compaction, cementation, recrystallization, precipitation, etc., will be expected to have smaller primary porosity values than clastic sediments. It has been reported that their porosity varies from 3 to 30%. Of course, in a nonstable sedimentary basin due to fracturing and chemical solution, the porosity increases to a certain extent depending on the degree of fracturing and solution. Later, in the life after the formation of a rock its porosity may increase due to either weathering which weakens the rock stability or tension cracks and fractures resulting from tectonic movements, release of overburden, as well as cooling or chemical solutions, creating voids and opening up the joints. In order to distinguish between the original voids present when the rock was formed and those that were developed as a result of later processes, the former voids are referred to as the primary porosity, n_p, and the latter as the secondary porosity n_s, which are related to the total porosity as

$$n = n_p + n_s \qquad (2)$$

The maximum ranges of primary porosity occur in clastic sediments. Table 1 represents expected primary porosity values for different types of sediments.

Igneous and metamorphic rocks have very low primary porosity, about 1 to 2%. However, fracturation of crystalline rocks gives 2 to 5% porosity, (Davis, 1969). The water-bearing capacity of these rocks is therefore related almost entirely to the secondary porosity. The weathering affects usually up to 100 m depth beyond which the fractures are closed by the weight of the overlying rock. The extent of fracturing depends on the rock type, geological history, and on the climate within the same rock type. It is almost impossible to generalize water-bearing capacities of these rocks. Sharp variations are possible within short distances and therefore detailed investigations are necessary before any conclusion concerning well yield can be made. Weathering of igneous and metamorphic rocks may increase the porosity up to 30 to 60% (Steward, 1964). Among the volcanic rocks, basalt may store significant

Table 1 Expected Porosity Values

Classes of sediments	Porosity (%)
Well sorted sand or gravel	25–50
Mixed sand and gravel	20–35
Glacial till	10–20
Silt	35–50
Clay	33–60
Limous	35–50
Argiles	45–50
Marnes	47–50
Limestones	0.5–17
Craie	15–45
Calcaireoolitique	3–20
Marble	0.1–0.2
Shist	1–10
Gres	4–26
Dolomite	3–5
Granite	0.02–15
Gypsum	3–4
Basalt	0.1–12

amounts of water due to its vesicular openings and shrinkage cracks which gives rise to porosity values up to 12%.

Last but not least, the porosity gives the capability of medium to store water. Hence, the total amount of ground water, V_t, that can be stored in a porous medium reservoir is calculated roughly as

$$V_t = nV_a \qquad (3)$$

in which V_a is the volume of saturation zone within the reservoir.

IV. SPECIFIC YIELD AND RETENTION

From hydrological point of view, the porosity of any formation is a measure of its water-bearing capacity. However, all of this water cannot be drained or yielded to the surface by springs and wells. Hence, a portion of water will be held in the void space by molecular and surface tension forces. The ratio of total drainable water volume to the bulk volume of a medium is referred to as the specific yield, S_y, and in the same way retained volume percentage is known as the specific retention, S_r. Among engineers the specific yield is known as the effective porosity. Usually the groundwater specialists divide water stored in the voids into the part that will be drawn under the influence of gravity and the remaining part that is retained as a film on internal rock surfaces or kept in very small openings. As a result the total porosity is composed of two parts as

$$n = S_y + S_r \qquad (4)$$

Depending on the material type, the contributions of S_y and S_r to n vary. The major role in the ratio of such a contribution is played by the particle surface area. The smaller the average grain size the larger is the surface area of the medium. Larger surface areas attract more water and accordingly their specific retention values are greater. Consequently, as a general rule the finer the grain sizes the greater the specific retention. It is tantamount to saying that there exists a reverse proportionality between the grain size and specific retention. It has been explained already in the previous section that the similar proportionality exists between the porosity and the average grain size. In light of these proportionalities and Equation 4, one can conclude that the specific yield is relatively small for very coarse and fine materials and attains its maximum value somewhere in between. Empirical studies have resulted in a graph similar to the one shown in Figure 2.

It is important that the specific yield of an aquifer is always less than the porosity, the difference being the specific retention. The absolute maximum specific yield corresponds to the range of sand and this is one of the reasons why most prolific aquifers in the world are found in sedimentary formations. The ranges of specific yield in percentage are given for sedimentary formations in Table 2.

In hydrogeological studies the specific yield concept is much more important than the porosity since the latter does not give any clue for water abstraction from the aquifers. For instance, in igneous rocks porosity is not more than 2% but they yield all the available water in their voids up to approximately 100%. On the contrary, clays have the maximum porosity but they yield less than 5% of stored water.

If a granular material with an average porosity of 30% exists below the water table, its voids are saturated, containing 30% water per unit volume. If the water table is lowered the soil will eventually give up about two thirds of its water, equivalent to 20% of total volume.

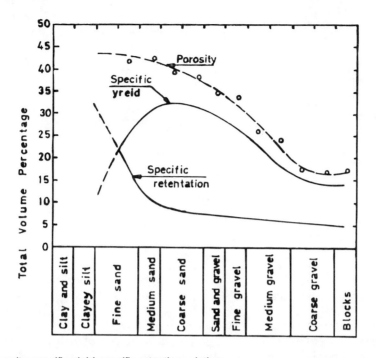

Figure 2 Porosity specific-yield-specific retention relation.

The remaining 10% water by volume is held by surface tension. In hydrological terminology one says that the specific yield is 0.2 and specific retention is 0.1.

Specific yield gives the amount of water available for man's use and the specific retention tells how much water remains in the rock after it is drained by gravity. Thus,

$$S_y = \frac{V_d}{V_t} \quad (5)$$

and

$$S_r = \frac{V_r}{V_t} \quad (6)$$

in which V_d and V_r are the drained and retained water volumes, respectively.

V. STORAGE AND SPECIFIC STORAGE COEFFICIENT

The ability of an aquifer to store water is one of the most important hydraulic properties. Groundwater at rest is under the influence of gravity and hydrostatic pressure. The former is

Table 2 Expected Specific Yields

Formation material	Specific yield (%)
Clay	5
Fine sand	10–20
Medium sand	20–25
Coarse sand	15–30
Gravely sand	16–28
Fine gravel	15–25
Medium gravel	14–24

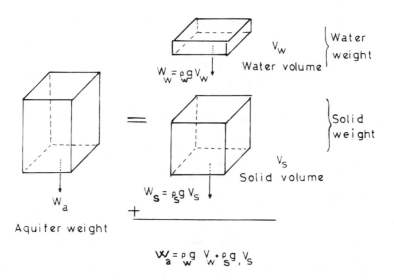

Figure 3 Bulk weight components.

the actual combined weight of solid and water above a certain horizontal area at any depth in the aquifer. As shown in Figure 3 the bulk weight is the sum of weights of solid and water.

On the other hand, it can be expressed as $\gamma_b V_t$ where γ_b is the bulk density and V_t is the volume of parallelepiped considered. The bulk weight per unit area can be expressed as

$$G = \gamma_b g h \qquad (7)$$

where h is the depth of saturation zone above that unit area as in Figure 4.

The direction of gravity force is always downward. In Equation 7 the only variable is h and consequently its increase gives rise to increase in total gravity force. Therefore, it is obvious that deep sediment is compacted under the force of gravity. The maximum compaction occurs when the voids are not filled with water.

On the other hand, the hydrostatic pressure, p, of water at the same depth is

$$p = \gamma_w g h \qquad (8)$$

in which γ_w is the water density. The direction of this pressure is upward provided that the bottom of the representative volume is horizontal. The net pressure affecting the bottom is

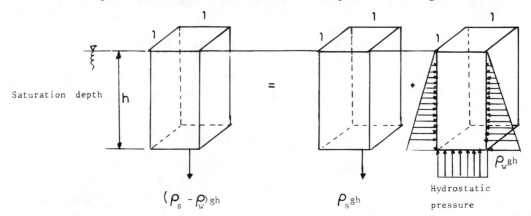

Figure 4 Bulk weight per area.

$$p' = (\gamma_b - \gamma_w)gh \qquad (9)$$

Since, γ_b and γ_w are constants any small increment in the pressure is directly related to the small increments in h as

$$dp' = (\gamma_b - \gamma_w)gdh \qquad (10)$$

Furthermore, $\gamma_b > \gamma_w$ and consequently the gravity force always greater than the hydrostatic pressure. However, the hydrostatic pressure at a point is equal at every direction with the same magnitude (Figure 5). This means that upward pressure works as a lifting force opposite the gravity force whereas downward it is combined with the gravity force and hence these two resultant forces try to compact the formation material. It is obvious from the same figure that the effective pressure acts as a compression stress on the solids as well as on the water in the voids.

As a result both solids and water will be compressed to a certain extent. Although the compressibility of water is smaller compared to the solids in large areas, it adds up to a significant volume. Equation 9 indicates that change in pressure is related directly to change in the hydraulic head.

In light of the above-mentioned discussion one can conclude that the storage ability of a reservoir depends on the effective pressure. If there were no pressure, the porosity would help to evaluate the available water volume in a certain reservoir, the specific yield would help in finding the volume of abstractable water. Such a situation is valid for unconfined aquifers with rather thin thicknesses. However, the specific yield concept becomes invalid in the reservoirs where the groundwater is stored under pressure. This is due to the aquifer material and water compressibilities.

The most distinctive characteristic of the confined aquifer is that as the water is withdrawn the aquifer remains fully saturated. The overburden is supported partly by the solid grains

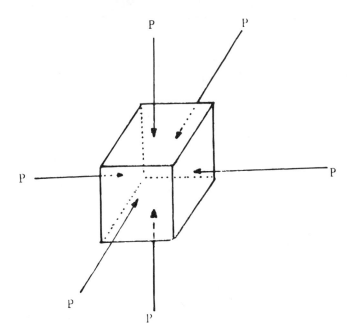

Figure 5 Hydrostatic pressure.

and partly by the pressure of the water (see Figure 6). Removal of water from the aquifer occurs at the cost of pressure drop and therefore more overburden must be taken by the solid grains which results in its slight compression leading to consolidation of the layer. Meanwhile, a slight expansion of water takes place. To express the overall effect in terms of water withdrawal, the storage coefficient parameter is defined as the volume of water that an aquifer releases or takes into storage per unit surface area of the aquifer per unit change in head. Figure 7 represents an imaginary columnar saturated portion of confined and unconfined aquifers with unit areas. The water released from reservoir parts in Figure 7 gives the storage coefficient. The value of the storage coefficient depends on whether the aquifer is confined or unconfined. If the aquifer is unconfined, the predominant source of water is from gravity drainage. In an unconfined aquifer, the volume of water derived from expansion of the water and compression of the aquifer is negligible. However, in the confined aquifer, the water is released from storage due to expansion of water and compression of the aquifer material under pressure release. Expansion of water is relatively smaller than compression of the aquifer and therefore the storage coefficient is smaller than an unconfined aquifer. Field experiences have shown that in an unconfined aquifer S varies between 0.3 and 0.1 whereas in confined aquifers S has a range from 10^{-6} up to 10^{-2}. In the case of an unconfined aquifer, the volume of water taken into or released from the reservoir as a result of compressibility is negligible in comparison with the water involved in draining or filling the pore space. Therefore, for practical purposes, the storage coefficient of an unconfined aquifer is taken as equal to the specific yield, i.e., $S \cong s_y$.

On the other hand, the storage coefficient can be defined as the ratio of dewatered volume, V_a, of the aquifer to the volume of water, V_w, abstracted from the aquifer as

$$S = \frac{V_a}{V_w} \tag{11}$$

Physically high storativities cause less drawdowns within the aquifer. A common point in the definition of storativity is that it is based on the whole saturation thickness of the aquifer. The division of the storativity by the saturation thickness is referred to as the specific storage:

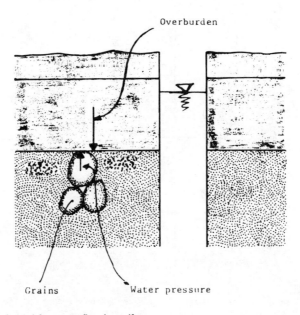

Figure 6 Water withdrawal from confined aquifer.

AQUIFER PROPERTIES

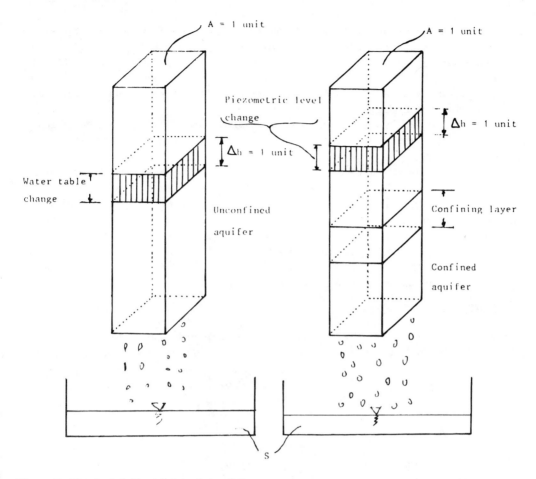

Figure 7 Simple definition sketch of storativity.

$$S_s = \frac{S}{m} \quad (12)$$

It has the dimension of inverse length. In fact, it means physically the amount of water that is stored or expelled from per unit volume of a saturated formation due to compressibility of the solids and water. The concept of specific storage is very useful in groundwater flow analysis especially in extensively deep aquifers.

Instead of being constant as assumed in many analytical models in Chapters 8 through 10, the storage coefficient sometimes varies to a considerable extent. The variations are caused by a lag in elastic compaction or in unconfined aquifers a lag in capillary-fringe drainage tends to level off after moderate duration of water abstraction. Errors from such a cause can be minimized by abstracting water for long periods before adaptation of field data for aquifer parameter determinations.

VI. FRACTURE SIZE DISTRIBUTION

A fracture is a plane along which rock material has lost cohesion. Fractures dissect the host rock mass into intact fragments which are referred to as block or also interchangeably as matrix. Figure 8 indicates simply blocks and fracture sets in a rock mass. Fractures occur usually in the form of one or more sets of nearly parallel planes. For instance, in the same figure there are three sets of fractures. Rock outcrops have plenty of fractures that give way to economic mining, quarrying, and groundwater movement but are unstable bases for foundations. Joints, faults, and fissures are special types of fractures, depending on the relative movement of adjacent rock blocks.

Fractures whose blocks have moved apart are called fissures; if there has been an appreciable movement of blocks then the fracture is called fault; without any appreciable movement they are joints (Figure 9).

Individual fracture parameters refer to the intrinsic characteristics, such as fracture aperture, length, penetration extent, closeness or openings of the fracture, type of filling material, and fracture wall roughness. However, quantitative analysis of fracture spacings is achieved by scan-line measurements either on the outcrop or in an excavation on fresh faces or along a bore hole (Deere, 1964; Şen and Qazi, 1984). The spacing between two successive fractures is the main variable that helps to have some quantitative idea about the block size distribution in an area. In addition, matrix block shapes, distribution, and density are among the important factors in any quantitative study of fractured reservoirs.

A. FRACTURE POROSITY

Fracturation of any rock increases the volume of voids in the parent rock. As explained in Section III the voids due to fractures give rise to secondary porosity. In fact any fractured

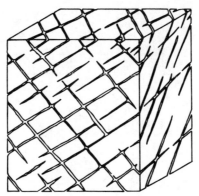

Figure 8 Fractures and blocks.

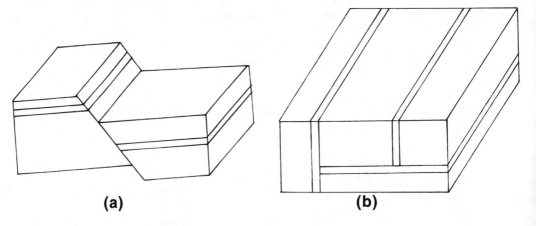

Figure 9 Fracture types: (a) fault; (b) joints.

reservoir can be treated as a double porosity system, one system in the host rock, i.e., matrix, and the other in the fractures. Fractures in clastic sediments are instantaneously sealed but in brittle rocks they remain open after formation thus giving rise to fracture porosity. Fracture porosity is a difficult parameter to measure and analyze because fractures range from microscopic to cavernous in size; they are difficult to catch in cores.

The governing factors of fracture porosity are fracture aperture, spacing, surface roughness, area, and filling. The aperture is the average distance between the walls of a fracture. This varies considerably for different rocks. According to aperture, a, fractures may be classified into different categories as shown in Table 4.3.

From the hydrogeological point of view, fracture porosity, n_f, is an important parameter of fracturation degree in calculating the available volume for water storage in fractured reservoirs. The fracture porosity is defined as the ratio of fracture volume to the volume of rock specimen. Hence,

$$n_f = \frac{l_i a_i}{F} \qquad (13)$$

in which l_i is the length of ith fracture, a_i is the aperture of ith fracture, and F is the surface area of the rock under investigation. In general, the degree of fracturation decreases with depth. It is observed that the fracture porosity ranges from 0.5 to 6% (Snow, 1969).

On the other hand, the fracture porosity, n_f, can be calculated in terms of average spacing between parallel fractures, D, and the average effective width of fractures, e, as

$$n_f = \frac{e}{D + e} \qquad (14)$$

Relatively small increases in fracture porosity give rise to immense changes in the easiness of water movement parallel to fractures. Fracture porosity is comparatively a smaller number than the matrix porosity which is equivalent to the primary porosity.

VII. HYDRAULIC CONDUCTIVITY

Water storage capabilities of reservoirs have been discussed through different parameters as porosity, specific yield, and storage coefficient, but these do not provide any information about the water transmission properties of the rocks. Whatever the water storage features of a rock they may not provide practically significant resources if the water cannot move rather easily through the lithological facies. Under normal conditions, as was noted in Chapter 3, the finer the reservoir medium, the more will be the resistance of the grains against the groundwater movement and the less will be the groundwater velocity. In general, conductivity is commonly defined as the ability of rocks to let the water through under any hydraulic gradient. In quantitative terms, the rock's ability to transmit water is related to the groundwater velocity (specific discharge). The hydraulic conductivity is equal to the specific discharge

Table 3 Fracture Classification

Fracture type	Aperture, a (mm)
Micro fissure	$a < 0.1$
Macro fissure	$0.1 < a < 1$
Micro fracture	$1 < a < 10$
Macro fracture	$a > 10$

under unit hydraulic gradient as shown in Figure 10. In light of this definition the unit of the hydraulic conductivity is physically the same as the velocity, i.e., $[L/T]$. The hydraulic conductivity depends on the pore and fracture sizes within the rock mass. The larger the pores and/or fracture apertures the more easily the water moves through these openings and therefore the hydraulic conductivity is high provided that these openings are interconnected.

In terms of the hydraulic conductivity values the rocks can be categorized into three distinctive groups:

1. Permeable rocks such as pebbles, gravel, sands, etc., for which the hydraulic conductivity values are relatively high. In other words, the groundwater moves easily with small resistance. Rocks with hydraulic conductivities of 1 m/day or more are regarded generally as permeable and they are likely to form good potential aquifers.
2. Semipermeable rocks such as argillaceous sand, sandy loam, loess, and shales where the hydraulic conductivity values are rather moderate.
3. Impermeable or practically water-resisting rocks such as clays, tight peat, and crystalline and some sedimentary rocks intact of fractures are characterized by negligible or zero hydraulic conductivities. These rocks might have high water storage capacity, but, unfortunately, their water-yielding abilities are negligible. This is due to the fact that during the water transmission through the tiny pores there appears a very strong resistance against the water flow. Rocks with hydraulic conductivities of less than 10^{-3} m/day would generally be regarded as impervious. However, all of these hydraulic conductivity values are relative and should be evaluated depending on the nature of the problem. For instance, in groundwater explorations for a hydrogeologist any media with the hydraulic conductivity of 0.1 m/day or less might mean an impervious layer but the similar media might be regarded as highly permeable in engineering works such as dam construction where very large hydraulic heads and consequently gradients are involved.

On the other hand, in homogeneous, isotropic, and saturated medium, the hydraulic conductivity is defined as the volume of the groundwater that will move in unit time under unit hydraulic gradient through a unit cross-sectional area perpendicular to the streamlines as in Figure 11. On the basis of such a definition the hydraulic conductivity has a unit of volume per time per area. Groundwater hydrologists usually use this definition with $L^3/T/L^2$, whereas engineers are more concerned with the seepage problems and therefore, they prefer to express the hydraulic conductivity in terms of velocity unit usually as centimeters per second. Representative hydraulic conductivity and storage coefficients are given in Table 4.

In general, development of the secondary porosity as a result of rock fracturation increases the hydraulic conductivity due to the number and size of the fracture apertures. As the minerals are dissolved from the rock, the fractures are enlarged; as a result the hydraulic conductivity

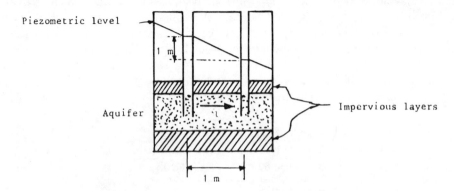

Figure 10 Hydraulic conductivity as velocity.

AQUIFER PROPERTIES

increases. Bedding planes in the sedimentary rocks may also cause an increase in the hydraulic conductivity. However, the chemical composition may fill the fracture apertures and decrease the overall hydraulic conductivity of the rock mass.

It is a general rule that in the fractured rock masses the fracture hydraulic conductivity is always more than the rock blocks. According to the transmissivity value aquifers are classified into five classes as shown in Table 5.

A. FRACTURE HYDRAULIC CONDUCTIVITY

The planar nature of the fracture surfaces gives an easy way for water to travel in the fracture network. The ease of water to move through rock is referred to as its permeability or hydraulic conductivity. The overall hydraulic conductivity of a rock mass is determined by the permeability of the rock matrix and of the fractures. In most of the practical cases the hydraulic conductivity of the fractures is considerably greater than that of the rock matrix so that it is justifiable to neglect the matrix conductivity. The fractures are finite in size, so they do not extend indefinitely within the same plane. As a result, a degree of interconnection between fracture clusters is a critical feature that contributes to the hydraulic conductivity of the whole rock mass. The fracture aperture determines its individual permeability. It is possible

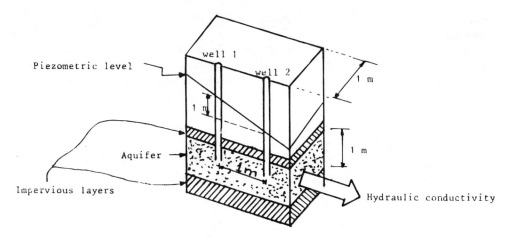

Figure 11 Hydraulic conductivity as volume.

Table 4 Representative Values

Reservoir rock		Hydraulic conductivity (m/day)	Storativity
Unconfined aquifer	Fine gravel	60	0.1
	Coarse gravel	20	0.10–0.30
	Fine sand	10	0.30–0.35
Confined aquifer	Porous	—	10^{-3}–10^{-6}
Aquitards	Sand and silt	2	>1
Aquicludes	Clay	$<10^{-3}$	—

Table 5 Aquifer Potentiality

Transmissivity (m²/day)	Potentiality
$t > 500$	High
$500 < t < 50$	Moderate
$50 < t < 5$	Low
$5 < t < 0.5$	Weak
$t < 0.5$	Negligible

to say that the fracture hydraulic conductivity is a function of fracture size, density, location, aperture, and orientation. It is obvious that, since each one of these fracture features is haphazardly distributed within the rock mass, the fracture reservoir is considered as a heterogeneous and anisotropic water-bearing layer. It is for this reason that available formulations for porous medium hydraulic parameters cannot be used directly for the fractured reservoir. It is a common rule that the number of fractures and their apertures diminish with depth due to overburden as well as the secondary sedimentation, especially in the form of clay minerals. It is expected, therefore, that the hydraulic conductivity reductions in rocks (especially igneous and metamorphic) vary with depth having a maximum value at the top and a minimum value at the bottom of the aquifer.

Early studies concerning the water flow in fractures were all confined to a single fracture which was idealized as an aperture between two parallel plates (Lomize, 1951; Snow, 1965; Louis, 1969). The laminar flow in a fracture with smooth walls, i.e., without roughness, complies with the classical Darcy law and the hydraulic conductivity, K_f, of the fracture with an aperture a is given as

$$K_f = \frac{g}{12\nu}(2a)^3 \qquad (15)$$

in which ν is the kinematic viscosity. Hydraulic conductivity is proportional with the cube of fracture aperture.

In fractured rocks the hydraulic conductivity depends on the density and aperture of the joints. The fractures may either be filled with fine material or else the aperture increases. In the crystalline rocks the fractures are very often sealed by clay, calcite, or silica deposits. Fractured crystalline rocks may vary in hydraulic conductivity from 10^1 to 10^{-3} m/d. On the other hand, fractured basalt is highly permeable. During the cooling period a basalt layer creates a dense network of vertical joints that divide the basalt mass into contiguous prisms and the hydraulic conductivity may be as high as 10^4 m/d. Weathering and deposition of minerals on the fracture plane reduce severely the fracture hydraulic conductivity and subsequent pumping of water causes hydraulic gradients to develop in the fracture and in the matrix blocks.

B. KARSTIC FORMATION

No doubt, initially during the original deposition of limestones and dolomite the primary porosity is the only void space and therefore the hydraulic conductivity is similar to the porous medium. Later, due to tectonic activities, they are fractured and the hydraulic conductivity of the fractures start to play a major role in the whole rock mass. The CO_2 dissolved by rainwater in the atmosphere and H_2CO_3 dissolve limestone deep in the aquifer thus enlarging the fractures in a random pattern as shown in Figure 12. This finally leads to a fully developed karstic formation and the fractures may locally become very large and form an underground system of chambers, tunnels, pipes, and siphons. In such a case the hydraulic conductivity does not apply. However, not all the karstic terrain is fully developed and the dissolution of limestone may create a network of open fractures with hydraulic conductivities in the range of 10^2 to 10^4 m/d.

VIII. COMPOSITE HYDRAULIC VARIABLES

These parameters are dependent on two or more basic aquifer parameters, such as the porosity, hydraulic gradient, specific discharge, and the geometric dimensions. Their first

AQUIFER PROPERTIES

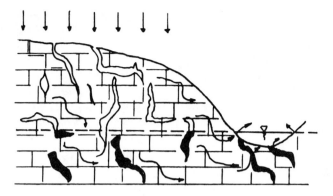

Figure 12 Fissures and enlargement due to water circulation.

appearance in the hydrogeological sciences was due to the convenience in definition of some global physical phenomenon.

A. TRANSMISSIVITY

This parameter characterizes the ability of the aquifer to transmit water. Its definition as it stands in the groundwater literature falls into one of the following categories:

1. The rate of flow under unit hydraulic gradient through a cross section of unit width over the whole saturated thickness of the aquifer (Bear, 1979). Bear also stated that, in three-dimensional flow through the flow media, the transmissivity concept is meaningless. As will be shown in Chapter 12 the transmissivity definition is not valid if the groundwater flow abides with a nonlinear flow law. By definition transmissivity is valid for Darcy (linear) flow law only.
2. The product of the thickness of the aquifer and the average value of the hydraulic conductivity (Hantush, 1964; Davis and De Wiest, 1966; Bouwer, 1978; Freeze and Cherry, 1979).
3. The ratio at which water of prevailing density and viscosity is transmitted through a unit width of an aquifer or confining bed under a unit hydraulic gradient. It is a function of the properties of the liquid, the flow media, and the thickness of the media (Fetter, 1980; Todd, 1980).

The overall capacity of an aquifer to transmit water is dependent on the saturation thickness and hydraulic conductivity of the aquifer material. To appreciate the definition and physical meaning of transmissivity let us formularize the amount of water that aquifer allows to pass through its saturation cross section which is perpendicular to the flow direction as schematized in Figure 13.

The discharge, Q, of water movement through cross section, A, can be written in general as

$$Q = Aq \qquad (16)$$

where q is the specific discharge. If the groundwater flow is assumed to have a linear regime (laminar flow), then the Darcy law $q = Ki$ becomes applicable and hence,

$$Q = AKi \qquad (17)$$

in which i is the hydraulic gradient and K is the hydraulic conductivity. Equation 17 takes its simplest form for a rectangular cross section with width, w, and saturation thickness m as

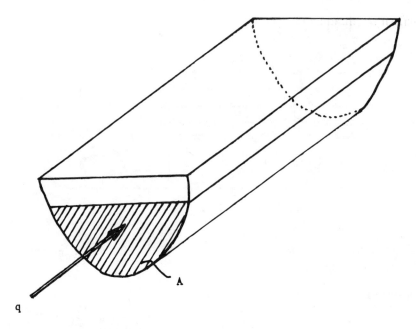

Figure 13 Transmissivity definition.

$$Q = wmKi \tag{18}$$

In fact the classical definition of transmissivity is based on this last expression. By definition, the transmissivity $T = mK$ and hence Equation 18 yields

$$T = \frac{Q}{wi} \tag{19}$$

This equation constitutes the basis of transmissivity and it reads in words that the transmissivity is the amount of water that can be transmitted horizontally by the whole saturated thickness of the aquifer from unit width and under unit change of hydraulic gradient. This definition of transmissivity is based on two very restrictive conditions: (1) that the flow is laminar and (2) that the cross section of groundwater flow is rectangular, i.e., the saturation depth is uniform. It is, therefore, advised to employ the transmissivity in groundwater problems with precaution unless these two conditions are satisfied. The second condition can be avoided provided that the irregular shape of the cross section is converted into an equivalent rectangular section by keeping the aquifer width the same. In such a situation the equivalent rectangular cross-section depth, m, will be

$$m = \frac{A}{w} \tag{20}$$

Then the aforementioned definition of transmissivity should be modified by considering the equivalent rectangular cross-section depth with the whole saturated thickness of the aquifer. If the groundwater flow is not linear, Equation 18 is no longer valid (see Chapter 12).

In light of the above-mentioned discussions it is possible to conclude that the transmissivity definition as it appears in Equation 19 is valid neither for the fractured nor karstic reservoirs but for porous medium only.

B. HYDRAULIC RESISTANCE

In classical leaky aquifers hydraulic resistance is a measure of the resistance of an aquitard to vertical flow movement. Any groundwater reservoir shows resistance to the movement of water resulting in a retarding effect. Physically, the length of path traveled will have a direct effect on this resistance, whereas the hydraulic conductivity of the medium has an inverse impact (see Figure 14). Because $K < K'$ the medium in Figure 14b will have less resistance than the medium in Figure 14a. Quantification of the medium's resistance is achieved by the ratio of distance traveled, D, to the hydraulic conductivity, K. Hence, the hydraulic resistance, R_h is defined as

$$R_h = \frac{D}{K} \qquad (21)$$

It is obvious that R_h has the dimension of time. It is noted that for impervious medium $K = 0$; hence, theoretically, $R_h \to \infty$. It may characterize the resistance of the semipervious layer to upward or downward leakage in a leaky aquifer. For tracer tests in aquifers the hydraulic resistivity corresponds to the time required by the tracer to travel a given distance.

In general, fractured or karstic reservoirs have less resistivity than porous medium and the water particle does not experience great resistance along the fractures.

The reciprocal of hydraulic resistance, called leakage by Jacob (1946) or specific leakage by Hantush and Jacob (1955), is defined as the quantity of water flowing across a unit area of the boundary between the aquifer and aquitard if the difference between the hydraulic head in the aquifer and that in the source bed is unity.

C. LEAKAGE FACTOR

Physical occurrence of leakage in groundwater studies is possible provided that there are two successive layers with different hydraulic conductivities as shown in Figure 15. If these two layers have the same hydraulic conductivities the leakage is referred to as seepage. For the leakage to take place easily, the source layer must have small hydraulic resistance. Otherwise, as was discussed in the previous section, for large hydraulic resistances the source layer behaves as an impervious layer preventing the leakage. Therefore, any factor definition for leakage quantification must be proportional with hydraulic resistance, R_h. On the other hand, small hydraulic resistance of the source layer is not sufficient for leakage occurrence. The leakage water must be conveyed away by the recipient layer with high transmissivity.

As will be discussed in Chapter 8 the necessary calculations for a leaky aquifer lead to a convenient definition of leakage factor, L, as

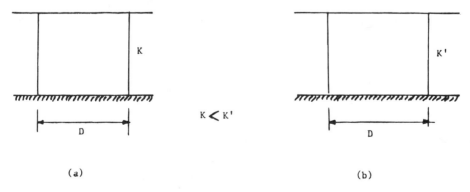

Figure 14 Resistance and water movement.

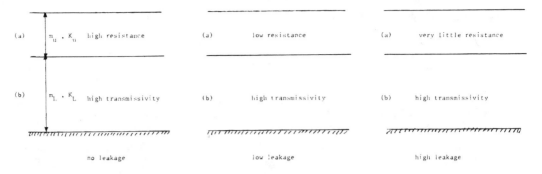

Figure 15 Leakage-hydraulic resistance and transmissivity.

$$L^2 = \frac{K_U m_U m_L}{K_L} \qquad (22)$$

in which m_U, m_L, K_U, and K_L are the thicknesses and hydraulic conductivities of the recipient and source layers, respectively. High values of L indicate a great resistance of source strata to flow as compared with the resistance of the layer itself. The factor L has the dimension of length. It is obvious from Equation 22 that the thickness of layers plays a direct role in the leakage factor evaluations. For practical guidance the overall leakage factor can be classified according to the value of L as shown in Table 6.

IX. GEOMETRIC PROPERTIES OF AQUIFERS

The significant role of flow net geometry on groundwater resources evaluations has been discussed in Chapter 3. This geometry is directly related to such geologic formations' dimensions as thicknesses, grain sizes, fracture sizes and distributions, dips, strikes, etc. Therefore, it is essential to have a good grasp of the aquifer geometric properties that are dependent on the geological setup of the reservoirs.

A. HOMOGENEITY AND ISOTROPY

Any physical phenomena that evolve in nature have sophistication and therefore their quantification is difficult if necessary simplifying assumptions are not set forward. These assumptions render the real events into idealized counterparts that can be grasped quite easily. In the earth sciences especially, the basic physical quantities discussed in this chapter vary spatially within a geological formation. Measurements of these quantities in a multitude of points during a field study show invariably their spatial variations either randomly or systematically or directionally.

These variations pose two exclusive questions: (1) should the researchers apply flow laws to each point separately and solve the problem or (2) should they make simplifying assumptions and write flow laws for the whole domain once and subsequently find the solution for all? Of course, the former approach should be preferable for finer solutions but it is almost

Table 6 Leakage Classification

Leakage factor (m)	Aquifer potentiality
$L < 1000$	Potential leakage
$1000 < L < 5000$	Moderate leakage
$5000 < L < 10{,}000$	Low leakage
$L < 10{,}000$	Negligible leakage

impossible to solve the multitude of equations. However, in practice, the latter approach is favored due to its easy treatment. To follow this path of action, a sequence of assumptions is in order. In heterogeneous aquifers the vertical and/or horizontal lithologies of most geological formations vary to a certain extent.

First, to avoid the variation from one point to another, the geological formation is assumed homogeneous. If any of the above-mentioned hydrogeological variables is dependent on the position within the aquifer, the formation is heterogeneous. In a homogeneous porous reservoir, pore sizes are evenly distributed even though randomly connected. Although a porous medium of sandy aquifer intercalated with relatively impermeable lenticular clays is heterogeneous, it may be regarded equivalent and homogeneous. Such an equivalence is not valid for an aquifer of distinctive layers with different hydraulic gradients as will be explained in Chapter 5.

A reservoir medium is regarded as an anisotropic domain if its basic hydraulic properties are dependent on direction. Anisotropy is a common property of fractured rocks. The movement of water is easier along the fractures than normal, provided that the principle directions of anisotropy are known. The medium concerned can be transformed into an isotropic system by coordinate change. For instance, the existence of flat minerals such as mica in sedimentary rocks as a result of overburden pressures flat surfaces become horizontal. Such a feature causes horizontal property different from the vertical counterpart at the same point. If the hydraulic property is independent of the direction of measurement at a point in a reservoir, the reservoir is isotropic at that point. The existence of such an isotropy at every point within the aquifer makes it spatially isotropic. In the majority of groundwater problem solutions by analytical methods, the aquifer is assumed as homogeneous and isotropic. Until stated otherwise, in this book homogeneity and isotropy refer to the hydraulic conductivity. Sedimentary and alluvial aquifers tend to be anisotropic, the major axis of permeability is along the bedding planes, and the minor axis at right angles to them. In locustrine or deltaic sediments anisotropy may be much more marked.

Most groundwater equilibrium or unequilibrium equations are based on the assumption that water-bearing formations are homogeneous and isotropic. In nature isotropy (anisotropy) and homogeneity (heterogeneity) appear in pairwise combinations, leading to isotropic-homogeneous, isotropic-heterogeneous, anisotropic-homogeneous, and anisotropic-heterogeneous medium types as shown in Figure 16.

B. EXTENSIVENESS AND UNIFORMNESS

These properties are related to the geometrical dimensions of a groundwater reservoir. To evaluate groundwater movement and volume, it is necessary to know the areal and cross-sectional geometry of an aquifer. Extensiveness of a reservoir depends on nondisturbance of groundwater flow phenomenon by any changes in the geological facies or structures. If the recharge affected area is not disturbed by any lateral boundary, the aquifer is said to be extensive during the analysis of recharge phenomenon (see Figure 17). For recharge calculations in Figure 17a the aquifer is extensive but in Figure 17b the same aquifer is nonextensive because there is boundary effect from the right bank of the channel.

As will be explained in Chapter 8 during water abstractions from a well, if the depression cone is not disturbed by any other geologic or hydrologic boundaries, then the aquifer is extensive; otherwise, for instance, when the well is close to a fault within the reservoir, the aquifer should be regarded as a bounded medium.

On the other hand, uniformity of an aquifer is related to its depth variations along a cross section. An aquifer is uniform if the saturation depth is the same along any cross section in which a groundwater phenomenon takes place. Figure 18 shows the uniform depth concept with recharge phenomena. As mentioned earlier, uniformity assumption entitles one to use the transmissivity concept throughout the uniform range as in Figure 18.

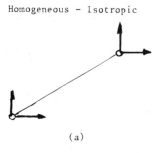
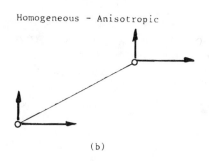
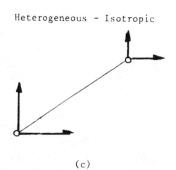
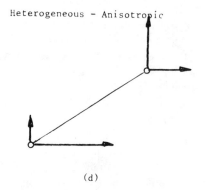

Figure 16 Homogeneity and isotropy.

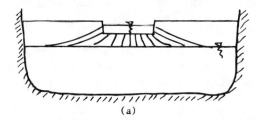
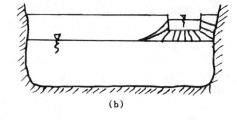

Figure 17 (a) Extensive and (b) bounded aquifers.

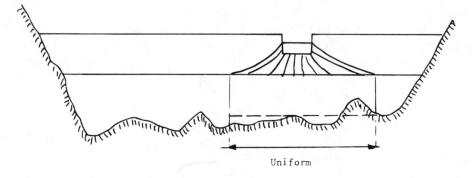

Figure 18 Uniformity of an aquifer.

REFERENCES

Deere, D. U., 1964. Technical description of rock cores for engineering purposes, *Rock Mech. Eng. Geol.*, 1, 17.
Davis, S. N., 1969. *Flow through Porous Media.* DeWeist, R. J. M., Ed., Academic Press, New York.
Jacob, C. E., 1946. Radial flow in a leaky artesian aquifer, *Trans. Am. Geophys. Union,* 27, 183.
Hantush, M. S. and Jacob, C. E., 1955. Nonsteady radial flow in an infinite leaky aquifer, *Am. Geophys. Union,* 36(1), 95.
Hantush, M. S. and Jacob, C. E., 1955. Non-steady radial flow in an infinite leaky aquifer and non-steady Green's functions for an infinite strip leaky aquifer, *Trans. Am. Geophys. Union*, 36(1), 95.
Lomize, G., 1951. Filtratsita v treshchinovatykh porodakh, Gosenergoizdat, 91.
Louis, C., 1967. Stromungsverganga in kluftigen Madian und ihre Wirkung auf die Standsicherheit von Bauwerken und Boshungen im Fels, Dissertation, University of Karlsruhe, Germany.
Bear, J., 1979. *Hydraulics of Groundwater,* McGraw-Hill, New York.
Bouwer, H., 1978. *Groundwater Hydrology,* McGraw-Hill, New York.
Davis, S. N. and De Wiest, R. J. M., 1966. *Hydrogeology,* John Wiley & Sons, New York.
Driscoll, F. G., 1986. *Groundwater and Wells,* Johnson Division, St. Paul., MN.
Fetter, C. W., 1980. *Applied Hydrogeology,* Charles E. Merrill, Columbus, OH.
Freeze, R. A. and Cherry, J. A., 1979. *Groundwater,* Prentice-Hall, Englewood Cliffs, NJ.
Hantush, M. S., 1964. Hydraulics of wells, in *Advances of Hydrosciences,* Vol. 1, Chow, V. T., Ed., Academic Press, New York 281.
Snow, D. T., 1965. A parallel plate model of fractured permeable media, Ph.D. thesis, University of California, Berkeley.
Snow, D. T., 1969. Anisotropic permeability of fractured media. *Water Resour. Res.*, 5(6), 1273.
Şen, Z. and Qazi, A., 1984. Discontinuity spacing and RQD estimates from finite length scanlines, *Int. J. Rock Mech. Min. Sci. Geomech.*, Abstr., 21, 203.
Todd, D. K., 1980. *Groundwater Hydrology,* John Wiley & Sons, New York.

CHAPTER 5

Groundwater Flow Laws

I. GENERAL

In previous chapters the groundwater phenomenon was presented from the geological and hydrological points of view, in addition to some basic physical parameter definitions. These initial concepts help to assess in a region the reservoir material nature, aquifer, and flow types. It is possible to evaluate the available groundwater storage, flow directions, and recharge and discharge areas. All this information is essential for a comprehensive understanding of any groundwater flow phenomenon. However, it is not yet enough to quantify the groundwater motion problems, and especially their relation to the basic aquifer parameters.

As explained in Chapter 3, the motion of water requires energy that is measured by the hydraulic head difference between two or more points. Furthermore, the reservoir must be permeable to allow groundwater movement similar to the water flow through pipes and channels.

In previous discussions very important agents that dominate the water movement were not mentioned: the solid boundary conditions of the flow domain. In the channel or pipe flow the boundaries are geometrically well defined. They are man made and therefore even the contractions and expansions in the flow domain are given simple geometrical shapes and cross sections to simplify the water motion equations. Existence of these contractions and expansions along the flow direction affects the flow velocity, kinetic energy, and potential energy through the hydraulic head changes.

On the other hand, imagination of the actual groundwater flow domain shows irregular void space, fracture, or solution cavity networks that are full of numerous randomly distributed contractions and expansions. If one tries to describe the groundwater movement along all of the individual flow paths within aquifer, it may seem that it is beyond human grasp. Fortunately, one must not be discouraged because of the following reasons:

1. It is not necessary to know the flow motion along the individual flow path but rather the total flow resulting from groundwater movement through the combination of all these paths.
2. Kinetic energy changes along any flow path are negligible due to the fact that groundwater generally moves very slowly. Consequently, as explained in Chapter 3 the total energy head consists of the geographical elevation from a datum and the pressure head.

This chapter is devoted to the basic laws governing the groundwater motion within the aquifers.

II. BASIS OF FLOW LAWS

These are the experimental or empirical relationships that relate the basic flow variables as the specific discharge and hydraulic gradient to different aquifer parameters, i.e., the hydraulic conductivity, fracture aperture, and roughness. Some do not have a theoretical basis but their validity has been verified through the laboratory and field studies. These laws are the basic cornerstones in any large-scale groundwater problem solution. In general, losses of energy in terms of hydraulic gradient, i, within pipes of any given cross section are described by the Darcy-Weisbach formula as

$$i = \lambda \frac{1}{H_r} \frac{q^2}{2g} \tag{1}$$

where λ is the friction coefficient, H_r is the hydraulic diameter, $q^2/2g$ is the kinetic energy relative to the unit weight, and q represents the specific discharge. There is an inverse relationship between λ and the Reynolds number. Different researchers have tried to determine a general relationship in (Re, λ) plane. The conclusion is that, whatever the material is, there exists an inverse relationship between Re and λ. The schematic slope of this relation is given on double logarithmic paper in Figure 1. The initial portion of the graph appears as a straight line representing the linear flow law, whereas for the moderate and large numbers it attains gradually and non-linearly to an asymptotic value. The Re domain on the horizontal axis is divided into two parts: linear transition and nonlinear flow zones on the basis of threshold Reynolds numbers as 1 and 10.

On the contrary, the hydraulic gradient-specific discharge relationship is an interesting function with similar linear and nonlinear flow domains as shown in Figure 2.

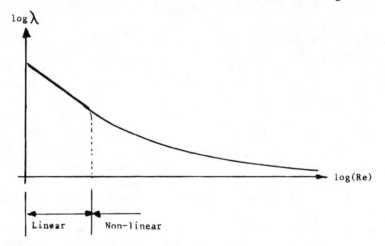

Figure 1 Re-$\lambda\nu$ graph of flow domains.

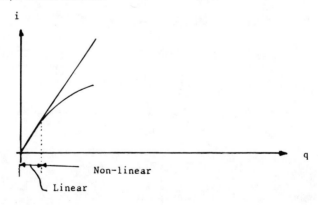

Figure 2 *i-q* Graph of flow domains.

III. POROUS MEDIUM

The fundamental characteristic of a porous medium is that the groundwater moves continuously in a manner governed by established hydraulic principles under the influence of the aquifer's inherent and geometric features. There appear various basic laws given by different researchers whereby the hydraulic properties are related to aquifer parameters.

A. DARCY'S LINEAR FLOW LAW

Darcy (1856) published an appendix to a groundwater report about the town of Dijon, which consisted of experimental results with the aim of determining the laws of water flow through sands. His simple experimental setup consisted of a cylinder in which there was a certain length, L, of sand column supported by a filter. Water flows through the cylinder at a steady rate, Q, measured at the outlet as shown in Figure 3. Two manometers located at a distance L apart along the cylinder measure the hydraulic head difference, Δh, between the top and bottom of the sample. The sole objective of this experiment is to determine the porous medium hydraulic conductivity, K, which is a constant but uncontrollable and unknown value. The only independent and controllable variable is the flow rate (discharge), whereas Δh is a resultant uncontrollable variable dependent on both the flow rate, Q, as well as on the medium property. Darcy observed that for the same sample the flow rate was directly proportional to the head difference and inversely proportional to the sample length in the direction of flow. By definition the rate of flow is directly proportional to the sample cross-sectional area perpendicular to the flow direction (Equation 16 in Chapter 4). Hence, from the experimental results Darcy deduced that

$$Q = KA \frac{\Delta h}{\Delta L} \qquad (2)$$

in which K is a constant of proportionality depending on the permeability of the sample. In practical terms, Darcy's law is a statistical relationship in which heterogeneities are averaged

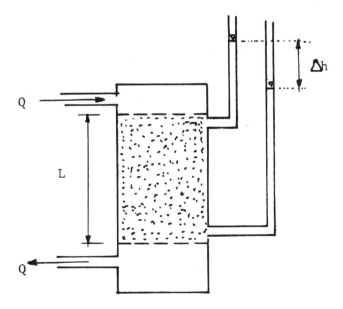

Figure 3 Darcy apparatus.

out between the manometer reading by fitting a straight line to the scatter of points. After dividing both sides of Equation 2 by A and considering Equation 16 in Chapter 4 and Equation 9 in Chapter 3 it is possible to write

$$q = Ki \qquad (3)$$

The coefficient of proportionality, K, appearing in Equation 3 is called hydraulic conductivity of the porous medium which has been explained in Chapter 4. In the Re-λ plane this linear relationship appears as

$$\lambda = \frac{C}{Re} \qquad (4)$$

where C is a constant. It is obvious from this equation that the Darcy's law is a linear expression and therefore it is sometimes referred to as the linear flow law. However, this concept must not be confused with the linear (laminar) flow where the streamlines are parallel straight lines as mentioned in Chapter 3.

The hydraulic gradient in Equation 2 can be written in alternative forms depending on the groundwater problem at hand. For instance, in any field work it is more convenient to write the linear flow law as

$$q = K \frac{h_2 - h_1}{\Delta L} \qquad (5)$$

where h_1 and h_2 are the elevations of water level from any datum (such as the mean sea level) in the two wells that are a distance, ΔL, apart horizontally as shown in Figure 4. Furthermore, it is very convenient to write the same law for the steady and unsteady flow conditions as

$$q = K \frac{dh}{dL} \qquad (6)$$

and

$$q = K \frac{\partial h}{\partial L} \qquad (7)$$

respectively. Here, the ordinary differential form of the hydraulic gradient, dh/dL, implies that

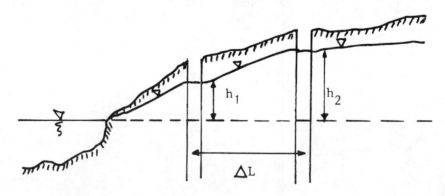

Figure 4 Field setup for hydraulic gradient.

the hydraulic head varies with the distance only, that is, the flow is in steady state, whereas in the unsteady-state case the hydraulic head varies with time also.

1. Validity Range of Darcy's Law

As explained before Darcy's law neglects the kinetic energy of water which is small for velocities normally encountered in most field problems but increases in the hydraulic gradient, for instance, especially adjacent to a well, resulting in high velocities. In addition, if the medium is coarse grained, fractured, or karstic, the groundwater flows comparatively much easier than fine sands and hence the groundwater velocity can be significant. Another drawback in the original form of the Darcy equation is the absence of any explicit variable for both the time and the fluid density which are important variables in the case of unsteady-state flow. Therefore, it is obvious that the Darcy law is an approximation for the unsteady-state flows. Laushey and Popat (1978) have shown that Darcy's law is incomplete for laminar unsteady flow through porous media. The generalization of Darcy's law is an oversimplification in the groundwater context but it helps to understand the more complex reality, the meaning of permeability, and the factors that affect it.

Dudgeon and Yuen (1970) state that, in the linear regime, laminar flow is present throughout and inertial effects begin to become significant and the velocity gradient required to cause a given flow velocity is greater than predicted by Darcy's law. At higher velocities still eddy shedding occurs in the wake of individual grains and the rate of head loss becomes even higher. Finally, as the velocity is increased turbulent spots appear and eventually the whole flow becomes turbulent with laminar sublayers around the particles except for extreme velocities or roughnesses.

Experimental data from different sources (Dudgeon, 1964; Kutilek, 1969; Slepicka, 1961; Swartzendruber, 1962; Wilkinson, 1956) have helped to evolve a general consensus that there is an upper and lower limit beyond which Darcy's linear law does not hold. Basak (1978) has combined the work of all the researchers and arrived at five zones as shown in Figure 5.

1. No flow zone—The groundwater flow is possible only after a certain hydraulic gradient that is greater than a threshold gradient. In other words, for the groundwater motion to start, the hydraulic head difference between two points must be great enough to counteract the surface forces. The finer the medium the greater the value of this threshold value.
2. Prelinear non-Darcian laminar zone—Swartzendruber (1962) noticed that the surface forces arising from the solid-fluid interaction due to strong negative changes on the clay particle surfaces and dipolar nature of water molecules cause a nonlinear and thus non-Darcian flow in the turbulent flow domain.
3. Darcian laminar flow—Almost all of the natural porous and finely fractured media exhibit this zone to a certain extent. The inertial forces are comparatively negligible against the viscous forces and the Darcy law is applicable confidently in this case.
4. Postlinear non-Darcian laminar zone—This zone is the transitional range from the laminar to turbulent flow during which due to gradual increase in the inertial force makes the flow deviate from linear flow.
5. Turbulent zone—Herein the turbulent flow starts and the substantial part of energy becomes dissipated in overcoming the inertial forces and comparatively to the other zones the slope of curve is smaller.

2. Application of Darcy's Law

One of the most important applications of Darcy's law is in predicting the groundwater flow to a well. This law is used to understand the natural flow through an aquifer cross section. For such a calculation it is necessary to know the saturation thickness, m, cross-

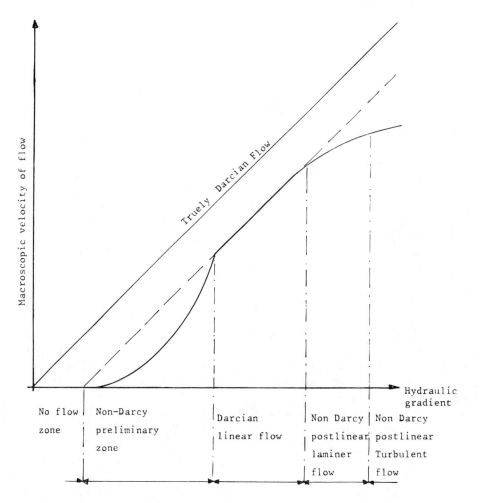

Figure 5 Different flow zones.

section width, w, hydraulic conductivity, K, and the hydraulic gradient, i, in the vicinity of the cross section. The hydraulic gradient can be determined from the difference in water levels in the two piezometers located at points A and B as shown in Figure 6. This difference is H so the hydraulic gradient is $i = H/L$ where L is the distance between piezometers. The discharge through the cross section can be calculated either from Equation 18 in Chapter 4 or as $Q = Twi$. Obviously, the effectiveness of a rock stratum as an aquifer depends not only on its hydraulic conductivity but on the geometric dimensions of cross-sectional area perpendicular to the flow direction. In a rectangular cross section such as in Figure 6a the geometric dimensions are the thickness and width. In irregular cross sections as shown in Figure 6b the cross-sectional area is converted into an equivalent rectangular shape by keeping the width of the saturation zone the same. In the unconfined aquifers, complications arise because of the water table slope thickness is not constant. The practical solution is that either an averaging procedure has to be used prior to the application of Darcy's law or this law has to be applied to successive reaches of the aquifer, perpendicular to the flow direction.

Another significant point in Darcy's law application is that in homogeneous aquifers the transmissivity definition as $T = mK$ is perfectly valid and straightforward. In real field studies, however, most of the aquifers consist of a combination of different geological layers of varying hydraulic conductivities as well as thicknesses as in Figure 7. It is physically plausible to

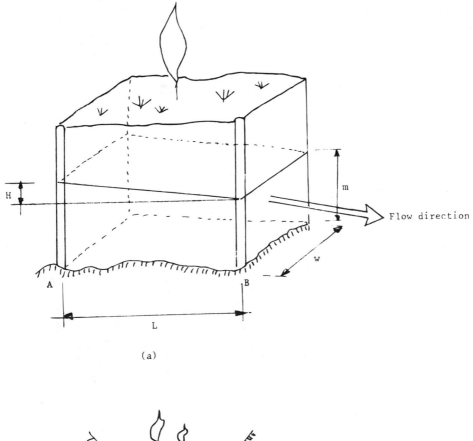

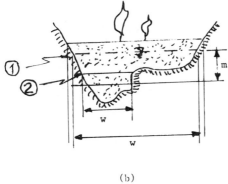

Figure 6 Hydraulic gradient calculations. (a) Regular geometry; (b) irregular geometry.

think strictly that the transmissivity must be worked out by calculating the contribution of each layer. Is it a solution to calculate an average hydraulic conductivity by dividing the total transmissivity by the total thickness? Does such a calculation depend only on the geology or is the flow direction also important?

Let the thickness and hydraulic conductivities be $m_1, m_2, m_3, \ldots, m_n$ and $K_1, K_2, K_3, \ldots, K_n$, respectively, where n is the number of different layers in a multiple aquifer such as in Figure 8. First the horizontal flow with total discharge Q is considered. Since no water is gained or lost in passing through the various layers, the principle of mass conservation can be used. In that, the total discharge is equal to the summation of discharges $Q_1, Q_2, Q_3, \ldots, Q_n$ in the individual layers. In such a configuration for horizontal flow the hydraulic gradient

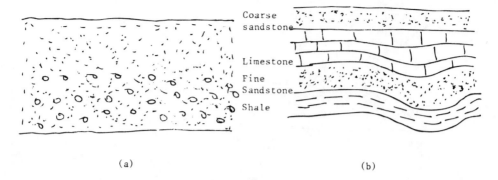

(a)   (b)

Figure 7 Various geological layers.

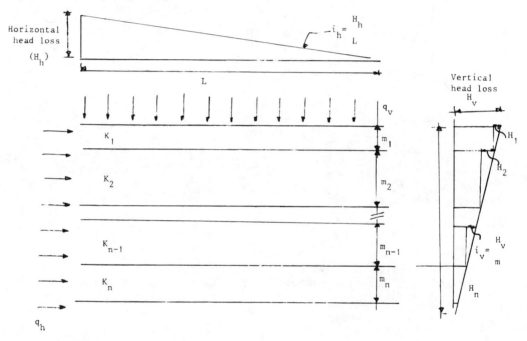

Figure 8 Multiple aquifer.

has the same value for each layer. Therefore,

$$Q_H = Q_1 + Q_2 + \cdots + Q_n \tag{8}$$

On the other hand, for unit width ($w = 1$) of the aquifer cross section the individual discharges become from Equation 18 in Chapter 4

$$\begin{aligned}
Q_1 &= K_1 m_1 i_h \\
Q_2 &= K_2 m_1 i_h \\
Q_3 &= K_3 m_3 i_h \\
Q_n &= K_n m_n i_h
\end{aligned} \tag{9}$$

It is obvious from Figure 8 that horizontal flow in each layer has the same hydraulic gradient.

GROUNDWATER FLOW LAWS

Substitution of Equation 9 and Equation 18 in Chapter 4 into Equation 8 leads to

$$Q_H = (K_1 m_1 + K_2 m_2 + \cdots + K_n m_n) i_h \tag{10}$$

By definition, the specific discharge in the horizontal direction is $q_H = Q_H/A = Q_H/(m_1 + m_2 + m_3 + \ldots + m_n)$ and therefore the substitution of Equation 10 yields

$$q_H = \frac{K_1 m_1 + K_2 m_2 + \cdots + K_n m_n}{m_1 + m_2 + \cdots + m_n} i_h$$

which means that the horizontal hydraulic conductivity, K_H, for the section considered is

$$K_H = \frac{K_1 m_1 + K_2 m_2 + \cdots + K_n m_n}{m_1 + m_2 + \cdots m_n} \tag{11}$$

This last expression implies that the horizontal hydraulic conductivity is the weighted average of the individual hydraulic conductivities with layer thickness being the weights. It is interesting to notice that if the layers have the same thicknesses then Equation 11 takes the form

$$K_H = \frac{K_1 + K_2 + \cdots + K_n}{n} \tag{12}$$

in which n is the number of layers. This last expression implies that only in the case of equal layer thicknesses is the average hydraulic conductivity equivalent to the arithmetic average of hydraulic conductivities; otherwise, this statement is never valid even approximately for unequal layer thicknesses. In fact, if the thicknesses are not equal the arithmetic averaging leads to under-estimation.

On the other hand, in the cases of vertical flow the overall hydraulic gradient is equal to the summation of the individual hydraulic heads divided by the total thickness (see Figure 8).

$$i_v = \frac{H_v}{m} = \frac{H_1 + H_2 + \cdots + H_n}{m_1 + m_2 + \cdots + m_n} \tag{13}$$

Each layer allows passage of the same total vertical discharge Q_v, and therefore the specific discharge in each layer, q_v, is the same. The application of Darcy's law in each layer gives the individual head losses as

$$H_1 = \frac{m_1}{K_1} q_v$$

$$H_2 = \frac{m_2}{K_2} q_v$$

$$H_3 = \frac{m_3}{K_3} q_v \tag{14}$$

$$\cdots$$

$$H_n = \frac{m_n}{K_n} q_v$$

Substitution of Equation 14 into Equation 13 yields

$$q_v = \frac{m_1 + m_2 + \cdots + m_n}{\dfrac{m_1}{K_1} + \dfrac{m_2}{K_2} + \cdots + \dfrac{m_n}{K_n}} i_v$$

which is tantamount to saying that the vertical hydraulic conductivity, K_v, is

$$K_V = \frac{m_1 + m_2 + \cdots + m_n}{\dfrac{m_1}{K_1} + \dfrac{m_2}{K_2} + \cdots + \dfrac{m_n}{K_n}} \qquad (15)$$

Most frequently multiple layers occur in the sedimentary rocks. In fact, they are often anisotropic with respect to hydraulic conductivity because they contain grains that are not spherical but elongated in one direction. During the deposition these grains settle with their longest axes more or less horizontally (see Figure 9) and this usually causes the horizontal hydraulic conductivity to be greater than vertical conductivity in a single layer. However, when many layers are considered, the bulk hydraulic conductivity of sediments is usually much greater than the vertical counterpart. In order to prove this point from the previous equations by considering two layers only the ratio K_H/K_v can be written as

$$\frac{K_H}{K_V} = \frac{m_1^2 + \left(\dfrac{K_1}{K_2} + \dfrac{K_2}{K_1}\right) m_1 m_2 + m_2^2}{m_1^2 + 2 m_1 m_2 + m_1^2} \qquad (16)$$

The hydraulic conductivity ratio terms in the denominator are always greater than two, because if x is any number, $x + (1/x) > 2$. Consequently, always $K_H > K_v$. It is also true for alluvial deposits that are usually constituted by alternating layers or lenses of sand and gravel on occasional clays.

B. NONLINEAR FLOW LAWS

As mentioned earlier Darcy's law does not cover all the groundwater problem solutions in nature. Deviations from the linear flow are due to either aquifer material composition such as coarse grains, fractures, and solution cavities or flow velocities leading to turbulent flow. However, in any case the head difference or more specifically the hydraulic gradient increases proportionally with the specific discharge but in a nonlinear manner. This nonlinearity causes greater deviations from the linear law especially at high velocities even though the medium is porous in which the linear law is valid at low specific discharges only. The deviations may occur along either a or b curves as shown in Figure 10. Theoretical considerations suggest that the relationship should begin as a straight line for a short duration and gradually curved upward as the curve b.

Darcy (1856) observed during his experimental studies that most of his tests obeyed the linear specific discharge hydraulic gradient law but in the meantime he realized that there

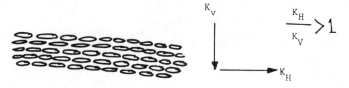

Figure 9 Heterogeneity.

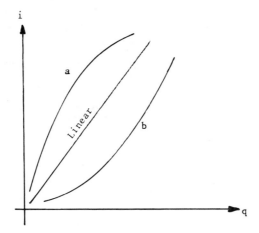

Figure 10 Porous medium nonlinear flow laws.

was an upper limit in terms of the specific discharge value approximately as 0.1 m/s for his materials. Because we know that the specific discharge is proportional physically with the hydraulic gradient and conductivity, any increase in these hydraulic parameters may lead to a nonlinear (turbulent) flow within the whole or part of an aquifer. For instance, in coarse-grained porous and fractured media due to relatively high hydraulic conductivity the groundwater flow may be turbulent as was shown by Givan (1934), Engelund (1953), and Basak (1977). In the fractured media a nonlinear relationship has been observed between the specific discharge and the hydraulic gradient by Louis (1969), Snow (1968), and Maini (1971). The nonlinearity has been attributed to various factors such as the kinetic energy effects, nonlinear pressure flow laws, leakage packers, and increase in the fracture aperture. Suggestions for nonlinearity can be attributed to

1. Kinetic energy increases.
2. Nonlinear pressure flow laws.
3. Increase in fracture aperture, etc. Louis and Maini (1970) gave the relationship between the hydraulic gradient and the discharge as shown in Figure 11. Initially for small velocities both normal and deformable rocks behave according to the linear flow but at large velocities each one shows distinct patterns of behavior. In this figure for normal rock there are four distinctive portions:

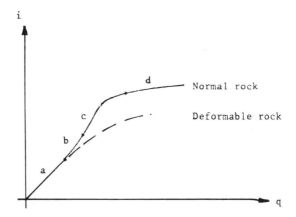

Figure 11 Fractured medium nonlinear flow laws.

1. Laminar flow domain where the Darcy law is valid
2. Turbulence effect.
3. Turbulence offset by fissure expansion.
4. Full fissure expansion. In the last three portions nonlinear flow laws are valid.

Nonlinear flow has been recognized for various field situations due to

1. Turbulent flow development where the hydraulic gradient and/or the pores are large.
2. Flow through coarse-grained alluviums and sandstones.
3. Very porous formations such as cavernous limestones and dolomites.
4. Extremely fine-grained materials (Kutilek, 1969).
5. Where sands are not completely saturated with water.

The nonlinear flows laws in Figures 10 and 11 have engaged researchers for many years and they tried to fit empirical curves to the experimental data obtained in the laboratory. Despite the increased efforts of research on the non-Darcian flow, a number of problems still await solutions in determining flow rates and hydraulic gradients outside the validity range of Darcy's law. Among these are the problem of defining the upper limit at which this law ceases to apply, the relationship for the particular granular material under consideration, and computing nonlinear flow nets.

Some of the basic definitions such as the transmissivity and the analytical or graphical procedures that are applicable to linear flow situations can be applied to the non-Darcian type of flows after modification as will be the main topic of Chapter 12. A large number of proposed nonlinear laws for groundwater motion, resulting from various theoretical and/or experimental investigations, fall into two main categories, polynomial and power laws as explained below.

Many studies were done in order to justify the validity of Darcy's law both experimentally and theoretically. These can be summarized as follows:

1. The application of Navier-Stokes equations indicated that, when the initial terms become significant, an upper limit appears beyond which the Darcy law is not valid.
2. The analogy with flow in capillary tubes suggested that the invalidity of the linear flow occurs with the commencement of turbulence implying an upper limit of validity.
3. The dimensionless analysis performed by Slepicka (1961) denied the unequivocal validity of the linear flow for the entire vast extent of hydraulic possibilities.
4. The application of the statistical methods indicated that the flow through porous medium can be modeled by the use of random walk approach (Scheideger, 1954). A simplified method has allowed the derivation of a linear expression equivalent to the Darcy law. Otherwise, deviations from the linear flow were observed.

All of the aforementioned theoretical studies showed that there appears an upper limit beyond which the linear law is no more valid. These theoretical studies, although sound in their approaches by combining some physical properties of the flow and the medium in addition to various simplifying underlying assumptions, do not yield quantitative value for the upper limit of validity. This drawback in the theoretical studies has been alleviated by the experimental investigations carried out in laboratory conditions. A number of researchers have attempted to find out a universal nonlinear flow instead of linear law. It has been customary to employ the Reynolds number for distinction between linear (laminar) and nonlinear (turbulent) flows. In practice, the linear flow is valid as long as the Reynolds number is less than a

GROUNDWATER FLOW LAWS

Table 1 Published Critical Reynolds Numbers

Original proposer	Porous medium material	Porosity	Critical Reynolds number
Zunker (1920)	Spheres	36.9–39.1	4
Ehrenberger (1928)	Homogeneous sand	38	5
Lindquist (1933)	Spheres (plumb)	37.1–39.1	4
Fancher et al. (1933)	Consolidated and unconsolidated media	11.9–38.5	1–10
Hickox (1934)	Uniform river sand	36.5–49.8	2
Givan (1934)	Uniform spheres (plumb)	35.7–42.1	10
Bakmeteff and Feodoroff (1937)	Uniform spheres (plumb)	36.9–45.9	5
	Uniform sand and gravel	32.7–45.6	5
Klinge (1940)	Uniform spheres (steel and glass)	39–41	10
Veronese (1941)	Uniform sand and gravel	38	5
Rose (1945)	Uniform spheres (plumb)	40	10
Rose and Rizk (1949)	Divers	40	10
Engelund (1953)	Uniform sand (calcarious and silicious)	39.5	3.55
Schneebeli (1955)	Uniform spheres (glass)	39	5
	Uniform granite pieces	47	2
Indri (1958)	Uniform and mixed sphere	38	5
Karadi and Nagy (1961)	Uniform spheres (steel, sand, and gravel)	40	5
Dudgeon (1964)	River gravel	37.2–51.5	5
Chauveteau (1965)	Consolidated sand uniform grain size	33.8–35.9	2–3

critical value. Table 1 presents the chronological list of experimental tasks with the aim of determining this critical number.

It is clear from this table that the critical Reynolds number may have any value between 1 and 10 and hence there is no unique number for any type of material and furthermore the porosity cannot be related to the critical number. A striking empirical rule can be deduced from the last column in Table 1 in that the critical Reynolds number has an approximate average value, mode, and median all equal to about 5. In statistical terminology the equivalence of these three central values implies a symmetrical frequency distribution function of the critical Reynolds number. In fact, for all the practical purposes it can be assumed to have a Gaussian distribution function with the standard deviation value equal to 2.8.

1. Polynomial Flow Laws

The first of the nonlinear expressions was suggested by Forchheimer (1901) for high flow rates as

$$i = aq + bq^2 \tag{17}$$

where a and b are dependent on the medium and fluid properties and their units are $[T/L]$ and $[T^2/L^2]$, respectively. In fact, a and b are also dependent on the velocity distribution and subsequently on the boundary configurations of the actual flow path and on the Reynolds number. However, they can be taken as constants within experimental accuracy over a wide range of Reynolds number from 1 to 50. It is clear from Equation 17 that for low flow rates $aq >> bq^2$; therefore, it approaches the linear flow law. However, for high flow rates $bq^2 >> aq$; therefore, it becomes equivalent to the Escande (1953) power law as $i = bq^2$. Hence, the Forchheimer law is general in the sense that depending on the flow conditions it may converge to the linear or power flow laws. A schematic classification of the Forchheimer flow law behavior is represented in Figure 12.

As stated by Şen (1986) a can be referred to as laminar flow coefficient whereas b is the turbulence factor. Their values for an extensive list of different types of porous media are

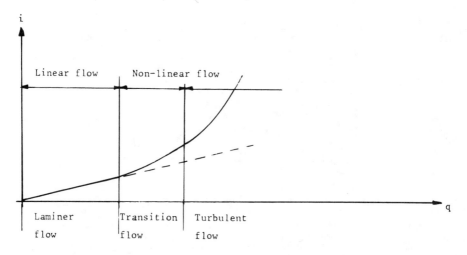

Figure 12 Flow classification.

given by Basak (1976). In addition, Tuin (1960) gave a and b values for gravel and boulders as shown in Table 2. The susceptibility of a medium for nonlinear flow to occur can be measured by a parameter, $c = b/a^2$ which has values less than 0.1 for fine or medium sand, 0.1 to 0.2 for coarse sands, and more than 2 for gravel. The value of c is an inverse function of the velocity at which the flow becomes turbulent. However, regardless of the c values, if q is low enough, then the flow is laminar; however, if q and Re are high enough the flow is turbulent. Other alternative polynomial nonlinear flow laws are summarized in Table 3.

2. Power Flow Laws

The main difference of this class from the previous ones is that they do not include any linear portion as schematically shown in Figure 13. Various power equations are presented in the literature to describe nonlinear flow law, in general, as

Table 2 a and b Coefficients

Material	Average grain diameter (mm)	a (s/m)	b (s²/m²)
Gravel A	7.05	12.30	305
Gravel B	14.31	2.50	244
Gravel C	28.53	0.04	107.70
Gravel D	15.72	5.00	288.50
Gravel E	11.40	6.30	443
Gravel F	22.13	4.38	225
Gravel G	25.65	4.33	117
Gravel H	62	0.19	69
Basalt boulders (10–60 kg)	225	0.13	15.30
Basalt boulders (60–200 kg)	385	0.21	3.63

Table 3 Polynomial Flow Laws

Proposer	Equation	Constants	Remarks
Forchheimer (1901)	$i = aq + bq^2$	a, b	Semiempirical
	$i = aq + bq^2 + cq^3$	a, b, c	Empirical
Rose (1945)	$i = aq + bq^{1.5} + cq^2$	a, b, c	Empirical
Poluborinova-Kochina (1962)	$i = aq + bq^2 + c\dfrac{q}{t}$	a, b, c	Empirical
Muscat (1946)	$i = aq + bq^n$	a, b, n	Empirical

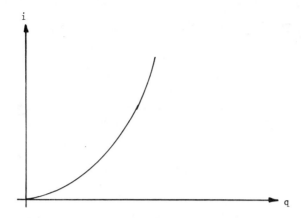

Figure 13 Power law graphs.

$$i = aq^n \tag{18}$$

where a and n take different forms and values as presented in Table 4.

On the other hand, Smreker (1878) is the first one to suggest a nonlinear relationship between i and q by imagining that the hydraulic conductivity, K, is dependent on the groundwater velocity and hence on the time. His expression was an exponential type as specified in Table 4 chronologically along with other researchers' proposals.

A common property to all of these flow laws is that the exponent term assumes values between 1 and 2 except in the case of Escande formulation which is valid for big boulders of basalt only.

IV. FRACTURED MEDIUM LAWS

The groundwater flow in an individual fracture is quite different from the porous medium flow but rather similar to the pipe flow. Contrary to the porous medium the moving water particle is not subjected to resistance at every point but only at the intersections of two or more fractures. Especially in rarely fractured medium the fracture network is similar to pipe network of water supply system within a city. Accordingly, there is a distinct piezometer

Table 4 Power Laws

Proposer	Nonlinear flow equation	Exponents and coefficients	Remarks
Smreker (1878)	$i = Knq^n$ $(1 < n < 2)$	K, n	Semiempirical resistance law
Izbach (1931)	$q = Mi^n$ $(1/2 < n < 1)$	M, n	Empirical
Escande (1953)	$q = (Bi)^n$ $n = 1/2$	B, n	Empirical B depends on grain size
Missbach (1967)	$i = aq^n$	a, n	Empirical
Slepicka (1961)	$q = K_f i^n$	K_f, n	Dimensional analysis
White (1935)	$i = cq^{1.8}$	c	Empirical for rocks
Wilkins (1955)	$i = cq^{1.85}$	c	Empiricial for rocks
Parkins (1962)	$i = cq^{1.86}$	c	Empirical for rocks
Anandakrishnan and Varadarajulu (1963)	$q^n = k'i$ $(1 < n < 2)$	n, k'	Semiempirical Dimensionless analysis

pattern as shown in Figure 14. In the piezometer network of a fractured medium the following points are worth noticing:

1. The piezometric line network is similar to the fracture network.
2. Each fracture has its own piezometric level.
3. Continuous head losses occur along each fracture.
4. Discontinuous (or local) head losses occur at the fracture intersections.

With the increase in the number of fractures the medium approaches the conventional porous medium and therefore there appears a piezometric surface with continuous head losses only. In the fractured rocks the continuous head losses are comparatively smaller than the porous medium; however, existence of sudden head losses at fracture intersections give rise to important overall head losses. In light of the above-mentioned discussions depending on the fracture density there are two different methods for tackling fractured medium flow laws either in an individual fracture or an equivalent porous medium.

A. INDIVIDUAL FRACTURES

In individual fractures the flow laws were investigated by Louis (1974) for laminar and turbulent flow by assuming a uniform flow velocity over the total aperture of the fracture which produces the same flow as the real one. In any single fracture there may appear one- or two-dimensional flow because the third dimension is negligibly small. One-dimensional flow happens in the form of either parallel or nonparallel pattern as shown in Figure 15. When

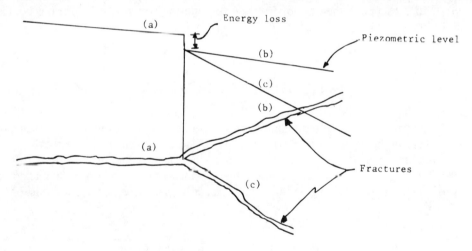

Figure 14 Piezometric levels along fractures.

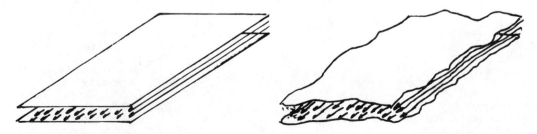

Figure 15 One-dimensional flow in fractures.

the streamlines are not parallel, a two-dimensional flow appears within the fractures as represented in Figure 16.

In addition to the Reynolds number in fractures, the relative roughness of the fracture walls plays a dominant role in flow regime determination and its validity. In general, fracture relative roughness, R_r, is referred to as the ratio of mean irregularity height, η, (see Figure 17) to the hydraulic radius, H_r, of the fracture cross section as

$$R_r = \frac{\eta}{H_r} \quad (19)$$

On the other hand, the hydraulic radius is the ratio of fracture cross-section area to its perimeter in saturated fractures. For a wide fracture with the aperture of $2a$, the hydraulic radius is $H_r = 2a$. In general the relationship between the hydraulic gradient and the average velocity of water in a pipe is given through the Darcy-Weisbach relation as in Equation 1, which is applicable to any conduit provided that the hydraulic radius and the friction coefficient, λ, are known. Friction coefficient (loss coefficient) is dependent on Re and relative roughness R_r. Many experimental investigations have been conducted by Louis (1974) just for determining the relationship between λ and Re. Depending on the R_r value the fracture roughness acts either as a hydraulically smooth for $R_r < 0.032$ or rough for $R_r > 0.032$. On the other hand, the critical Reynolds number for water flow within the fracture is 2300 which separates laminar flow from turbulent. On the basis of critical Re and R_r values and the laboratory experiments, five different flow domains emerge as schematically shown in Table 5. To convert equations in this table into practically usable forms in terms of specific discharge and hydraulic gradient each one of the friction coefficients λ is substituted into Equation 1 with the introduction of Reynolds number for fractures explicitly as

$$Re = \frac{H_r q}{\nu} \quad (20)$$

in which ν is the kinematic viscosity of water. The following flow laws result for different flow regimes in the fractures:

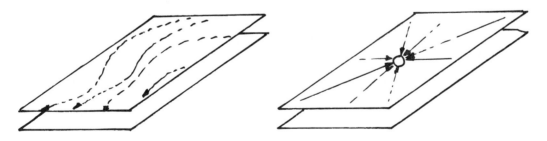

Figure 16 Two-dimensional flow in fractures.

Figure 17 Fracture roughness.

Table 5 Different Flow Domains

$\lambda = \dfrac{96}{Re}(1 + 8.8\,R_r)$	$\dfrac{1}{\sqrt{\lambda}} = -2\log\dfrac{R_r}{1.9}$
Rough laminar (Poissuille, 1846)	Very rough turbulent (Louis, 1974)
	$\dfrac{1}{\sqrt{\lambda}} = -2\log\dfrac{R_r}{3.7}$
	Rough turbulent (Louis, 1974)
$\lambda = \dfrac{64}{Re}$	$\lambda = -0.316\,Re^{-1/4}$
Smooth laminar (Blasius)	Smooth turbulent (Nikuradse)

1. For smooth-laminar flow regime a linear flow law is valid as

$$q = \frac{ga^2}{12\nu} i \qquad (21)$$

which is similar to Darcy law in porous medium.

2. In the rough-laminar flow regime the flow law in q-i plane becomes

$$q = \frac{ga^2}{12(1 + 8.7\,R_r^{1.5})\nu} i \qquad (22)$$

This is also a linear flow law but the fracture relative roughness plays significant role in the flow behavior. In fact, for zero relative roughness Equation 22 turns into Equation 21.

3. In the smooth-turbulent flow regime the flow law is of nonlinear type as

$$q = c i^{4/7} \qquad (23)$$

in which c is a constant dependent on the fracture aperture only as

$$c = \left[\frac{g}{\nu}\left(\frac{2a^5}{0.079}\right)^{1/4}\right]^{4/7} \qquad (24)$$

In its form Equation 23 is similar to the power laws of the porous medium with a significant difference that the value of power is less than 1.

4. In the rough-turbulent flow regime the flow law is also nonlinear as

$$q = c i^{1/2} \qquad (25)$$

where

$$c = 4(ag)^{1/2}\ln\left(\frac{3.7}{R_r}\right) \qquad (26)$$

5. Equation 25 is also valid for the very rough-turbulent flow domain with a slight change in the constant which becomes

$$c = 4(ag)^{1/2}\ln\left(\frac{1.7}{R_r}\right) \tag{27}$$

After all the above-mentioned flow laws it is clear that linear and power laws are the only alternatives in the fracture flow and the value of power is always less than one. These laws were determined in the laboratory on joint models with artificially created roughness and, therefore, they do not reflect necessarily the behavior of natural fractures. To understand their validity Sharp (1970) performed tests on the natural joints with four different roughness configurations as shown in Figure 18. In a series of tests on the roughness pattern as in Figure 18a the joint walls were fixed in direct contact with each other, whereas in other cases apertures of 0.25, 0.50, and 1.25 mm were maintained. The tests consisted of measuring the hydraulic gradient change with a dimensionless flow rate variable which is defined as q/q_1 where q_1 is the flow rate under unit hydraulic gradient. The test results are shown on a double logarithmic paper in Figure 19. It is seen from this figure that the test results appear in the form of power law and they are reproduced relatively well by the rough regime flow laws. However, the test results for configuration in Figure 18a show considerable deviation from the flow laws. This deviation may be due to the fact that in the mean aperture and hydraulic radius calculations the contact points appear as singular points with zero values of these quantities. As a general conclusion, however, it can be assumed that the flow obtained from the artificial fractures is approximately valid for the natural joints. One important point to remember is that all of the discussions on fracture flow laws were obtained under the steady-state flow condition similar to Darcy's law.

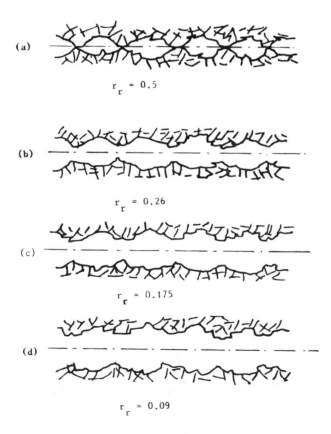

Figure 18 Natural fracture roughnesses (after Sharp, 1970).

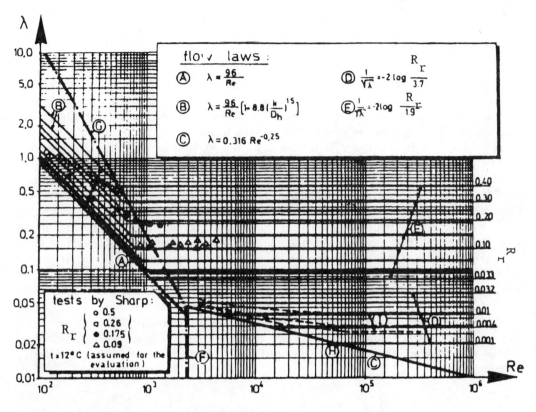

Figure 19 Natural fracture test results (after Sharp, 1970).

B. EQUIVALENT POROUS MEDIUM

Rock masses with the multitude of fractures in various directions and different apertures with more or less the same properties with no superiority of any fracture over others are broken up into blocks of irregular size and shape as represented in Figure 20. Barenblatt et al. (1960) assumed that any small volume of rock consists of a large number of porous blocks as well as fissures, both of which are randomly (size, shape, orientation, etc.) distributed within the rock mass. Therefore, within the same rock unit, there exists two different media: fractures and blocks. As explained in Chapter 4 their hydraulic behaviors are different. This difference becomes distinct only in the unsteady-state flow cases. Otherwise, it will be explained in Chapter 10 that in the steady-state case the fractures and blocks act together as one aquifer unit.

In the unsteady flow case, because fracture hydraulic conductivity is greater than the block hydraulic conductivity and fracture storativity less than block storativity, it is necessary to

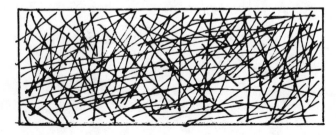

Figure 20 Fractured media.

consider the whole rock mass as consisting of two different but coexisting porous media with different hydraulic heads. Under such a consideration there will appear three different flows, namely, in the fractured medium, the porous medium, and from the blocks to the fractures. The flow law in both media can abide by any of the previously mentioned flow laws, linear or nonlinear. However, the same laws cannot be used for block-to-fracture flow.

1. Block-to-Fracture Flow Laws

The simplest of these laws is given by Barenblatt et al. (1960). They assumed that the water exchange between blocks and fractures depends only on the difference between the head in the fracture, h_f, and the average head in the block, h_b. No geometrical consideration is given to the porous blocks. With these assumptions and the theory of dimensionless analysis they proposed.

$$q = \alpha(h_b - h_f) \tag{28}$$

where q is the specific discharge from block-to-fracture and α is a parameter depending on the geometry of the fractured rock and it has the unit of inverse time [l/L]. The flow law in Equation 28 is based on the assumption of pseudo-steady-state flow. Another law for block-to-fracture flow is based upon the regular block shapes either in the form of cubes or parallelepipeds as will be discussed in Chapter 10.

The major assumption is that the prismatic blocks have low hydraulic conductivity. The flow within these blocks is of linear type perpendicular to the fracture, as mentioned in Chapter 3. Furthermore, the fractures have constant apertures and they are equally spaced. The suggested flow law relates the temporal and spatial variation of block hydraulic head as

$$K_b \frac{\partial^2 h_b}{\partial z^2} = S_{sb} \frac{\partial h_b}{\partial t} \tag{29}$$

subject to

$$h_b = h_0 \text{ (initial head) for } t = 0$$

$$h_b = h \quad \text{for } z = 0$$

$$\frac{\partial h_b}{\partial t} = 0 \quad \text{for } z = 0$$

in which S_{sb} is the block-specific storage coefficient and z is the vertical axis with its origin at the block-fracture boundary.

V. KARSTIC MEDIUM FLOW LAWS

Karst aquifers are very heterogeneous reservoirs in which water is transferred in a network of interconnected cracks, caverns, and channels. In short, all these secondary features will be referred to as the conduits. The flow in conduits is similar to the pipe flow. Hence, it is possible to say that the karstic media flow laws will be more or less similar to the fractured medium flows. However, in karstic conduits the water may not fill the whole cross section and therefore it resembles flow in pipes with partial fill, i.e., the water may not be under pressure.

The presence of large solution cavities and caves in karstic aquifer gives rise to laminar or turbulent flow regimes and Darcy law in granular rocks is no more valid. It is also possible by the passage of time that the karstic features become larger and as a result the flow regime might change.

REFERENCES

Anandakrishnan, M. and Varadarajulu, F. H., 1963. Laminar and Turbulent Flow of Water through sand, *J. Soil Mech. Foun. Div.,* 5, 1.

Bakmeteff, B. A. and Feodoroff, N. V., 1937. Flow through granular media, *J. Appl. Mech.,* 4A, 97.

Barenblatt, G. E., Zheltov, I. P., and Kochina, I. N., 1960. Basic concept in the theory of seepage of homogeneous liquids in fissured rocks, *J. App. Math. Mech.,* 24(5), 1286.

Basak, P., 1976. Steady non-Darcy seepage through embarkment, *J. Irrig. Drain. Div.,* 102(IR4), 435.

Basak, P., 1977. Non-penetrating well in a semi-infinite medium with non-linear flow, *J. Hydrol.,* 33, 375.

Basak, P., 1978. Analytical solutions for two-regime well flow problems, *J. Hydrol.,* 38, 147.

Chauveteau, G., 1965. Essai sur la Loi de Darcy et les Ecoulement Laminaires a Perte de Change Non-Lineare. Thesis presented to the Universite de Toulouse, Toulouce, France, in partial fulfillment of the requirements for the degree of Doctor of Philosophy.

Darcy, H., 1856. *Les Fontaines Publique de la Ville de Dijon,* Victor Dalmond, Paris.

Dudgeon, C. R., 1964. Flow of Water through Coarse Granular Materials, Water Research Laboratory, Report No. 76, University of New South Wales, Australia.

Dudgeon, C. R. and Yuen, C. N., 1970. Non-Darcy Flow in the Vicinity of Wells, Proceedings of Groundwater Symposium, The University of New South Wales, Australia, p. 13.

Ehrenberger, R., 1928. Versuche Uber die Ergiebigheit von Brunnen und Bestimmung der Durchlassighkeit des sandes. Zeitschrift des Osterreicher Ingenieur und Architektunvereins, Heft 11/12, Austria (in German).

Engelund, F., 1953. On the Laminar and Turbulent Flows of Ground Water through Homogeneous Sands, Bulletin of the Technical University of Denmark, Copenhagen, No.4.

Escande, L., 1953. Experiments Concerning Infiltration of Water through a Rock Mass, Reprint from Proc. Minnesota Int. Hydraul. Convention.

Fancher, G. H., Lewis, J. A., and Barnes, K. B., 1933. Soil Physical Characteristics of oil sands. State of Pennsylvania Mineral Industries Exploration, State College Bull. 12, College Park.

Forchheimer, P. H., 1901. Wasserbewegung Durch Boden. Zeitschrifft Ver. Deutch. Ing., No. 49, 1736.

Givan, C. V., 1934. Flow of Water Through Granular Materials—Initial Experiments with Lead-Shot, Am. Geophys. Union, 15th Annual Meeting, 572.

Hickox, G. H., 1934. Flow through Granular Materials, Trans. Am. Geophys. Union, 15th Annual Meeting, 567.

Indri, E. 1958. Experience sur L2ecoulement de L'eau a Travers des Ames de Materiaux Spheriques. Acqua 3 (in French).

Izbach, S. V., 1931. O Filtracii V Kropnoaernstom Materiale. Izv. Nauchnoissles, Inst. Gidrotechniki (NIIG), Leningrad.

Karadi, G., and Nagy, I. V., 1961. Investigations into the Validity of the Linear Seepage Law, Proc. 9th Convention, Int. Assoc. Hydr. Res., Dubrovnik, Yugoslavia, 556.

Klinge, W., 1940. Was ist Wasserwirtschaft? Die Wasserwirtschaft, 39, Heft 4 (in German).

Kutilek, M., 1969. Non-Darcian Flow of Water in Soils (Laminar Region), 1st IASH Symp. Fundamentals, Israel Elsevier, Amsterdam, 327.

Laushey, L. M. and Popat, Y., 1978. Darcy's Law during Unsteady Flow, Int. Assoc. Sci. Hydrol., General Assembley of Bern, Ground Water,

Lindquist, E., 1933. On the Flow of Water through Porous Soil, 1er Congress des Grandes Barrages, 5-18 (in French).

Louis, C., 1969. A Study of Groundwater Flow in Jointed Rock and Its Influence on the Stability of Rock Masses, Rock Mech. Res. Rep. No.10, Imperial College of Science and Technology, University of London.

Louis, C., 1974. Introduction, in *Hydraulique des Roches,* Bureaux de Recherches Geologiques et Minieres, Orleans, France,

Louis, C. and Maini, Y. N., 1970. Determination of In Situ Hydraulic Parameters in Jointed Rock, Proc. 2nd Congress of Int. Soc. of Rock Mech., Belgrade, 1.

Maini, Y. N., 1971. In Situ Hydraulic Parameters in Jointed Rock. Their Measurement and Interpretation, Ph.D. thesis, Imperial College of Science and Technology, University of London.

Missbach, A., 1967. Non-Linear Flow through Porous Materials, Stark, K. P. and Volker, R. E., Eds., Res. Bull., No. 1, April, Dept. Civil Engineering, University College, Townsville, Australia.

Muskat, M., 1946. *The Flow of Homogeneous Fluid through Porous Media,* 2nd ed., McGraw-Hill, New York.

Parkins, A. K., 1962. Rock-Fill Dams with Inbuilt Spillways. I Hydraulic Characteristics, Dept. Civil Engineering, University of Melbourne, Australia.

Poiseuille, J., 1846. Recherches Experimentales sur le Movement des Liquides dans les Tubes de tres Petit Diametre, Mem. Savants Etrange, Vol. 9.

Poluborinova-Kochina, P. Ya., 1962. *Theory of Groundwater Movement. Gostekhizdat, Moskow,* Princeton University Press, Princeton, NJ.

Poluborinova-Kochina, P. Ya., 1962. *The Theory of Ground-Water Movement,* Princeton University Press, Princeton, NJ.

Rose, H. E., 1945. An investigation into the laws of flow of fluids through beds of granular materials, *Inst. Mech. Eng. Proc.,* 153, 511.

Scheidegger, A. E., 1954. Statistical hydrodynamics in porous media, *J. Appl. Phys.,* 25(8), 271.

Scheidegger, A. E., 1954. The physics of flow through porous media, *J. Appl. Physics,* 25(8).

Schneebeli, A. E., 1955. Experience sur la Limite de Validite de la Loi de Darcy et L'application de la Turbulence dans un Ecoulement de Filtration. La Houille Blanche, 2 (in French).

Şen, Z., 1986. Volumetric approach to non-Darcy flow in confined aquifers, *J. Hydraul. Div.,* 87, 337.

Şen, Z., 1989 Non-linear flow towards wells, *J. Hydraul. Div.,* ASCE, 115(2), 193.

Sharp, J. C., 1970. Fluid Flow through Fissured Media, Ph.D. thesis, Imperial College of Science and Technology, University of London.

Slepicka, F., 1961. The Laws of Filtration and Limits of Their Validity, Int. Assoc. Hydraul. Res. Proc. 9th Convention, 383.

Smreker, O., 1878. Entwicklung eines Gesetaes fur den Widerstand dei der Bewegung des Grundwassers, *Zeitschrifft des Vereines Deutscher Ingenier,.*

Snow, D., 1968. Rock fracture spacings, openings and porosities, *J. Soil Mech. Foun. Div.,* 94(1), 73.

Swartzendruber, D., 1969. The applicability of Darcy's law, *Soil Sci. Soc. Am. Proc.,* 32.

Tuin, V. D. H., 1960. La Permeabilite et les Applications Pratique des Materiaux Gross. Hydraul. Souterraine, Soc. Hydrotech., France, 17.

Veronese, A., 1941. Interpretazione della prove di Permeabilita nelle Sabia et Ghiaie. L'ingeniere, 15, 463 (in Italian).

White, A. M., 1935. *Trans. Am. Inst. Chem. Eng.,* 31, 390.

Wilkins, J. S., 1955. Flow of water through rock fill and its application of the design of dams, *N. Z. J. Eng.,* 5, 382.

Wilkinson, J. K., 1956. The Flow of Water through Rock Fill and its Application to the Design of Dams, Proc. 2nd Australia-New Zealand Conf. on SMFE.

Zunker, F., 1920. Das Allgemeine Grundwasserflissgesetz. Gasbeleuchtung und Wasserversorgun, (in German).

CHAPTER 6

Water Wells

I. GENERAL

Any vertical excavation on the earth's surface is a water well provided that the purpose is to extract groundwater from the zone of saturation for domestic, agricultural, industrial, or similar uses. A water well is designed to get the optimum quantity of water from a saturated geological formation. Wells should have shapes that simplify the groundwater movement toward the well storage. The choice of well shape and dimensions depend on the topography, piezometric level, subsurface geological conditions of the surrounding aquifer, regional climate, rainfall, and recharge possibilities as well as quantities in addition to the water demand from the well itself. On many occasions economics and politics play an important role in well location determination. Technological aspects in a well excavation are outside the scope of this book, however, they are discussed elsewhere (Driscoll, 1987; Anderson, 1967).

Prior to any well excavation it is advisable to study the hydrological, subsurface geological, economical, and other aspects of the region from the available data that should be supported by reconnaissance field surveys. It is important to gain some insight into the general topography, geomorphology, geology, hydrology, and hardness of the soil layers. In such a study, geographical, topographical, and geological maps are indispensable means that can be assisted by aerial photographs, satellite images, drilling logs, and existing water supply structures.

In light of all the available data decision variables about the well cross-section shapes, dimensions, penetrations, and depths should be determined. The main theme of this chapter is not the well drilling, design, or development but the explanation of the interaction of the well penetration, shape, and diameter with the actual groundwater flow in addition to the expected objectives of well measurements in the field. Hence, this chapter is confined to practical aspects of well excavation and its mutual interaction with groundwater movement and aquifers.

II. REGULAR CROSS SECTION

Cross section of a water well is important for different functions to have the optimum performance during planning, design, and operation stages of water supply. In calculating discharge or in identification of any aquifer parameters, the geometric shapes of flow and equipotential lines in the vicinity of the well provide restrictive conditions. For instance, if the well boundary is smooth without any discontinuity (corners, fractures,) then the flow lines will end up without any dilemma. Existence of corners on the well periphery gives rise to complications in the stream and equipotential lines which may cause extra energy loss around the well vicinity as shown in Figure 1a. The geometric shape of equipotential lines become irregular, giving rise to difficulties in the analytical treatments of groundwater problems.

In regular cross sections the equipotential and flow lines take simple regular shapes. For instance, in Figure 1b the equipotential lines are concentric circles, whereas the flow lines are straight along radial directions but in Figure 1a they are irregular. As explained in Chapter 4 the calculation of discharge depends on the equipotential line length which is rather difficult to measure or visualize in irregular cross sections. The longer the stream lines the greater will be the energy loss.

For practical purposes, it is necessary to have a smooth surface at the well face in order to avoid any extra energy losses.

A. CIRCULAR CROSS SECTION

An ideal water well should have circular cross section which is advantageous from the following points of view:

1. The groundwater pressure at any point along the well circumference on the horizontal plane is radial and therefore the tangential stresses are all equal. As a result, the pressure distribution is radially uniform. This creates compression strength on the well casing and screen without any tensile strength. The geological formations are more resistant to compression stress than tensile stress and consequently collapses on well face are reduced to a minimum in a circular cross section, particularly in unconsolidated formations.
2. For any given cross-sectional area the circular cross section has the minimum circumference area and therefore during the well drilling the friction between the drilling bit and the hole lateral surface will be the minimum. In addition, the casing is cheaper. In fact, all the drilling techniques lead to circular shapes. Other shapes are impossible, especially in deep well drilling.
3. Flow lines are radially straight lines and the circular circumference is a potential line. This means that the flow net geometry is completely known in circular-cross-section wells. The calculation of this case is the simplest among many others as will be seen in Chapter 8.

1. Large-Diameter Wells

For many centuries groundwater has been withdrawn from the aquifers by means of large-diameter wells. These wells are used mainly for individual exploitations and their excavation

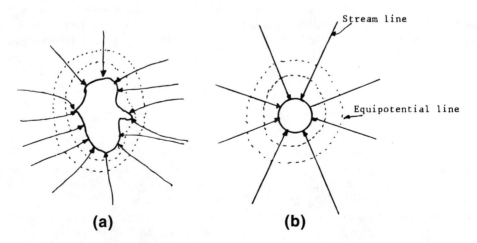

Figure 1 Effect of cross section on flow nets.

requires some minimum conditions and that the position of the water table should be close to the earth surface. The wells are generally circular in cross section, but rectangular or square shapes are not uncommon in some areas. In practice, this distance varies between 1 and 30 m. Large-diameter wells are convenient to extract large quantities of water from a low-permeability aquifer such as unconsolidated glacial and alluvial deposits of shallow depths. Performance of a large-diameter well poses special problems of analysis and interpretations.

The main instruments used in the excavation of a large-diameter well are the picks and shovels. Mostly they are dug by hand and therefore known as hand-dug wells. The minimum size of diameter is about 1 m which gives comfort for one person to work in the well digging. The larger the diameter the more will be the storage of well and therefore at the time of need large quantities can be abstracted initially with no difficulty. These wells cannot be deep and the maximum depth observed in the practice are about 50 m. They are mostly dug in unconfined aquifers and the actual depth of a large-diameter well at any time is a function of the groundwater table elevation. Initially, when the well reaches the groundwater table the aquifer is penetrated for about 1 to 2 m and the water is pumped out. A regional drop in the water table may require additional digging as necessary. In this way, the maximum depth is limited by the underlying impermeable nonfractured crystalline rocks. Applications of large-diameter wells in the Middle Eastern countries and in India indicate that mostly they are dug in Quaternary deposits of wadi alluviums that are underlain by weathered rocks and fractured media before penetration into solid rocks. Figure 2 shows a typical large-diameter well. A large-diameter well pierces from the aquifer a volume in which the storage of water is directly

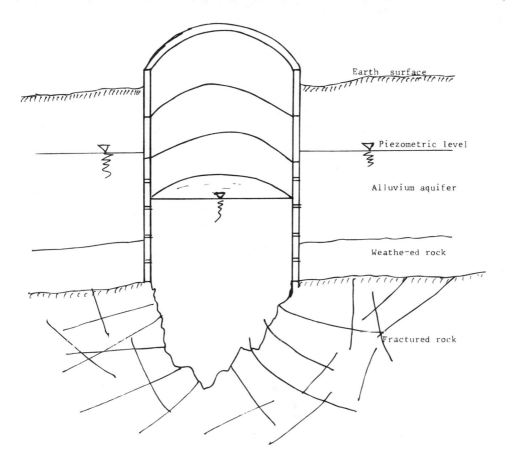

Figure 2 Large-diameter well.

ready for exploitation. Therefore, during a short time period the water is abstracted from well storage only. This gives rise to a difference between the water levels in the aquifer and well and, consequently, groundwater flows toward the well at small rates. It is logical to conclude that the larger the diameter the smaller is this difference and smaller the groundwater flow rate. The time lag between the aquifer response and abstraction tends to be long. This point is taken into consideration in calculations related to both aquifer and/or the well. Large-diameter wells have the following advantages:

1. Water already derived from the aquifer storage prior to abstraction is available in the well storage for use at any time.
2. The excavation process does not require sophisticated implements apart from simple picks and shovels.
3. Skilled personnel are not required for excavation, construction, and operation of well.
4. The well can be extended easily after the water table drop by either digging or boring a few vertical bores at the bottom.
5. Because the elevation difference between the earth and water surface in the well is small, simple centrifugal pumps can be employed for pumping.
6. Economically they are rather cheap and provide good solution for domestic, ranch, small agricultural land, or private estates.

It is not surprising to see that every day more and more hand-dug wells are being constructed for providing the rural population of the world with water. Large-diameter wells also have disadvantages that may abolish their use:

1. Susceptible to contamination or pollution.
2. Occupy large space for the well and excavated material.
3. Dangerous especially for small children.
4. Susceptible to filling in from flood sediments if excavated in flood planes.
5. The well may dry up in drought years.
6. Not suitable for high potential aquifers where the transmissivity is more than 50 m^2/d.
7. Subject to water table fluctuations for 1 year.
8. Deep aquifers cannot be tapped economically.

The larger the diameter the longer is the time necessary for reaching a quasi-steady-state flow after discharge of water from the well. Figure 3 exposes the general layout of a properly designed large-diameter well. The bottom plaque is necessary for the well stability and it protects the movement of particles in the bottom of the excavation. In hard rocks, large-diameter wells intersect more numerous fractures compared to small-diameter wells.

2. Small-Diameter Wells

Industrialization has brought with it an enormous increase in water demand so that shallow-dug wells become unsatisfactory for water supply. Advanced technology gave ways to drill holes in the earth at very great depths for water, oil, mineral resources, or/and subsurface geological explorations. Hence, another type of well, i.e., drilled, is used extensively in practice. They have small diameters and invariably have circular cross sections. The diameter is usually 12 in. (30 cm) and the well depth may reach down to 2000 m. It is, therefore, possible to abstract water from a multitude of aquifers along a hydrostratigraphical log. Because the well diameter is small the well storage for direct use is almost negligible. Even any small amount of water taken from the well storage creates large head differences between the water surface in the well and the piezometric level within the aquifer. Consequently, aquifer response to

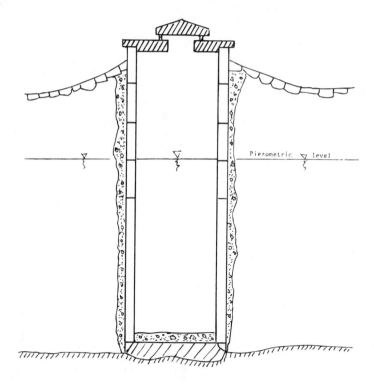

Figure 3 Large-diameter well.

water abstraction is rather instantaneous and no lag time between the two is assumed which simplifies the calculations concerning the well or aquifer hydraulics to a significant extent.

There are different methods in drilling a small-diameter well as explained by Driscoll (1987) and Huismann (1972). The most widely used methods are cable tool, hydraulic rotary, and reverse hydraulic rotary. A typical small-diameter well is presented in Figure 4. The main advantages of having a small diameter well are

1. It does not occupy large area of drilling.
2. Multiple aquifers can be drained easily.
3. Water supply can continue without stoppages even in drought years. This is one of the reasons why in arid regions with desert landscape small-diameter wells are drilled.
4. Great depths can be reached with quick constructions.
5. Groundwater is of better quality from the pollution point of view.
6. Small-diameter wells are very economical if deep-seated aquifers are encountered.
7. Flowing artesian wells may be hit so that the water elevation does not become a problem.
8. In high permeability aquifers the small-diameter wells discharge large amounts of water quickly.

On the contrary, the main disadvantages of small-diameter wells are

1. Well loss, which is the difference between water table within the well and the piezometric level in the aquifer but at the well face, is comparatively greater than a large-diameter well. This causes extra cost in hoisting water.
2. Initial capital on sophisticated drilling equipment and machinery is high.
3. Requires skilled personnel and great care to drill and operate the well.
4. Costly submersible pumps are required.
5. The maintenance of well is rather costly.

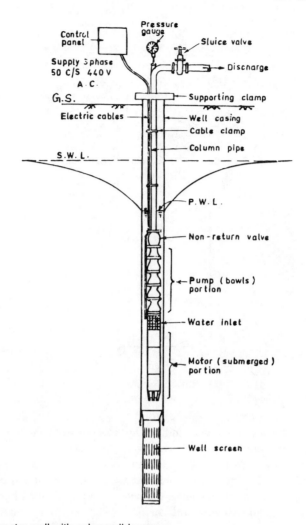

Figure 4 Small-diameter well with submersible pump.

In well type choice the following must be taken into consideration.

1. Space availability.
2. Hydrogeological and geological characteristics of the subsurface strata.
3. Hydrological considerations concerning natural recharge and seasonal fluctuations of water table level.
4. Cost of construction, operation, and maintainance.

B. RECTANGULAR CROSS SECTION

There are two practical reasons why rectangular cross-section wells are of interest. First, in an individual well the groundwater enters the well along its peripheral surface area. Hence, there is a direct relation between this area and the rate of flow from the aquifer into the well. It is, therefore, advisable to choose a well shape that will have the maximum perimeter for a given cross-sectional area. Under this condition, the circular well has the least perimeter which is larger for a square well and the largest case is in a rectangular well. In loose clastic rocks the cohesion is rather weak and they cannot bear even small tensile stresses. Because of benefits already mentioned in the previous section circular cross sections should be preferred in loose unconsolidated rocks. However, hard rocks can take tensile strength and therefore a rectangular well should be preferred. In fractured media especially, the longer the perimeter, the greater is the number of fracture intersections as shown in Figure 5. In circular cross-section, small-diameter wells there is a possibility of missing the fractures, fissures, and joints in hard rock areas resulting in many dry wells.

The second reason is that a rectangular array of wells may be drilled, in practice, so as to supply water to a city or in dewatering for foundation-laying purposes. Such a well-pumping scheme is applied to provide water to the Riyadh, capital of Saudi Arabia, as shown in Figure 6. If the positions of wells are close to each other, practically less than about 500 m, then the wells collectively act as an equivalent extensive rectangular well. For instance, in water supply to Riyadh, from the Wasia sandstone aquifer there are 40 wells arranged at 400 m apart from each other on 4 lines with 10 wells on each line. Such a well field is similar to an extensive rectangular well.

C. EXTENDED WELLS

In fractured media the existence of any major fracture provides an opportunity to have a maximum surface area by drilling a small-diameter well in the fracture itself or it is possible that the groundwater is abstracted from the fracture directly. Such situations are very convenient in vertical or almost vertical fractures. Of course, the fracture extensions on both sides of the well make the total periphery many orders greater than the actual drilled well surface. The open fracture is a planar production surface that becomes an integral part of the well (see Figure 7). The longer the fracture the larger will be the area of groundwater entrance surface that will support the productivity of the well. In such an extended well situation groundwater moves from two sides toward the fracture and then within the fracture toward the drilled well. In field situations, the groundwater moves easily within the fracture than the aquifer itself which may be fractured or porous medium.

In extensive planar fracture the flow lines in the aquifer toward the fracture become parallel to each other which implies that the flow is one dimensional. Extended wells have been used in practical applications by Jenkins and Prentice (1982). The main advantage of these wells is maximum production surface with minimum cost. The extended portions of the well do not need any casing or screen.

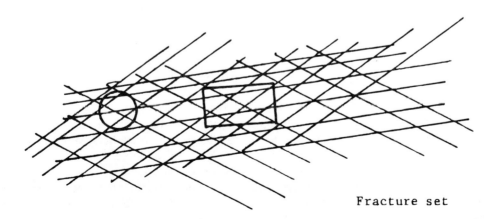

Figure 5 Rectangular well in fracture sets.

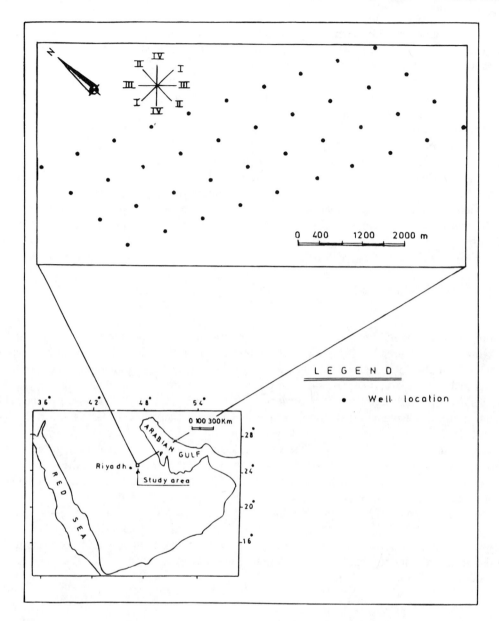

Figure 6 Rectangular array of wells.

III. COLLECTOR WELLS

These wells yield large supplies of water from relatively thin shallow aquifers. The advantages of these wells are (1) availability of large filtered sterilized water supply, (2) reduced operation and maintainance costs, (3) suitability in thin aquifers, and, (4) reduced wear of pumping due to the entrance velocities, resulting in sand-free water.

As a general setup, these wells have two main parts, a vertical and circular cross-sectional water storage volume acting as a collector and horizontally driven close to the collector base radial drains as groundwater conveyors from the aquifer to the collector (see Figure 8). The collector has groundwater seepage neither from its wall nor bottom but only from the radial

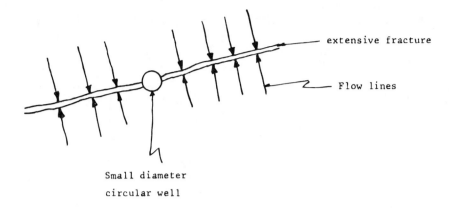

Figure 7 Extended well.

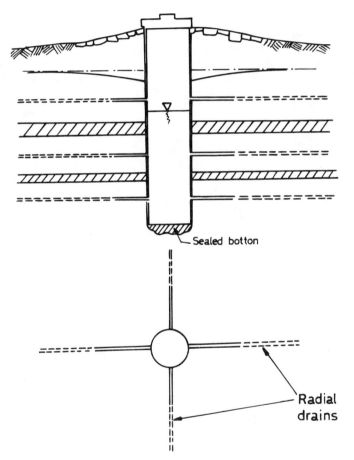

Figure 8 Collector well.

conveyors through gate valves that are operated from the top. The water collector is essentially a large-diameter, usually 4 to 5 m, watertight tank. After the completion of this tank from reinforced concrete with a sealed bottom by pouring a thick concrete plug heavy enough to resist the buoyancy, lateral pipes made of steel having 15 to 20% slot area are driven horizontally into the groundwater reservoir by special hydraulic jacks. There may be more than one layer of conveyor pipe, especially if the groundwater reservoir is dissected by a clay layer preventing free vertical movement of groundwater.

Collector wells are constructed in alluvial deposit groundwater reservoirs. They prove especially useful in getting large irrigation supplies with a permanent source for continuous recharge such as lakes, perennial streams, or waterlogged areas irrigated by large canal systems. Collector wells have the following differences from common well types.

1. The initial cost of a collector well exceeds that of a vertical well.
2. The maintenance cost of collector wells is less than other wells.
3. Collector wells have large yields under low pumping heads.
4. The maximum head difference between the water level in the well and in the adjacent aquifer occurs at the time of well construction completion.
5. Horizontal drains in various directions will have a chance to intersect more fractures than vertical borehole wells. Hence, the well productivity will increase significantly.
6. Hydraulic head losses, i.e., drawdowns around the collector well, will be smaller than the drilled wells. Therefore, the sanding problem will not occur in the collector wells.
7. The saline groundwater generally lies at great depths of the saturation thickness. Use of the collector wells will not give rise to groundwater quality deterioration due to the mixing of deep-lying saline water with overlying relatively fresh water.
8. Although the initial cost of collector wells is comparatively higher than vertical wells, maintenance is inexpensive and durability is very long. The risk of any problem in collector wells is negligibly small.

IV. KANAT

A system of wells connected together by a gallery that brings water from the foothills to the plains is called a "kanat" as shown in Figure 9. Kanats were originally devised in Persia by the year 800 B.C. The main groundwater carrying part of the system is the gallery which penetrates with a small gradient deep into the aquifer.

The first stage in the construction of a kanat is to dig large-diameter exploratory wells through the potential aquifer. These wells are about 300 to 400 m apart and may reach down

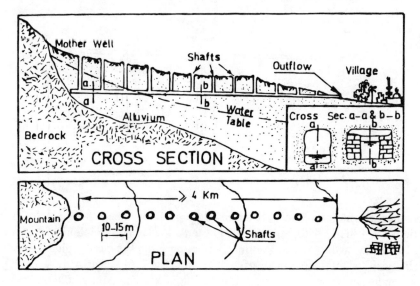

Figure 9 A schematic kanat diagram.

to 100-m depth and they have diameters of at least 75 cm. The gradient of a kanat is taken usually around 1:1500. Figure 10 indicates excavation stages of a kanat. Kanats appear to have various local names in different countries. For instance, in Saudi Arabia, they are known as iyn (or iyun in plural), shat-at-ir in Morocco, foggariur in Algeria, falaj (or aflaj in plural) in Oman, shariz in Yemen, and karez in Pakistan, Iraq, and Afghanistan.

Kanats are constructed as gently sloping tunnels in the alluvial fills or fans. The tunnel collects the groundwater seepage and carries it down the slope until it appears as surface water near the exploitation center above the water table. The widespread use of a kanat in the arid or semiarid regions is due to

1. The groundwater flows due to the gravitational force and thus there is no need for extra power source.
2. The evaporation losses are at the minimum level because the flow takes place completely in the subsurface.
3. It provides a dependable and sustainable water supply for local, domestic, and agricultural lands for long duration. For instance, the Iyn Zubaydah kanat system was constructed mainly for supplying water to the pilgrims in the holy places of Makkah circa 750 AC and it is functioning even today.
4. The groundwater flow is safely guarded against pollution.

How does the water reach the land surface in a kanat? The loss of hydraulic head incurred in tunnel flow is less than in the aquifer so that, if the aquifer somewhere has a water table above the level of the land to be irrigated, it can be exploited. The water flows into the kanat down the potential gradient that is set up during its construction. The equipotential lines around a kanat are shown in Figure 11. The flow in the tunnel is like the flow to a perennial river. There is a radial flow within the aquifer to the kanat, normal to the equipotential lines that pass under the kanat. Of course, the rate of flow increases with distance down to the outlet.

A detailed account of kanat construction in the arid zones is given by Amin et al. (1983) in addition to their hydrologic importance, maintenance, and specific properties.

V. IRREGULAR CROSS SECTIONS

Due to hydrological, geological, or economical reasons, some wells are given irregular shapes. In the arid regions of the world especially, alluvial deposits are recharged by infre-

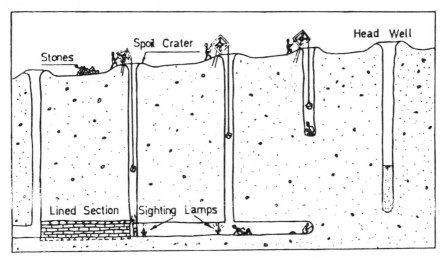

Figure 10 Excavation stages of kanats.

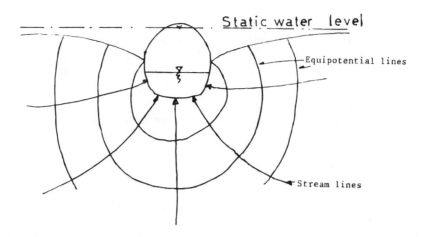

Figure 11 Flow net around a kanat.

quently occurring but intensive flash floods. On the other hand, flash floods damage or fill in the open wells with sediments in a destructive way. Therefore, rather than having regularly shaped wells, irregular shapes may become preferable under these circumstances for economic reasons. Furthermore, if the groundwater table is close to the surface, digging of wells with bulldozers reaches the groundwater table in a short time. Therefore, similar to a small lake those ditches provide groundwater for local irrigation or domestic uses. As the groundwater table is lowered by abstraction the depth of these ditches is increased by successive bulldozer diggings. Next flash flood fills all the ditch with sediment but the same procedure is repeated after the flood as long as there exists a demand for water (see Figure 12).

Geological reasons may appear in the fractured or karstic groundwater reservoir areas to have irregular shapes. It is well understood that, although there are some definite volcanic or chemical reasons for fracture and solution cavity evolutions, their distributions in any rock mass are rather irregular. Accordingly, irregular cross-sectional areas have a better chance to intercept more fractures or cavities than a regular section as shown in Figure 13. Besides, fractures and solution cavities appear in hard rocks which have better slope stability than unconsolidated rocks. As a result, there is no casing requirement. Sometimes wells tapping an alluvial aquifer are not cased due to economic reasons. This leaves the well bore face to have irregular shapes. In karstic terrains especially, the wells are bound to have different irregular cross sections along the well bore due to chemical precipitation as well as solution. Accordingly, the flow lines will have irregular shapes near the well which complicates the analytical calculations.

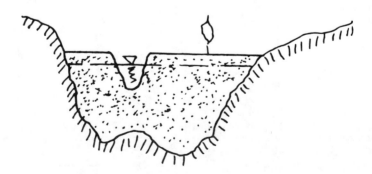

Figure 12 Swamps.

Figure 13 Wells in karstic formations.

VI. WELL PENETRATION

The common depth between the well and saturation zone of the groundwater reservoir is referred to as the well penetration. Determination of this depth depends on the reservoir lithological log at the well site. The amount of discharge is directly proportional with the penetration length. Furthermore, the shapes of stream and equipotential lines on a vertical cross section depend on the well penetration.

A. FULLY PENETRATING WELL

In a given aquifer the maximum quantity of water can be abstracted with a well that spans the whole saturation thickness as in Figure 14. In confined aquifers full penetration provides planar groundwater flow which facilitates calculation of discharges. The flow is two dimensional and radial.

In a vertical cross section the flow lines toward the well are horizontal and the equipotential lines are vertical. Even though the aquifer may have different horizontal and vertical transmission characteristics due to the radial horizontal flow, the vertical transmission does not affect the groundwater flow. It is, therefore, obvious that fully penetrating wells simplify the groundwater flow net geometry. The well need not penetrate the whole thickness of the aquifer, but one cannot expect to have the maximum well yield if the well penetrates partially.

B. PARTIALLY PENETRATING WELL

In a partially penetrating well the water-entering portion of the well is somewhere in between the upper confining layer or water table and the bedrock as shown in Figure 15.

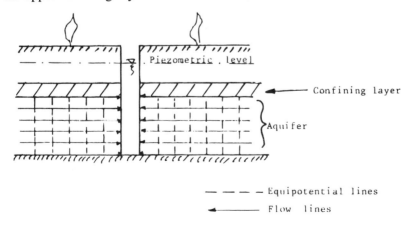

Figure 14 Fully penetrating well.

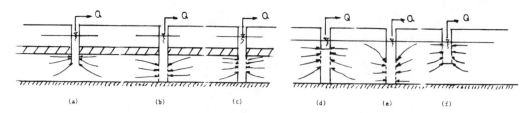

Figure 15 Partial penetration.

These wells are common in areas where the aquifer is relatively thick and the required discharge can be supplied safely by a well drilled only through a portion of the saturation thickness. There are various practical reasons in drilling a partially penetrating well.

In a partially penetrating well the flow lines are forced to converge vertically toward the well screen. As a result a water molecule has to travel longer distances than the fully penetrating well. Such a situation causes larger drawdowns than a fully penetrating well. In the vicinity of the well the flow lines have a three-dimensional radial rotational shapes. However, with increase in distance from the well the flow domain converges to two-dimensional radial flow (see Figure 15).

The majority of the hand-dug wells in alluvial fills of arid regions are partially penetrating. Compared to the fully penetrating wells partially penetrating wells are more economical but their hydraulic treatment for calculations poses some difficulties (see Chapter 8). For instance, the definition of transmissivity is not valid in the vicinity of partially penetrating wells.

C. NONPENETRATING WELL

It is a cavity well that is quite commonly used in confined aquifers due to its ease of construction and economy. It is completed by puncturing the upper impervious layer of the confined aquifer and hence withdrawing water through the cavity formed in the bottom as shown in Figure 16. These wells have been invented and run by the local people in northern

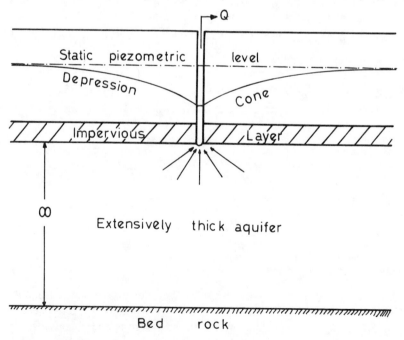

Figure 16 Schematic diagram of a cavity well.

Indian alluvial plains in the last 2 decades. Cavity wells do not require any screening or gravel packs. The equipotential lines appear as semispherical surfaces provided that the aquifer thickness is rather large. In convenient areas the cavity wells are recommended especially for preliminary studies because their initial investment capital is comparatively small.

VII. WELL PURPOSE

In groundwater resources evaluation studies wells are drilled either for groundwater abstraction or monitoring purposes. Abstraction is achieved by means of rather large-diameter wells such that either a bucket or a submersible pump can easily enter the well storage.

A. PUMPING WELL

They are interchangeably referred to as the main well, abstraction well, or pumping well. These wells are drilled, designed, developed, and operated to get the optimum quantity of water from an aquifer. The choice of well site and depth depends upon topography, subsurface geological composition, depth to the saturation zone, recharge, and demand of water. For instance, in shallow Quaternary alluvial fills, large-diameter wells and depths less than 50 m are preferred whereas in the sedimentary regions small-diameter wells with large depths are unavoidable. The well must be large enough to accommodate the pump with proper clearance for installation and efficient operation. The well diameter plays a primary role in the entrance velocity of water to well. The diameters should be chosen such that the entrance velocity does not exceed 3 to 6 m/s. Such a choice serves three purposes:

1. The well losses are reduced.
2. The finest sand particles are hindered from entering the well, i.e., the sanding problem is reduced.
3. The corrosion as well as the encrustation are prevented to a certain extent.

Any pumping well must provide an adequate intake area through which water flows into the well. This area depends on the well diameter and the thickness of aquifer it penetrates. A common rule is that increase in the well diameter will not necessarily mean proportionately large discharges. In fact, doubling the diameter of a well will decrease the drawdown by only about 10% for the same discharge; therefore, it is a waste of money to drill a larger-diameter well.

Usually the depth of a well is determined from the lithological logs and confirmed from electrical resistivity as well as the drilling time logs. Pumping well completion requires well screen design, screen diameter, material slot size, gravel pack composition, and prior to any water supply a proper development. These points are explained by Huisman (1972), Todd (1980), and Marino and Luthin (1982).

Pumping wells are indispensable in any field test in that water is pumped at a constant rate (aquifer test) or variable rate (well test), which are carefully controlled and measured.

B. OBSERVATION WELLS

These wells are also known as piezometers and they are the fundamental means for measuring groundwater head changes in an aquifer for evaluating the aquifer parameters or monitoring the groundwater level fluctuations. In fact, there is a practical difference between the piezometer and observation well in that the piezometer measures the pressure in a confined aquifer whereas the observation well measures the level in a water table aquifer. However, in practice, they are used synonomously to describe any vertical shaft for monitoring water levels. Changes in the piezometric surface are monitored by one or more observation wells. The observation well usually consists of a blank pipe with a screen at the bottom.

In a uniform, isotropic, and homogeneous medium any ordinary observation well as in Figure 17 gives representative data for determining aquifer parameters with techniques in Chapters 8 through 10.

On the other hand, if there are multiple layers or heterogeneity causing vertical water flows, the ordinary piezometers will indicate an average of the hydraulic head. To observe the actual situation two or more observation wells are required. For instance, when two aquifers are penetrated as in Figure 18 the observation well A indicates an average level that is representative of neither aquifers but observation well B represents the average level within aquifer B only.

More detailed information about the observation wells is provided by Kovacs (1981). The piezometers usually consist of a blank pipe with a screen at the bottom. The ordinary piezometers such as in the bore hole backfilled with clean sand as in Figure 19 are open to the entire aquifer. However, Figure 20 shows a true piezometer in a confined aquifer that is isolated within a specific zone of the aquifer. Although the ordinary piezometer gives the overall representative data in a uniform isotropic unconfined aquifer, the true piezometer gives the particular data for an individual aquifer in a multiple-aquifer situation. For instance, when two aquifers, confined and unconfined as in Figure 21, are present then the ordinary piezometer is representative of neither aquifer. To have the real picture of the groundwater flow two piezometers are required. It is interesting to have as many piezometers as conditions allow but at least three are recommended in practice. The number of piezometers depends on the rate of depression cone expansion and material heterogeneity. In densely fractured rocks response to water abstraction resembles that of single porosity homogeneous aquifer. In a single vertical fracture, the number and location of piezometers will depend on the fracture

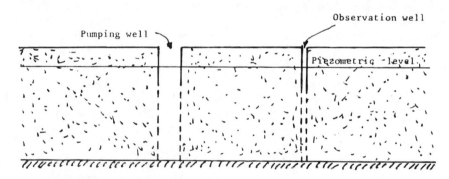

Figure 17 Single-aquifer observation well.

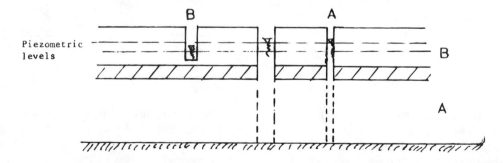

Figure 18 Two-aquifer observation wells.

WATER WELLS

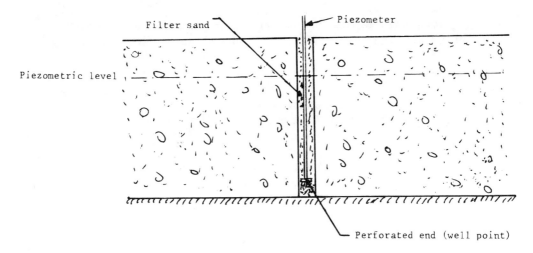

Figure 19 Ordinary piezometer.

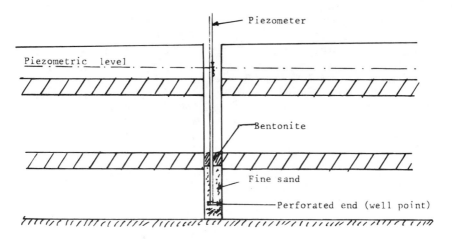

Figure 20 True piezometer.

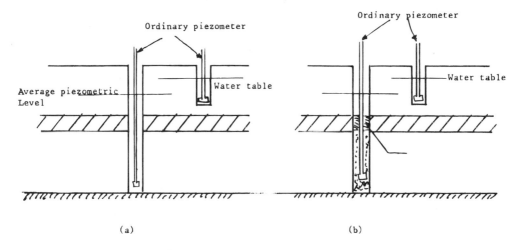

Figure 21 Ordinary and true piezometers.

orientation, transmissivity of host rock on opposite sides of the fracture, and openness or closeness of the fracture. In open fractures the hydraulic conductivity within the fracture is extremely high and it will resemble a canal whose water level is lowered suddenly. In a vertical fracture (extended well) it is preferable to arrange piezometers along a line perpendicular to the fracture. In addition, in order to see whether the fracture behaves as an "extended well" a few piezometers should be placed in the fracture itself.

There is no fixed rule in choosing the number of piezometers but it depends virtually on local conditions and piezometer placements within the radial range from 10 to 100 m to give satisfactorily reliable results. For thick confined aquifers, the range may change between 100 and 250 m. At least one piezometer must be placed outside the radius of influence to see the natural behavior of water level decline in the aquifer. These piezometers must be several hundred meters away from the main well. The lower end of a piezometer should end almost at the same elevation as with that of half the length of the well screen.

REFERENCES

Amin, M. I., Qazi, R., and Downing, T. E., 1983. Efficiency of infiltration galleries as a source of water in arid lands, *Symp. Water Resour. Res. Saudi Arabia*, 1, A-107.

Anderson, K. E., 1967. *Water Well Handbook*, Scholin Brothers, St. Louis.

Driscoll, F. G., 1987. *Groundwater and Wells*, Johnson Division, St. Paul, MN.

Huismann, L., 1972. Artificial Groundwater Recharge, Delft University of Technology, Dept. of Civil Engineering, Delft, The Netherlands.

Jenkins, D. N., and Prentice, J. K., 1982. Theory for aquifer test analysis in fractured rock under linear (non-radial) flow conditions, *Ground Water*, 20(1), 12.

Kovacs, A., 1981. *Subterranean Hydrology*, Water Resources, CO.

Marino, M. A. and Luthin, J. N., 1982. *Seepage and Groundwater*, Elsevier, Amsterdam.

Todd, D. K., 1980. *Groundwater Hydrology*, John Wiley & Sons, New York.

CHAPTER 7

Field Measurements

I. GENERAL

In this chapter, some common techniques are described for the measurement of water levels, drawdowns, and discharges. Any successful application of the methodologies presented in the following chapters depends on the precise and accurate field and laboratory measurements of the medium and flow properties. No doubt, in earth sciences the quantities vary within short distances and the aquifer properties may have as many different values as the number of samples taken in the field. However, because of the extensive flow domain their averages give satisfactory results for practical purposes. On the other hand, some of the parameters need to be measured with more care and accuracy than others. It is, therefore, a prerequisite to know the error limits of the measurements in arriving at meaningful, representative, and valid practical results. Although the sensitivity of any person to obtain precise results makes him collect as many data as possible to reduce the error band, it may not be practical to do so due to the cost of data and limited time available. It is necessary therefore to accept a certain error percentage in the results which should be less than 10%; concerning decision variables for design less than 5% is preferable. For instance, as mentioned earlier it is difficult to find pure steady state in nature. However, a substitute of this state as a quasi-state can be adopted. What would be the difference between two successive groundwater level measurements to accept quasi-steady state? If the two successive hydraulic head measurements in an observation well are h_i and h_{i+1} ($h_i > h_{i+1}$) then the relative error percent, α, can be evaluated as

$$\alpha = \frac{|h_i - h_{i+1}|}{h_i} \times 100 \tag{1}$$

provided that $\alpha < 5\%$, the flow is in quasi-steady state.

An efficient field work requires experienced personnel, funds, and time. Sometimes any one of these three requirements may put a restriction on the field activities. The best field work is the one during which the optimum amount of these sources is used to achieve a specific purpose.

II. MAIN WELL LOCATIONS

To collect the most representative field data for aquifer parameters, identification, and regional assessments, the locations of abstraction and observation wells must be selected such that the assumptions concerning the geometry, geological setup, and flow properties are satisfied to a great extent. The following points are important in deciding on relative well location sites.

1. To explore surface and subsurface geological composition of the study area. If possible, stratigraphic sequences, lithologic units, facies changes, fence diagrams, and hydrolithologic sequences must be known.
2. Position of hydrological (lake, river, surface depressions) and geologic structural barriers (faults, joints, cleavages, etc.) must be determined.
3. To know depth variations and areal extents of the aquifers, if possible, by using geophysical methods.
4. To have some idea about the regional groundwater flow directions and piezometric level variations.
5. To make economical evaluations.

The selection of main well locations is the most important decision because the observation well locations are relative to the main well. In making this decision the following points provide useful guidelines.

1. The area of the aquifer surrounding the well must be extensive.
2. This extensive area must be intact of geological and hydrological boundaries.
3. The depth of the aquifer at the well location must be great enough to have a large entrance area of groundwater to the well.
4. If possible one type of geological facies to exist within the extensive area. This is for the satisfaction of the homogeneity and isotropy assumptions.
5. If there is an already existing well the interference from the new well must be avoided.

Selection of well location in karstic formations is tricky. If the well remains within the main rock body without intersecting any conduit, it will not supply significant water. Therefore, the best well locations in karstic terrain is either on a fracture or solution cavity or even better at the intersection of two or more fractures. During the drilling, if the fracture is not reached prior to the abandoning of the well, it may be helpful to inject HCl acid to open existing fractures in the hope that they will eventually connect the well with a drain. Alternatively, one may try dynamite blasting so as to make artificial fractures. It is also possible to reach an open network of conduits (fractures) under the base flow level by deepening the well.

III. OBSERVATION WELLS AND PIEZOMETERS

They serve as points of piezometric level measurements within the aquifer. Their properties have been discussed in the previous chapter. Some piezometers are necessary for continuous monitoring of groundwater level fluctuation measurements but others are prerequisites for completing any aquifer test successfully and are located around the pumping well. Practically any type of vertical shaft penetrating the aquifer is suitable for the observation and measuring of groundwater. However, some restrictions are necessary in selecting the observation well for hydraulic head measurement purposes:

1. The observation wells should represent large areas of the aquifer. For instance, if the surface lithology of an aquifer includes 80% fine-grained rocks then the observation well in such a facies is preferable rather than in the remaining 20%.

FIELD MEASUREMENTS

2. One observation well must serve only one aquifer. This rule suggests that for a complete measurement the number of observation wells must not be less than the number of aquifer layers (see Figure 1).
3. If there is a vertical flow within the aquifer due to various reasons such as leakage, partially or nonpenetrating wells, gravity drainage, fractured media, etc., then on a single vertical there should be at least two observation wells to be able to distinguish different piezometric levels at the same point. Some examples are shown in Figure 2. In homogeneous permeable water table aquifers, it is sufficient if piezometers tap only the upper part of the water table aquifer, below the zone of water level fluctuations.
4. The observation well must not be used for any purpose other than for monitoring groundwater level fluctuations. Otherwise, the measurements may be biased.
5. Too large or small observation well diameters may affect the measurements within the well. For instance, a too-small-diameter well may have a capillary effect and therefore the groundwater elevations appear more than the true elevation. On the contrary, the larger the diameter the shorter the time necessary for the equalization of the piezometric levels within the aquifer and the water surface in the well (see Figure 3). In other words the difference between these two levels is small in large-diameter wells.

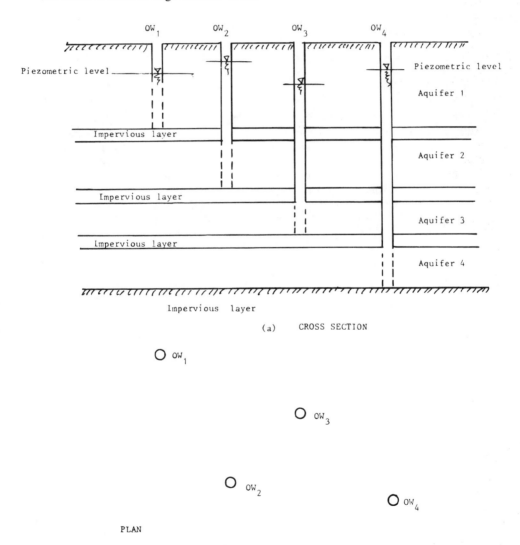

Figure 1 Observation wells in multiple aquifers.

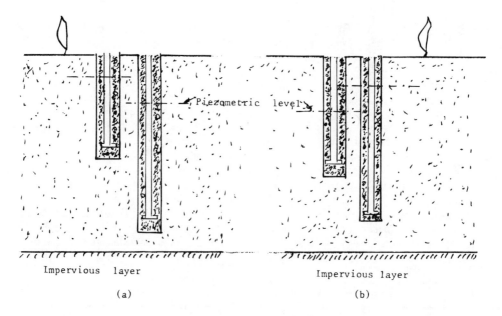

Figure 2 Piezometers for vertical flow: (a) upward, (b) downward.

6. The number of observation wells brings an economic restriction rather than any hydrogeological limit. If possible it is advisable to have at least three observation wells in monitoring groundwater drops during an actual aquifer test. In nonpumping durations, these three wells help determine the general groundwater flow direction. Otherwise, measurements from the observation wells provide potential data for hydraulic conductivity estimations from the steady-state flow and transmissivity as well as storativity estimation from a distance-drawdown plot as will be explained in the next two chapters. The observation wells around the main well must be at different directions and distances. Different directions are useful in obtaining information about the aquifer isotropy whereas a difference in distances leads to inferences about the aquifer homogeneity as well as the application of distance-drawdown formulas.

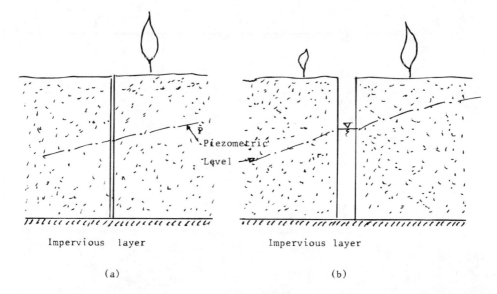

Figure 3 (a) Small- and (b) large-diameter observation wells.

7. The distances of observation wells to the main well necessitate the consideration of some practical and theoretical points. Practically, to any excitement the confined aquifer response to drawdown spread around the well is quicker than an unconfined aquifer. Field experiences guide us that average maximum excitement distance from the pumped well is about 250 m for a confined aquifer whereas it is about 100 m in an unconfined aquifer. For refinement of measurements or for the purpose of well design criterion direct measurement of well loss is possible by locating one observation well in the very close vicinity, in fact, within the gravel pack of the well. The distance between the center of the main well and observation well should not be less than twice the thickness of aquifer to avoid the partial penetration effects.

A quantitative way of observation well distance calculations can be obtained from the steady-state formulations after plotting the drawdown (the square of drawdown) vs. logarithm of the radial distance that appears as a straight line for a confined (unconfined) aquifer (see Figure 4). The necessary formulations for calculations are presented in Chapter 8.

At large distances there is uncertainty about applicability of the drawdown. Therefore, only the intermediate portions of the graph in Figure 4 appear as a straight line. Notice that the straight line portion in the confined aquifer case is longer than the unconfined aquifer. This intermediate portion is divided into a number of observation wells leading to equal distances on the logarithmic scale. For instance, eight observation wells are shown in Figure 4a. Hence, the actual field distances can be found from antilogarithms of the results along the horizontal distance axis.

The arrangements of observation wells is preferably done on a Cartesian coordinate system (practically N-S and W-E directions) as shown in Figure 5. Such a configuration of observation wells enables one to have a good picture about the equipotential and flow lines around the main well.

8. The actual distances of observation wells to the main well depend mainly on their number, the type of aquifer material, and its hydraulic characteristics. Short distances are preferable in detrital rocks and unconfined aquifers whereas long distances are recommended in the case of fractured and karstic aquifers. Table 1 indicates practical guides for selecting observation well distances to the main well.

IV. DRAWDOWN MEASUREMENTS

By definition the drawdown is the difference between the static and dynamic piezometric levels, both of which should be measured with respect to a common bench point. In unconfined aquifers especially, if possible the measurements should be made at periods when there is the least likelihood of heavy rainfalls which may lead to infiltration and deep percolation. On the

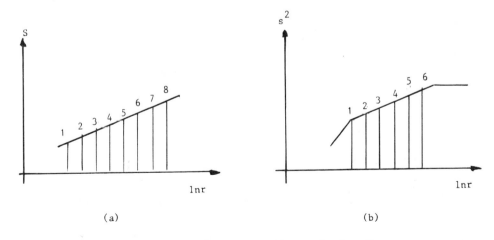

Figure 4 Observation well distance. (a) Confined and (b) unconfined aquifer.

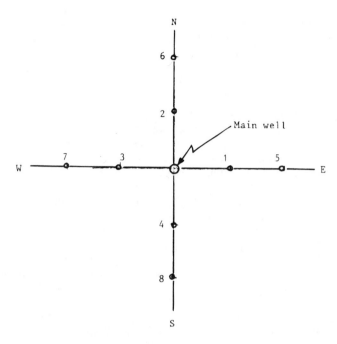

Figure 5 Plan view of observation well configuration.

other hand, if the aquifer is confined barometric changes may affect piezometric level. An increase in the barometric pressure may cause a decrease in the piezometric level. The measuring point at the pumping and observation wells should be selected and marked clearly with a permanent pointer and the elevations determined by altimeter or surveying. In field application drawdown definition needs piezometric head measurement changes with time. Static piezometric level corresponds to the initial piezometric level with no movement and hence prior to any other measurement the static piezometric level must be confirmed. To do this two successive piezometric level measurements are taken at a certain time and after some time period, e.g., half an hour. If these values are different from each other by more than 5% relative error calculated from Equation 1, then another period is waited and a new reading is taken. This procedure is continued until at least three consecutive readings are equal to each other within 5% error which indicates the static piezometric level. The measurements of piezometric level are schematically given in Figure 6.

Although there is only one static level, theoretically there are an infinite number of dynamic piezometric levels at each time instance during groundwater movement. With the start of water withdrawal from the well the piezometric level drops and this phenomenon may continue, depending on the hydrologic conditions and aquifer types from 2 to 3 h up to several weeks. Again theoretically in an aquifer with no external recharge the duration of dynamic piezometric level continues indefinitely. However, in practice confined aquifers have longer durations than

Table 1 Observation Well Distances

Number of observation wells	Hydraulic conductivity (cm/s)		
	High (>1)	Medium ($1 < k < 10^{-2}$)	Low ($<10^{-2}$)
1	1–3	3–5	5–10
2	3–10	10–20	20–50
3	10–50	50–100	100–200

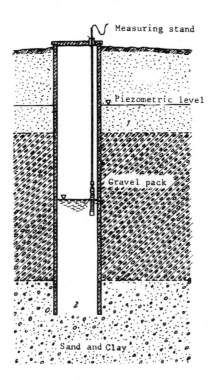

Figure 6 Piezometric level measurement.

any other type of aquifer and it may continue for 4 to 5 days or weeks whereas unconfined aquifers have 2 to 3 h or 1 day.

At what time intervals one should sample the dynamic piezometric level recordings during the pumping? The answer is rather subjective and one may take records at every second or minute generally with no harm but practically some of these data are superfluous and do not bring any benefit. In fact, the change of drawdown is proportional to the logarithm of time which indicates that the sampling time intervals should be small at the beginning but might increase at large times. Table 2 gives a suggestion based on the author's experience in the field. The time intervals suggested in Table 2 need not be followed exactly but they should be modified in response to local conditions.

All measurements concerning the dynamic piezometric level variations, well and aquifer configuration, discharge, and other particulars are written in a log book with the format given in Table 3. In the table T_P and T_R represent in daily hour intervals the beginning of pumping and recovery periods. The first reading in the piezometric level column is the static piezometric level. The drawdown column includes differences of the piezometric level from the first one. The test duration is continued until the last two successive piezometric level readings give less than 5% relative error by using Equation 1. This is tantamount to saying that quasi-

Table 2 Time Intervals in Piezometric Level Measurements

Time since pump start (min)	Intervals (min)
0–5	1
5–15	2
15–30	5
30–60	10
60–120	15
120–240	20
>240	60

Table 3 Field Test Data Sheet

Sheet number: Study area: Remarks: Date of test:
Type of test: Aquifer test
Location:
Performed by:
Aquifer type:
Well type:
Casing type:
Pump type:
Well diameter:
Casing diameter:
Distance to observation well:
Total depth of well:
Reference level:

	Time (hour)	Time since start (min)	Piezometric level (m)	Drawdown (m)	Discharge remarks (m³/min)
T_P		0	h_0	$s_0 = h_0 - h_0 = 0$	Q
		t_1	h_1	$s_1 = h_1 - h_0$	Q
		t_2	h_2	$s_2 = h_2 - h_0$	Q
		t_3	h_3	$s_3 = h_3 - h_0$	Q
		t_n	h_n	$s_n = h_n - h_0$	Q
T_R		t'_1	h'_1	$ms'_1 = h'_1 - h_n$	0
		t'_2	h'_2	$s'_2 = h'_2 - h_n$	0
		t'_3	h'_3	$s'_3 = h'_3 - h_n$	0
		t'_m	h'_m	$s'_m = h'_m - h_n \cong 0$	0

steady-state flow is reached and the changes in hydraulic heads are no more significant practically for calculations based on drawdown data. Techniques presented in Chapter 8 are valid only after the appearance of quasi-steady-state flow; otherwise, unsteady-state methods in Chapters 9 or 10 are used for aquifer parameter determinations.

After the pump is shut the piezometric level starts to retreat to its original position with time. This operation is called the recovery period and during its duration there appears two types of drawdown, the recovery drawdown, s_{rec}, which is the difference between the piezometric level measurements after the pump shut, and the maximum drawdown prior to pump shut. The other type is just the complementary of s_{rec} which is called residual drawdown, s_{res}, and it is the difference between the piezometric head measurement during recovery and the static piezometric level prior to pumping start. In fact, the summation of these two drawdowns is equal to the maximum drawdown of the pumping period.

Field experiences have shown that the recovery period is invariably longer than the pumping period. Initial recovery for the first hour or so is very rapid, but as time goes on the rate of water rise in the aquifer decreases. The accuracy of the recovery period piezometric head behavior is determined by the frequency of the observations and following recommendations given in Table 4 concerning the time intervals between measurements.

Because the recovery follows the pumping period, the recovery piezometric measurements can be continued on the same sheet for the pumping test as shown in Table 3.

A. WETTED TAPE

Many hydrogeologists and engineers alike prefer a steel surveyor's tape with a rough surface that can be chalked or dusted. There is a weight attached on its lower end and it keeps the tape stretched. The held point is noted and then the tape is removed and the water level is noted. The difference between the held point and the water mark on the tape will be the

FIELD MEASUREMENTS

Table 4 Recovery Time Intervals for Piezometric Levels

Time since pump shut (min)	Intervals (min)
0–5	1
5–20	5
20–60	10
60–240	30
>240	60

depth to water. A wetted tape is probably the most accurate method but its application is cumbersome especially in taking a series of rapid readings since the tape must be fully removed each time. This method is a very accurate way of measuring water depth up to about 30 m.

B. FLAT-BOTTOMED WEIGHT

Some of the engineers use a flat-bottomed weight at the end of a steel tape and listen for the splash when the weight strikes the water surface within the well. This method leads to rough measurements with poor accuracy. A refined method is to mount a heavy whistle open at the bottom and attached at the end of the tape. When it submerges into the water a proper whistle will give a "peep" as the air in the whistle is displaced. Such measurements can be made to within 1- to 2-cm error.

C. ELECTRIC TAPE

This instrument is also referred to as the electric sounder and it is battery operated with an electric cable marked off in centimeters wound on a spool as in Figure 7. When the

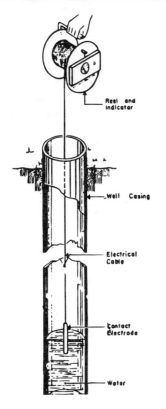

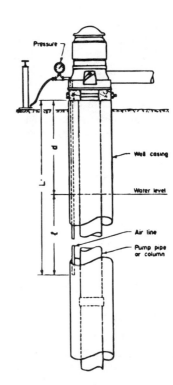

Figure 7 Electrical type measurement.

Figure 8 Air-line method.

electrode at the end of electric cable touches the water surface the circuit is closed and a flashing lamp indicates this situation. To improve the accuracy of readings the electrode should be kept hanging in the well for a series of subsequent readings. Hence, kinks and bends in the wires that may lead to errors are avoided.

The electrical sounders are very convenient and more suitable than any other device because they do not require removal from the well for each reading. Measurements with such a probe become difficult when there is oil on the water surface or water cascading into the well or turbulent water surface.

D. AIR LINE

This line consists of a small-diameter tube with sufficient length to extend from the top of the well to a point about 1 m below the lowest anticipated water level as shown in Figure 8. The tube and the necessary connections should be completely air tight and the exact length between a marked reference point at the well top and its end within the water must be measured as it is placed in the well. The upper end of the air line is connected through a valve to an ordinary pump. In addition, there is a pressure gauge near the valve. The operation of such a device is based on the principle that the air pressure required to push all the water out of the submerged portion of the tube equals the water pressure of a column of that height. The air pressure shown by the gauge increases until it reaches a maximum and then remains a constant value which means that all the water within the tube has been forced out of the air line. At such an instant the gauge reading shows the pressure necessary to support a column of water of a height equal to the distance from the water level in the well to the bottom of the tube.

Substitution of this gauge reading from the total length of the air line gives the depth to water below the chosen reference point. Of course, a measurement prior to any pump start indicates the static water level. Referring to Figure 8 the depth to water is calculated as $b = L - d$, in which L is the depth to the bottom of the air line and d is the gauge reading in meters.

Any change in water level is represented directly by a difference in pressure gauge. Drawdown during pumping periods can be recorded readily from the pressure readings. This method is not accurate enough for use in observation wells during an aquifer test but most practical for the measurements within the main wells. The depth to well can be determined usually within 10 cm of exact value.

E. MERCURY METER

The above-mentioned methods are not suitable at all for the flowing wells where the piezometric level is above the earth's surface. Manometers are U-shaped glass tubes mounted on a wooden board fixed with a graduated scale in between the two limbs of a U tube as shown in Figure 9. One end of the manometer is connected to the tap in the well cap by a plastic tube whereas the other end is open to the air. The manometer should be filled with an appropriate quantity of mercury so that it is level with the zero mark as shown in Figure 9a. When the tap is open the water enters one limb of the U tube (Figure 9b). After the mercury stabilizes, the difference between the mercury level in the limbs is read, h_m^0, on the scale, and the pressures on the left and right limbs are

$$p_L = w_w h_w + p = 62.4\ h_m^0 + p \tag{2}$$

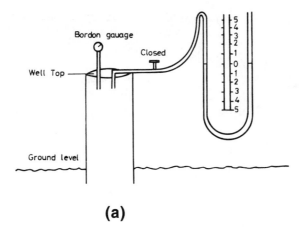

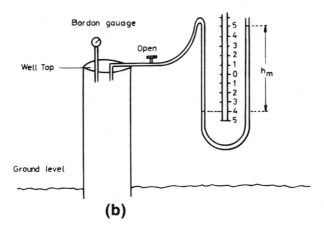

Figure 9 Mercury meter.

and

$$p_R = w_m h_m^0 = 13.6 \times 62.4\, h_m^0 \tag{3}$$

respectively. Herein, w_w and w_m are the specific weights of water and mercury, respectively, and p is the atmospheric pressure read from the pressure gauge. Due to the equivalence of pressure in the left and right limbs, i.e., $p_L^0 = p_R^0$, the piezometric head h^0 after the initial reading is calculated as

$$h^0 = 13.6\, h_m^n \tag{4}$$

Similarly, the nth reading of piezometric level at time t_n is

$$h^n = 13.6\, h_m^n \tag{5}$$

Hence, the drawdown, s^n, at nth reading is

$$s^n = h^n - h^0 = 13.6(h_m^0 - h_m^n) \tag{6}$$

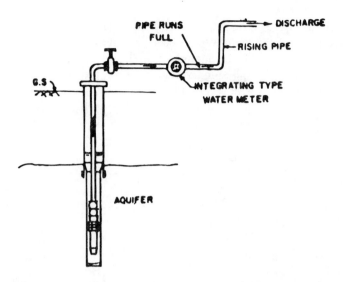

Figure 10 Well discharge measurement.

V. DISCHARGE MEASUREMENTS

In various analytical models to be explained in the following chapters, although there is no direct assumption concerning the drawdown, the discharge is postulated to remain constant throughout the tests. Therefore, it is necessary to satisfy this assumption as closely as possible by its frequent measurements in an actual field work and necessary adjustments are made to keep it constant. It is a controllable variable and its uncontrolled variations make the application of analytical models almost impossible. Well discharges are measured usually during an aquifer or well test at the main well. The commonly used methods consist of volumetric measurements by using containers or observing the head of water as it passes through a weir or pipe. Discharge should be measured at least once every hour. If necessary, the adjustment is done by a valve on the delivery pipe rather than by changing the pump speed.

There are different direct ways of measuring the discharge according to the available field facilities, some being more accurate than the others. Usually aquifer tests are conducted at low discharges but it is preferable to maintain a high discharge so that even in distant observation wells the drawdown is not masked by fluctuations due to external influences.

A. CONTAINER METHOD

The basic definition of discharge is the ratio of water volume to a certain time needed to fill this volume. Hence, simply the discharge, Q, can be calculated by determining the time, t, required to fill a container whose volume, V, is known as $Q = V/t$. The stopwatch is desirable for the time measurement. The containers may have various forms such as oil barrels, stock tanks, or rectangular or circular reservoirs. Small changes can be accurately measured by noting the time taken to fill a container of known volume. This method is suitable for low discharges up to 0.50 m^3/min.

B. WATER METER

This is a counter in a cubic meter of water that flows through the delivery pipe and it is located on this pipe as shown in Figure 10. The pipe should run full and in order to enable this a vertical bend is given after the meter. The meter records automatically the total volume discharged through the pipe. Hence, it is necessary to take two readings at certain time intervals and the difference between the two gives actually the volume of water. Then the ratio of this volume to the time interval yields the amount of discharge. A water meter has the advantage that its register will record any pumping variations when the system is not attended. Some water meters can be equipped with recording charts.

C. ORIFICE METER

The end of the delivery pipe is mounted with an orifice which is a perfectly round hole in the center of an iron plate as shown in Figure 11. The orifice should be from one half to three quarters of the size of the pipe. Such a design is necessary for guaranteeing the full run of water through the orifice. The pressure on the pipe wall is measured by a piezometer tube located at 24-in. (61 cm) distance from the orifice at the side of pipe. The difference between elevations of water in the piezometer and the center of the orifice gives the water pressure, h. The area, A, of the orifice and h are related to the discharge, Q, as

$$Q = kA\sqrt{2gA} \tag{7}$$

in which g is the gravitational acceleration (9.81 m/s²) and k is the discharge constant calculated as

$$k = \left(\frac{d_c}{d_0}\right)^2 \frac{x}{2\pi yh} \tag{8}$$

where d_c is the diameter of the jet at the contracted section; d_0 is the actual diameter of the orifice; x and y are any corresponding horizontal and vertical distances.

If the discharge water contains air or the pump is surging then pulsations in the water tube make the reading of h difficult. Commercially available orifice meters are supplied with a calibration chart.

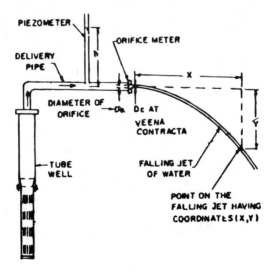

Figure 11 Orifice meter.

D. JET STREAM METHOD
1. Horizontal Pipes

The basic physical principle of this method is that a water jet leaving a horizontal pipe will, like a bullet from a gun barrel, follows a path that is determined by its exit velocity and the gravitational acceleration. One of the major assumptions in this method is that the velocity of a water particle at the jet surface is representative of the average velocity in the stream which is not precisely true. Nevertheless, the method gives reasonable estimations within 10% error.

In the field applications of the method, an open-ended straight and horizontal pipe is required with a length at least equivalent to eight pipe diameters after the last elbow, tee connection, or node as in Figure 12. The horizontal distance from the pipe end corresponding to the fall of 12 in. is measured and the estimated flow is read from Figure 13. The 12-in. fall must be measured from the inside of the pipe. If measured from outer surface then the pipe thickness, t, must be added to this figure in order to obtain correct results. It is advisable to have a ready L ruler for the measurements.

The jet stream method can also be used if the pipe runs partially full as shown in Figure 14. Herein, the horizontal distance to the 12-in. fall is measured from the free water surface in the pipe. In addition, the maximum height, y, from the free water surface to the inner surface of pipe is measured. Therefore, in practice, the horizontal x distance will be measured more easily along the pipe extension corresponding to a fall equal to $(12 + y + t)$ in. First the discharge corresponding to this value is obtained for a full pipe as explained above. The

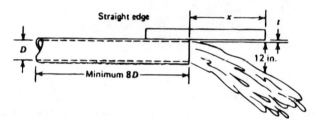

Figure 12 Horizontal pipe.

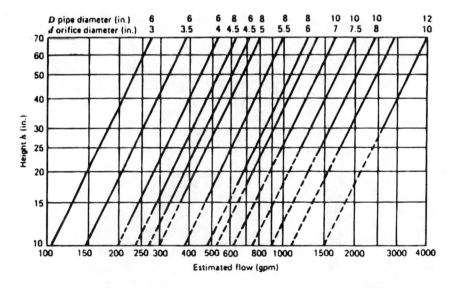

Figure 13 Curves for trajectory method.

FIELD MEASUREMENTS

estimated discharge from the partially full pipe is approximately proportional to the air space ratio $(D - y)/D$ percentage. For more precise estimates of the percentage of the full pipe the chart in Figure 15 can be used.

2. Vertical Pipe

This method depends on the height of a jet due to water exit from the end of a vertical pipe as shown in Figure 16. The approximate discharge from vertical pipes is a function of the maximum jet height, h, and the inside diameter of the pipe both measured in inches. The flow in liters per second is given in Table 5. This method is very suitable for flowing wells and the vertical final run is at least eight to ten times the pipe diameter. The final vertical run must be free of any hindrance such as elbows, valves, etc.

E. PARSHALL FLUME

Simple hydraulics structures are ideal pieces of instruments for discharge measurements if the water passes through them in critical conditions. If the discharge is to be measured permanently as in irrigation, drainage, or sewage networks, then construction of a suitable hydraulic structure saves time and capital in the long run. Most often these structures are of standard configurations. Standardization makes their cost considerably less.

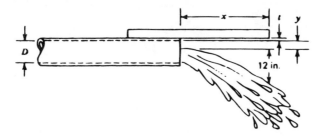

Figure 14 Partially full pipe.

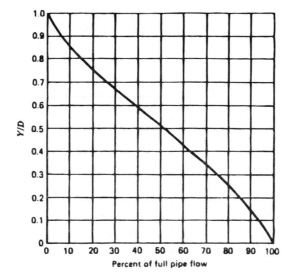

Figure 15 Percentage of full type flow.

The simpler and successful structure that is developed for the discharge measurements in small open channel flows is the Parshall flume as shown in Figure 17. Along its longitudinal direction there are three sections, a converging throat and a diverging section to give the flow a critical condition. For free flow in the throat section the discharge in the flume can be computed by measuring the upstream water depth, H_a, before the converging section. The approximate formula for the discharge, Q, is

$$Q = 4wH_a^{1.522w^{0.026}} \qquad (9)$$

where w is the throat width. The nomograph in Figure 18 facilitates the solution of the Parshall flume equation.

F. WEIRS

Weirs are hydraulic structures that render the irregular cross-sectional area of a small channel into regular sections over which the water flows. They are constructed with openings of simple geometric shapes such as triangular or rectangular shapes, the most commonly employed ones. Weirs allow water passage over the crest in a free fall. Under this condition they are good flow-measuring devices, especially if their wall at the crest and sides is thin.

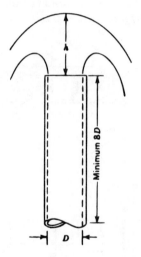

Figure 16 Vertical pipe.

Table 5 Vertical Pipe Discharge Measurement

Height (mm)	Nominal pipe diameter (in.)		
	2	3	4
50	2	4.3	7.4
75	2.4	5.3	9.1
100	2.8	6.1	10.5
150	3.4	7.4	12.8
200	3.9	8.8	14.8
250	4.4	9.6	16.6
300	4.8	10.5	18.1
350	5.2	11.4	19.6
400	5.5	12.2	21
450	5.9	12.9	22.2
500	6.2	13.6	23.4
550	6.5	14.3	24.6
600	6.8	14.9	25.7

FIELD MEASUREMENTS

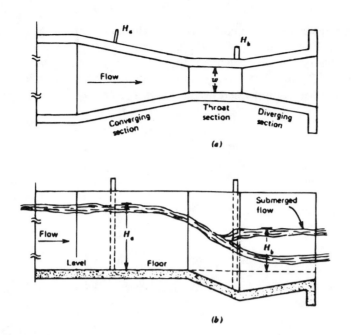

Figure 17 Parshall flume.

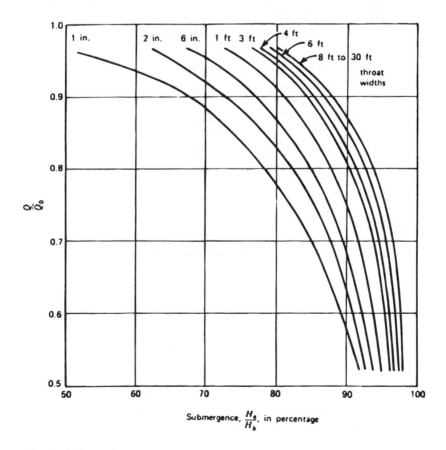

Figure 18 Parshall plume chart.

1. The V-Notch Weir

This is a reliable method of estimating discharges up to 1000 gallons per minute (gpm). Its shape is given in Figure 19. The weir can be constructed as a portable instrument and the water from the pipe end is discharged into it. Discharge Q is given by

$$Q = \left(\frac{8}{15}\right)\sqrt{2g}\,C_d \tan\left(\frac{\theta}{2}\right) H^{5/2} \tag{10}$$

where g is the gravitational acceleration, C_d is the discharge coefficient, θ is the angle between the sides of the notch, and H is the total head of water. Values of C_d for 90° thin plate weir with fully developed end contraction are given in Table 6. The limitations in applying this expression with notations in Figure 19 are that

1. Tail water should remain below the vertex of the notch h and y shall not be less than 5 cm or greater than 38 cm.
2. The vertex height, p, shall not exceed 45 cm and shall not be less than 10 cm.
3. H/p shall not exceed 1.2.

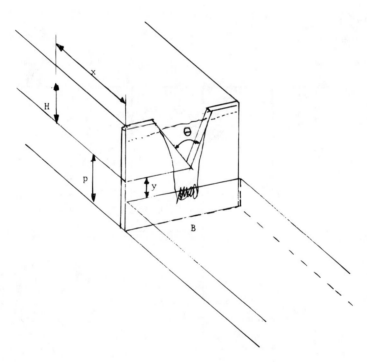

Figure 19 V-notch weir.

Table 6 V-Notch Weir Discharge Coefficient Table

Head, H (m)	Value of C_d
0.050	0.608
0.075	0.598
0.100	0.592
0.125	0.588
0.150	0.586
0.200	0.585
0.300	0.585

FIELD MEASUREMENTS

4. The width of approach channel, B, shall not exceed 90 cm.
5. H/p shall not exceed 0.2.
6. x should be more than 3 to $4 \times H$.

If the flow is led to a field channel in which a 90° V-notch is installed then the discharge can be estimated by determining the head, H, over the notch as

$$Q = 1.38H^{2.48} \tag{11}$$

or can be obtained from Table 7.

2. The Rectangular Weir

For flow rates larger than 1000 gpm the V-notch weirs are not suitable. Instead, the rectangular weirs can measure up to 5000 gpm discharge. It is similar to the V-notch weirs except that the triangular notch is now a rectangle. For the application the weir dimensions

Table 7 Discharge through a 90° V-Notch

Head, H (cm)	Discharge, Q (l/min)
2	4.7
4	26.5
6	73
8	149
10	255
12	414
15	725
20	1490
25	2580
30	4060

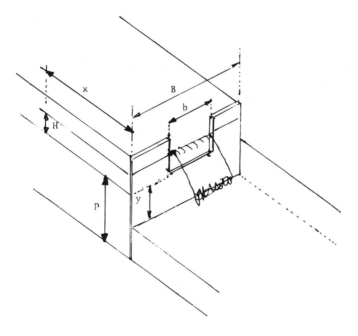

Figure 20 Rectangular weir.

should have the relationship to the head, H, as shown in Figure 20. Discharge, Q, is given by

$$Q = \left(\frac{2}{3}\right)\sqrt{2g}C_d w H^{3/2} \tag{12}$$

where w is the notch width and other terms are for V-notch weirs. In the rectangular weirs C_d is given by

$$C_d = C_1 + C_2 \frac{H}{p} \tag{13}$$

where p is the weir height and H is the gauged head both in meters. The limitations applying to the rectangular weir relationship are that

1. H/p shall not exceed 1.
2. H shall be between 3 and 75 cm.
3. The weir width, b, shall be at least 30 cm.
4. p shall not be less than 10 cm.

In Equation 13 the constants C_1 and C_2 should assume values according to Table 8. The following limitations should be considered in practical applications:

1. The recommended lower limit for H is 3 cm.
2. Upper limit for the ratio is 2 while p should be less than 15 cm.
3. To facilitate the aeration of the nape, the tail water level y should be at least 10 cm.
4. b should not be less than 15 cm.

VI. DISCHARGE CALCULATIONS

During field studies a common difficulty in discharge measurement is that the leading pipe from the well either branches into many small pipes supplying water to different agricultural lands or to a common reservoir from where the water is distributed by means of many uncontrollable outputs. Therefore, none of the above-mentioned direct discharge devices can be used. It is well known from the experience that, as the water is abstracted from a well initially, all the water comes from the well storage. During this period the relationship between the drawdown and time appears as a straight line. The physical consequences of this straight

Table 8 Rectangular Weir Discharge Coefficients

b/B	C_1	C_2
1	0.602	0.075
0.9	0.599	0.064
0.8	0.597	0.045
0.7	0.595	0.030
0.6	0.593	0.018
0.5	0.592	0.011
0.4	0.591	0.058
0.3	0.590	0.0020
0.2	0.589	−0.0018
0.1	0.588	−0.0021
0.0	0.587	−0.0023

FIELD MEASUREMENTS

line is that the pumping discharge comes from the well storage contribution only. It is possible to write this relationship as explained by Şen (1986) as

$$Q = \pi r_w^2 \frac{ds_w(t)}{dt} \tag{14}$$

where t is the time since pump start and r_w is the radius of the well. In this equation, $ds_w(t)/dt$ is equal to the initial slope of the time-drawdown curve. As a result, the initial drawdown measurements during a pumping test can be plotted against the corresponding times on a millimetric paper. The slope can be read from this plot and after the measurement of the well diameter the aforementioned equation gives the discharge value. In the absence of direct discharge measurements this method gives practically reliable results.

Example 1—This method is applied for five large-diameter wells in Wadi Qudaid, western Saudi Arabia. Alluvial deposits constitute the only potential aquifer in the wadis. The early drawdown measurements of these wells with their radii are given in Table 9. The plots of time-drawdown data for each well are shown in Figure 21. The best straight lines are drawn among the scatter points.

Table 9 Early Time-Drawdown

Time since pump start (min)	W1 $r_w = 1.30$ (m)	W2 $r_w = 0.98$ (m)	W3 $r_w = 1.08$ (m)	W4 $r_w = 1.19$ (m)	W5 $r_w = 0.95$ (m)
1	0.08	0.08	0.12	0.05	0.12
2	—	0.21	0.19	0.10	0.28
3	0.24	0.32	0.29	0.15	0.44
4	0.31	0.45	—	0.20	0.60
5	0.40	0.55	0.51	0.26	0.72
6	—	0.70	0.62	—	0.86
7	0.56	—	—	—	0.99
8	0.66	0.89	0.81	—	1.10
9	—	—	—	—	—
10	0.79	—	0.96	—	—

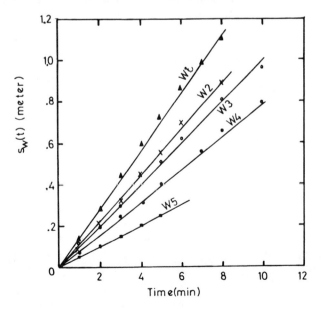

Figure 21 Early time drawdown plots.

Table 10 Comparison of Measured and Calculated Discharges

Well no.	Discharge (m³/min) Measured	Discharge (m³/min) Calculated	Relative error (%)	Slope (°)
W1	0.38	0.41	8	0.078
W2	0.31	0.33	6	0.113
W3	0.33	0.36	10	0.100
W4	0.24	0.22	−9	0.050
W5	0.44	0.44	0	0.240

The calculated discharge and direct discharge measurements in the field are presented in Table 10. Also, the relative errors are given in the same table and each individual relative error is less than 10% with an average of 3%.

VII. SPECIFIC DISCHARGE MEASUREMENTS

In some groundwater studies it is necessary to measure the groundwater movement velocity. Usually velocity is related to the hydraulic gradient, conductivity, and effective porosity. These quantities are difficult to measure simultaneously in a field study. The most practical specific discharge measurement in the field is through the use of tracers.

Prior to the use of tracers the direction of groundwater is established by water table or piezometric surface mapping. Two wells along the streamline are located and a tracer is injected from higher hydraulic head well. In practice, when there are uncertainties about the flow direction, different wells are observed around the injection well. For velocity calculations, the necessary time, t, of trace travel and distance, D, between the injection and observation wells are given approximately as (UNESCO, 1972)

$$t = \frac{D\theta}{Ki} \quad (15)$$

in which K is the hydraulic conductivity in i is the hydraulic gradient, and θ is the effective porosity. The specific discharge estimation is given by

$$q = \frac{D}{t} \quad (16)$$

If D is in meters and t in days, then q will result in meters per day.

VIII. GRAPHICAL DATA TREATMENT

After completion of any aquifer test before the field data analysis the time data should be converted to a single set of elapsed time unit (e.g., minutes) since pump start and water-level data to drawdown values of uniform unit. External influences on water levels must be corrected. Among these influences are natural effects, precipitations, evapotranspiration, atmospheric pressure changes, tides, and regional trends in groundwater levels. Accidental variations may appear as a result of earthquakes and subsidence. A simple impression about the external effect is gained by comparing water level measurements prior to pumping and after the recovery. If these two levels are practically equal to each other it can be safely assumed that the external effects did not influence the hydraulic head during the test. Otherwise, the water level is subject to changes and it must be corrected. Useful interpretations about the time-

FIELD MEASUREMENTS

drawdown data are obtained primarily through visual inspection from convenient graphical representation.

A preliminary step in determining aquifer parameter estimations from field measurements is to plot the observed data on various graph papers for interpretation purposes depending on the type of data available and aquifer-well configuration, hydrological conditions, and on the selected method of analysis. The responses of confined, unconfined, and leaky aquifers to pumping are different. Visual inspection of test data in graphical form is very essential to draw significant meaningful interpretations, perhaps qualitative prior to the application of any formal methods in the following chapters. For data evaluation in groundwater studies there are three types of graphical representations depending on the type of graph paper: arithmetic (ordinary), semilogarithmic, and double-logarithmic.

A. ARITHMETIC GRAPH PAPER

Changes in the piezometric level represent changes in the groundwater storage. These changes recorded in an observation well reflect natural recharge, tides, and surface water level fluctuations. Regional changes in piezometric level are depicted by groundwater maps showing the rise or fall of the groundwater level in a specified time interval at many observation wells. However, time-dependent changes in water levels in wells can be depicted by groundwater hydrographs which are defined as the change of groundwater level with time on an arithmetic paper (see Figure 22). A decline in the groundwater level implies water abstraction in excess of recharge to aquifer. In other words, during a decline period output is more than input into an aquifer: there is decrease in the groundwater storage. Otherwise, increase in the storage takes place. Fluctuations are related to climatic elements such as rainfall, evapotranspiration, atmospheric pressure, hydrological influence including surface body water fluctuations, and man-made causes among which are the water abstraction through wells, irrigation, and artificial recharge.

Natural recharge and discharge result in hydraulic head rise and fall, respectively. This rise or fall should be determined for the aquifer test period by interpolations from the main and observation well groundwater hydrographs. Supposing that at the start, t_0, of the test the hydraulic head elevation is h_0 and during any time, t_i, the hydraulic head is h_i; if there were no test performed, the head difference, $\Delta h = h_0 - h_i$ should either be added to s_i (natural recharge) or subtracted (natural discharge) from it. The corrected drawdown s'_i without any external effect due to water abstraction only is

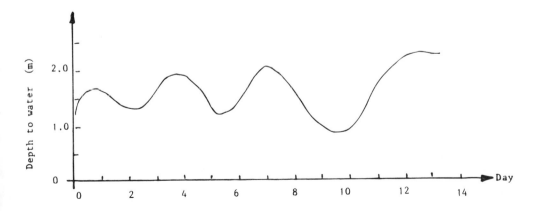

Figure 22 Groundwater hydrograph.

$$s'_i = s_i - \Delta h \qquad (17)$$

or

$$s'_i = s_i + \Delta h \qquad (18)$$

Corrections in hydraulic heads are also necessary if the water table in unconfined aquifers is close to the earth's surface being affected by diurnal variations due to evapotranspiration. Plant water usage causes groundwater hydrographs to appear in the form of rhythmic fluctuations. Groundwater hydrographs of main and observation wells for sufficiently long periods prior to water abstraction and posterior to recovery periods provide the necessary information for correction of water levels during a test period. Groundwater level in a well changes also due to the atmospheric pressure. For corrections in drawdown measurements it is necessary to have simultaneous records of groundwater hydrographs and atmospheric pressure change with time covering the pretest period. These two graphs help calculate the barometric efficiency of the aquifer defined by Jacob (1946) as the ratio of water level change, Δh, in a well to the corresponding change in atmospheric pressure, Δp. The plot of Δh vs. Δp on an arithmetic paper yields a straight line and its slope gives the barometric efficiency. The same graph provides a basis for determining hydraulic head change Δh_p, due to the atmospheric pressure only provided that the atmospheric pressure change is Δp during a test. Subsequently, field recorded drawdown, s, can be corrected as $s' = s \pm \Delta h_p$, plus and minus sign for falling and rising atmospheric pressure, respectively.

During abstraction of groundwater a decline occurs in water level in a well. Drawdowns calculated from the differences between static and dynamic water levels increase directly with

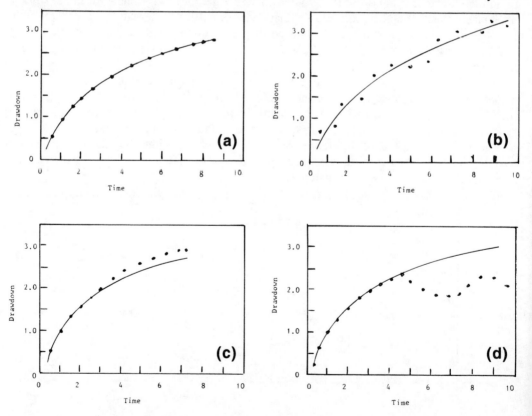

Figure 23 Arithmetic paper time-drawdown plot.

FIELD MEASUREMENTS

time since pump start. In general, time-corrected drawdown plots appear as a parabolic pattern on an arithmetic paper. These plots are essential in qualitative assessment of aquifer and flow properties prior to any numerical treatment of data.

If the points appear perfectly on a definite curve (Figure 23a) it implies that the aquifer medium is homogeneous and isotropic. Erratic deviations from a smooth curve as in Figure 23b mean that the aquifer material has nonsystematic heterogeneities. However, if the deviations appear in a systematic manner as in Figure 23c and d then significant changes in aquifer properties appear after a certain time and/or distance. A scientific account of these deviations is presented under the title of "hydrogeophysical concepts" in Chapter 9.

B. DOUBLE-LOGARITHMIC PAPER

The numerical determination of aquifer parameters requires theoretical mathematical models capable of representing the characteristics of the real aquifer in the best possible manner. The choice of theoretical model is a crucial step in the interpretation of field test data. This is due to a few possible representative theoretical models for the same problem. The uncertainty involved in proper model selection can be reduced by conducting more field study if financial sources are available. If the wrong model is chosen, the aquifer parameter estimations will not represent the real aquifer properties. A key tool in model selection is the double-logarithmic (log-log) plot of drawdown vs. time or distance. As explained in Chapter 9 all theoretical model outputs are represented on log-log graph papers. Hence, by knowing the behavior of different aquifer types on these graphs it is possible to identify visually the best representative model from field time-drawdown data plot on log-log paper. The key log-log plots of major aquifer types are given in Figure 24. Major assumptions for double logarithmic behaviors in this figure are that the well diameter is infinitesimally small, aquifer material is homogeneous and isotropic, water is abstracted with a constant discharge, and the well fully penetrates the aquifer and the flow law is linear (Darcian).

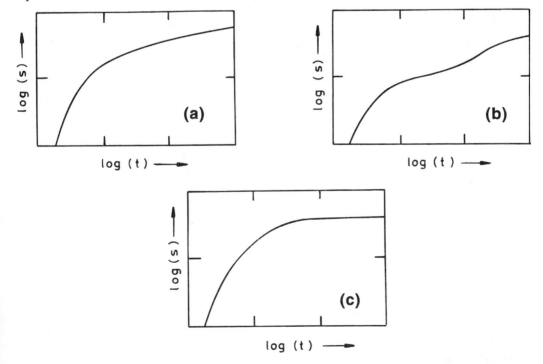

Figure 24 Double-logarithmic paper. (a) Confined aquifer; (b) unconfined aquifer; (c) leaky aquifer.

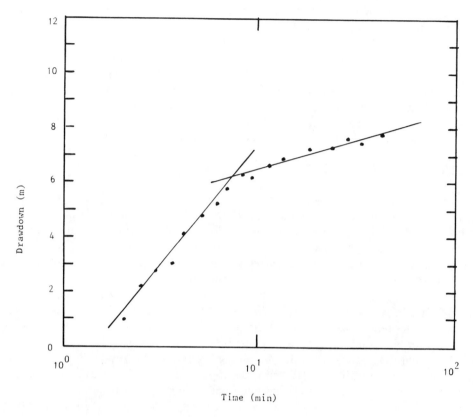

Figure 25 Semilogarithmic paper straight lines.

After plotting time-drawdown data on a double-logarithmic paper and inspecting a smooth curve through the scatter of points, it is possible to identify the aquifer type by comparing these curves with the ones in Figure 24. In such a comparison the following points are useful to consider:

1. In all graphs early times do exhibit nonlinear portions. This is a good indication that the well diameter is extremely small and all the abstracted water originates from the aquifer storage only. The appearance of a straight line at early times is due to a large-diameter well where the abstracted water comes initially from the well storage.
2. In Figure 24a the curve does not have any inflection point and there are drawdown increments, though insignificant, at large times. This curve implies confined aquifer type.
3. Existence of a flat portion during medium times as in Figure 24b indicates that the aquifer is unconfined provided that large time-drawdown behavior is similar to confined aquifer. Flatness implies further that there is a delayed recharge from the aquifer itself.
4. If there are no drawdown changes at large times the aquifer is said to be "leaky". Along this final horizontal segment the groundwater reached a steady-state flow. The initial segment in Figure 24c resembles the initial segment of the confined aquifer. At medium times more and more water from the aquitard(s) reaches the aquifer.

Double-logarithmic papers are used for showing the time-drawdown data scatter for the whole test period.

C. SEMI-LOGARITHMIC PAPER

Time-drawdown plots on these papers are useful for the following purposes.

1. To work with a straight line instead of a curve. In practice, semi-log papers are used for late time-drawdown data that appear theoretically around a straight line. In the case of quasi-steady-state time-drawdown (or distance) data use of semilogarithmic papers is effective.
2. To identify dominating flow regimes (see Chapter 12).
3. To determine the radius of influence from distance-drawdown data.
4. To determine impermeable and recharge boundaries.

Conventionally, drawdowns are shown in linear scale vertical axis and time-distance on a logarithmically scaled horizontal axis (Figure 25).

REFERENCES

Şen, Z., 1986. Discharge calculation from early drawdown data in large-diameter wells, *J. Hydrol.*, 83, 45.
Jacob, C. E., 1946. Radial flow in a leaky artesian aquifer, *Am. Geophys. Union*, 27(11), 198.
UNESCO, 1972. Ground Water Studies, Studies and Reports on Hydrology, No. 7, Genera, 4.

CHAPTER 8

Steady-State Flow Aquifer Tests

I. GENERAL

The steady-state flow toward or away from a well, fracture, or ditch has theoretical as well as practical significance and implications. Physically, steady-state flow occurs either after or before an unsteady-state flow. Initially the groundwater is in the steady state before pumping but after a long period of continuous pumping the groundwater flow approach is in a quasi-steady state depending on the geological environment as explained in Chapter 3. Mathematically, the steady or quasi-state provides relatively easier solutions which are special cases for the unsteady state. The practical significance of steady-state flow is due to the simplicity of equations, necessary field measurements, and evaluation of some basic decision variables relating to the discharge, well spacing, design and construction, and pump and piping system. Basic formulations were developed by Thiem (1906) and Muskat (1937) and later supplemented by many investigators. However, all formulations assume ideal aquifer and well behaviors. Computed values are therefore regarded as approximations for natural systems. It is suggested at this point that any result should be related to geology prior to its use in further planning, design, and operation of groundwater resources. The steady-state groundwater flow formulations show how basic variables regarding the well and the aquifer are combined together to determine the consequences of any decision variable. It is also possible to grasp the validity of the basic assumptions and the impact of the final decision variables if some of the assumptions were incorrect. These variables are the discharge, maximum drawdown, or radius of influence depending on the objectives.

The main purpose of this chapter is to present steady-state groundwater problem solutions with a minimum level of mathematics but detailed physical implications of assumptions and practical applications.

II. ASSUMPTIONS

Numerous analytical and graphical solutions for steady- and unsteady-state flows are available for determining the aquifer hydraulic parameters from available aquifer test data. Various formulae for the solution of groundwater flow problems are based on certain assumptions and generalizations concerning the hydraulic properties and geometrical features of aquifers and piezometric surface. It is, therefore, essential by all means to verify the assumptions on which a formula is based.

Applications of the continuity and the fundamental groundwater flow laws to real situations require a set of simplifying assumptions that are necessary for mathematical convenience. The more restrictive and numerous the assumptions, the more unrepresentative are the results obtained thereof. Avoidance of assumptions is not possible but their restrictiveness and numbers can be reduced by refined mathematical approaches. Mathematical simplifications imply sacrifices from the physical reality. However, they are necessary for refinement to achieve physically reliable and practically accurate results.

The assumptions are classified in this book under three categories concerning the aquifer, well, and the groundwater flow properties. Physical and mathematical reasoning of various assumptions are given in the following sections.

A. AQUIFER ASSUMPTIONS

These are essential to analyze the aquifer geological composition and the geometry so as to render the real problem into a mathematically tractable form. Four fundamental assumptions are:

1. That aquifer material is isotropic and homogeneous as explained in Chapter 4. It is tantamount to saying that the aquifer has a single layering (simplest stratigraphy), no change in the areal geological facies (single lithology), and no discontinuities (fractures, folds, faults, dikes). The physical consequence of isotropy and homogeneity is that the equipotential lines are perpendicular to the flow lines. Isotropy and homogeneity imply that fewer samples or field measurements are needed at different sites to arrive at meaningful conclusions.
2. That the aquifer is aerially extensive, has uniform thickness, and is horizontal. Extensiveness and uniformness have been discussed in Chapter 14. Definition of horizontal layer is different in confined and unconfined aquifers. In confined aquifers the upper and lower confining layer slopes indicate the horizontality. The piezometric level slope does not play any role in horizontality of an aquifer. Within the depression cone, the thickness variations of the aquifer can be estimated along four principle directions by considering dip and strike. If the variation gradient, i, is less than 0.05, which is frequently the case in practice, then the aquifer is said to be horizontal. Figure 1 shows different confined aquifer cross sections. Figure 1a is the ideal case where confining layers are horizontal even though the piezometric level is not. However, in Figures 1b through d either one or both of the confining layers are not horizontal. In these cases the aquifer is assumed as horizontal, provided only that both confining layer gradients are less than 0.05. To confirm aquifer horizontality at least two hydrogeological cross sections in different directions are necessary. Otherwise, the decision based on one direction might be misleading as in Figure 2, although the aquifer is horizontal along the E-W cross section but not along the N-S direction. On the other hand, in an unconfined aquifer slopes of the impervious layer and water table govern the aquifer horizontality. If the relative difference between the slopes is less than 0.05, the aquifer is considered horizontal. In Figures 3a and b the aquifer is horizontal but in Figures 3c, d, and e nonhorizontal aquifers appear. The horizontality in confined aquifers is determined by geology only, but in unconfined aquifers additionally the hydrogeological factors (seepage, infiltration, recharge, etc.) affecting the groundwater table are also important. Hence, the same unconfined aquifer may be horizontal or not at different times. Horizontality of leaky aquifers is determined by geological information only similar to confined aquifers. In horizontal aquifers the groundwater flow is also horizontal provided that the well is of the fully penetrating type.
3. That the aquifer material is fine granular and hence the flow domain is of porous medium type.
4. That the aquifer parameters are independent of temporal and spatial changes which is not true in a micro scale but approximately valid at macro scales.

B. GROUNDWATER FLOW ASSUMPTIONS

The assumptions in this section are concerned with groundwater velocity, discharge, hydraulic conductivity, and flow modes. These are

1. That the state of flow is steady or quasi-steady. Complete steady state implies an external recharge, the amount of which is equal to water abstracted from the aquifer. The drawdown

STEADY-STATE FLOW AQUIFER TESTS

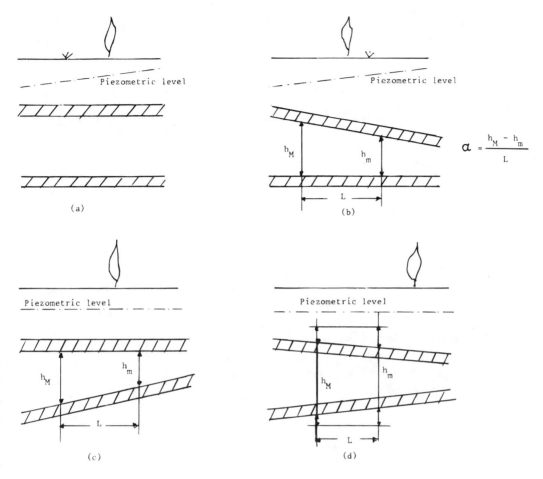

Figure 1 Horizontality of confined aquifer.

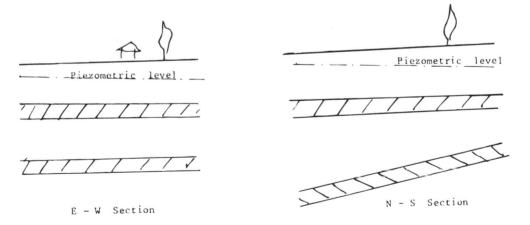

Figure 2 Hydrogeologic cross sections.

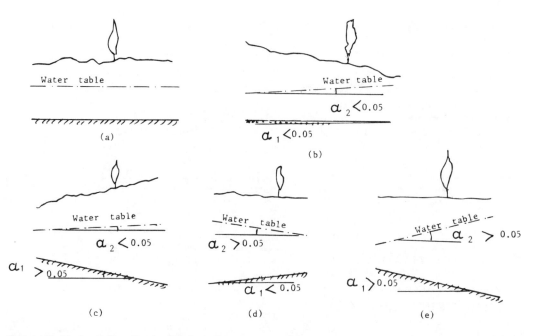

Figure 3 Horizontality of unconfined aquifer.

varies with the distance only and, therefore, all the partial derivations in groundwater movement equation become ordinary derivatives.
2. That the flow is laminar and the Darcy law is applicable. Physically, this is tantamount to saying that the kinetic energy of water is not significant, the specific discharge is small, and that the dominating forces are due to the viscosity and friction.
3. That the pumping discharge is constant.

1. Dupuit-Forchheimer Assumptions

Quantitative treatment of confined aquifers is easier than unconfined aquifers because in confined aquifers the saturation zone is completely defined by impervious layers and its geometry does not change with time. The difficulty in unconfined aquifers comes from spatial and temporal changes of saturation thickness. Temporal changes render the flow not only into a three-dimensional type near the well but also equipotential lines become curves leading to decreases in saturation thickness toward the well as in Figure 4. Thus, the groundwater velocity

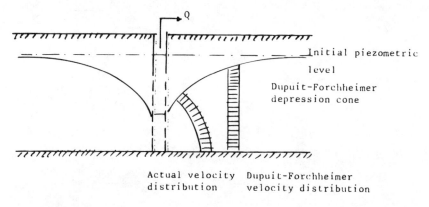

Figure 4 Flow net around the well.

STEADY-STATE FLOW AQUIFER TESTS

direction is not horizontal at every point but inclined especially in the well vicinity near the water table.

The hydraulic gradient also changes from one point to another within the flow domain. The inclinations and variations cause difficulties in the groundwater movement quantification. To avoid such difficulties Dupuit (1863) proposed the following assumptions that were later advanced by Forchheimer (1930):

1. That the water is homogeneous and in every direction it has the same physical properties.
2. That the flow lines are horizontal and, accordingly, the equipotential lines are vertical. This assumption implies horizontal groundwater velocity and its uniform distribution along the whole saturation thickness.
3. That the hydraulic gradient at every point along a vertical line is equal to the slope of the free surface. The velocity of flow is proportional to the sine and not the tangent of the angle of the water table, and the flow is horizontal and uniform near the water table.
4. That the capillary zone is negligibly small.
5. Thiem assumed that in unconfined aquifers transmissivity is constant provided that the variations in saturation thickness are small compared to the initial saturation thickness.
6. That the aquifer material and water are incompressible.

It is obvious that these assumptions neglect completely the vertical flow component. However, the practical value of Dupuit-Forchheimer assumptions lies in the fact that they reduce a three-dimensional flow into a two-dimensional type that is easy to deal with.

C. WELL ASSUMPTIONS

Apart from providing an ample space for pump submergence into the groundwater storage, the well affects the geometry and dimensions of groundwater flow net especially in the vicinity of the well as explained in Chapter 6. The following set of presumptions are necessary for simple analytical treatment of groundwater flow toward wells:

1. That the well cross section is circular. It provides mathematical convenience in calculating the aquifer discharge as the volume of water crossing the lateral surface of concentric cylinders. The groundwater flow takes place along radial streamflow lines.
2. That the well diameter is small. It means that the well storage does not play any significant role in the continuity equation. In other words, from the beginning of pumping until the end, the pump discharge comes from the aquifer storage only. This assumption implies instantaneous aquifer response that is not valid for large-diameter wells (see Chapter 6).
3. That the well fully penetrates the aquifer from the top to the bottom through the whole saturation zone. A detailed amount of this situation is given in Chapter 6.
4. That there is no well loss. This assumption treats the maximum drawdown in the aquifer as equal to drawdown within the main well. Schematic demonstration of well losses is shown in Figure 5. It implies further than the depression cone has steep hydraulic gradients in the

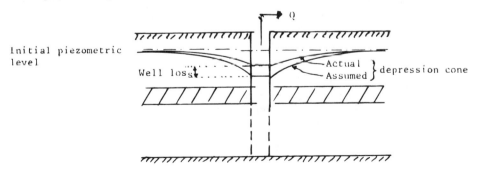

Figure 5 Aquifer well loss.

vicinity of well. Consequently, in calculations the specific discharge near the well becomes more than actual velocity. Further implications and evaluations of well losses will be discussed in Chapter 11.

III. LIMITING CONDITIONS

It is possible to obtain general solutions for groundwater movement by considering the assumptions listed in the previous section but specific solutions of any problem require additional knowledge about the initial and boundary conditions. Unique solutions of groundwater movement equations for a specific flow system are obtained after knowing relevant initial and boundary conditions. Otherwise, each equation has an infinite number of possible solutions. The reliability of final solution depends essentially on the accuracy of these conditions. Prior to any problem solution identification of these conditions is important on the basis of available geological, hydrological, and geophysical information or past experiences. Since different boundary and initial conditions lead to different solutions, the correct determination of the conditions is a prerequisite in groundwater problem solutions.

A. INITIAL CONDITIONS

These conditions indicate the state of piezometric surface of groundwater at rest or in movement at some initial time. For instance, in the steady-state case the possible initial conditions are

$$q(x,y,z,0) = C_1$$
$$s(x,y,z,0) = C_2$$

or

$$i(x,y,z,0) = C_3$$

in which $q(x,y,z,0)$, $s(x,y,z,0)$, and $i(x,y,z,0)$ are the specific discharge, drawdown, and hydraulic gradient, respectively, in space at a point with coordinates (x,y,z) and time, t, $(t = 0)$. The constants C_1, C_2, and C_3 depend on the physical conditions of the problem at hand. In most of the steady-state cases in this book, with no loss of generality, these constants are taken as zero.

B. BOUNDARY CONDITIONS

They indicate spatial properties of groundwater flow at any time depending on geological and hydrological information that give possible facies changes, discontinuities, geological structures, or recharge or discharge sources within the flow domain.

1. Impermeable Boundaries

Any equipotential on the impervious boundary is called a constant potential. There is no cross flow and hence the specific discharge component perpendicular to the boundary or the hydraulic gradient is equal to zero. The mathematical expression of this statement is

$$i(x,y,z,t) = \frac{\partial h(x,y,z,t)}{\partial n} = 0$$

in which n is the normal to the impervious surface at point (x,y,z). Examples are confining layers, impervious barriers or lenses, or groundwater table if there is no cross flow due to recharge. Figure 6 represents the most commonly existing impermeable boundaries in and around the well.

2. Well-Aquifer Continuity

Consideration of continuity equation for a well implies simply that the difference between abstraction, Q, and aquifer delivery rates to the well, Q_a, is equal to well storage change. For infinitesimally small diameters the well storage is negligible and the abstraction rate is equal to the aquifer discharge which assumes its maximum value at the well face. Mathematical formulation of this physical boundary condition is

$$Q = \lim_{r \to 0}[2\pi m r q(r,t)] \tag{1}$$

The right hand side (r.h.s.) term gives aquifer discharge where $q(r,t)$ is the specific discharge, m is the aquifer thickness, and r is the radial distance. On the other hand, if a well has a large diameter the same continuity equation leads to

$$Q = \lim_{r \to r_w}[2\pi m r q(r,t)] - \frac{ds_w(t)}{dt} \tag{2}$$

The second term on the r.h.s. shows the well storage effect and $s_w(t)$ is the drawdown measured in the well at time t. Practically, either for small r_w values and/or large times, the quasi-steady-state term converges to zero and Equation 1 becomes valid even for large-diameter wells.

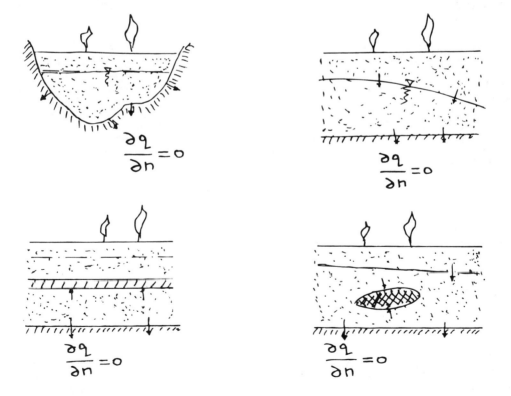

Figure 6 Impervious boundaries.

Another continuity condition is the assumption that the drawdown in the well is equal to the maximum drawdown, $s(r_w,t)$ in the aquifer and hence

$$s_w(t) = s(r_w,t) \tag{3}$$

These conditions of well-aquifer continuity are valid at all times.

3. Extensiveness Condition

In nature most often aquifers are finite, but simple derivations of analytical equations require the assumption of infinite areal extent. In practice, however, aquifers are considered as extensive if they extend over several kilometers and the abstraction period is not too large. Extensiveness of an aquifer has been explained qualitatively in Chapter 4. In extensive aquifers the drawdown at large distances is equal to zero. This boundary condition can be expressed mathematically as another limiting case:

$$\lim_{r \to \infty}[s(r,t)] = 0 \tag{4}$$

in which $s(r,t)$ is the drawdown at any point within the depression cone and time, t. This condition is not valid for bounded aquifers.

IV. FULLY PENETRATING WELL IN AN AQUIFER

We will consider in this section different combinations of aquifer and flow types to arrive at practically useful formulations of radial groundwater flow toward wells.

A. CONFINED AQUIFERS

The very first solution of groundwater flow toward wells was presented by Thiem (1906). Figure 7 illustrates a finite radius, r_w, well that fully penetrates a confined aquifer of permeability K and thickness m. The well is discharging at a constant rate, Q.

The question is how to derive an expression for the drawdown distribution in the aquifer. In the following, the derivation will be presented from physical, mathematical, field implications, economy, and practical points of view. It is hoped that these reasonings will give some incentive to the reader and hence the mathematical concepts presented herein will be employed in other sections of this book.

Since the flow is in steady or quasi-steady state the rate of input, Q_a, across any imaginary concentric cylinder in the aquifer is equal to pump discharge,

$$Q = Q_a \tag{5}$$

On the other hand, Darcy's law can be written for a steady-state radial flow as

$$q(r) = -K\frac{ds(r)}{dr} \tag{6}$$

where $q(r)$ and $s(r)$ are time-independent specific discharge and drawdown at radial distance, r, in the aquifer with hydraulic conductivity K. Considering the specific discharge definition from Equation 15 in Chapter 3 together with the cylindrical surface as $2\pi rm$ and their

STEADY-STATE FLOW AQUIFER TESTS

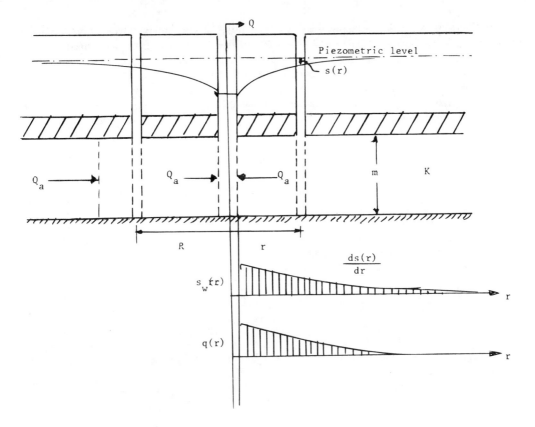

Figure 7 Confined aquifer radial flow.

substitution into Equation 6 gives

$$Q = -2\pi Tr \frac{ds(r)}{dr} \qquad (7)$$

in which T is the transmissivity. Using the method of separation of variables and then taking the integration the general solution of this first-order linear ordinary differential equation is found as

$$s(r) = \frac{Q}{2\pi T} Lnr + C \qquad (8)$$

Determination of integration constant C is possible only after some measurement of drawdown, $s(R)$, through an observation well at any radial distance R. With this information at hand Equation 8 becomes

$$s(r) - s(R) = \frac{Q}{2\pi T} Ln\left(\frac{R}{r}\right) \qquad (9)$$

Here, $R > r$ and correspondingly $s(R) < s(r)$. It implies an inverse relationship between the radial distance and the drawdown. These inequalities compel one to think that at a certain distance from the main well the drawdown should be equal to zero. Such a distance is referred to as the "radius of influence", R_0. Equation 9 is referred to as the Thiem equation of groundwater flow. Equation 9 can be used in practice for different purposes as follows:

1. Physically, during a continuous water abstraction the depression cone deepens and expands which implies that in Equation 9 $s(r)$ and $s(R)$ increase with time because in confined aquifers drawdown variations are never zero. After some time the depression cone starts to deepen uniformly implying that the difference $s(r) - s(R)$ remains constant. This is a graphic case of quasi- (transient) steady-state flow.
2. Considering the definition of the radius of influence, $s(R) = 0$ for $R = R_0$, the drawdown in the aquifer becomes

$$s(r) = \frac{2.3Q}{2\pi T} \log(R_0/r) \qquad (10)$$

which shows a straight line on the semilogarithmic paper with the distance r on the logarithmic axis. By measuring the drawdowns at least in three observation wells at different distances from the main well center, it is possible to construct a straight line on a semilogarithmic paper through the field points as in Figure 8.
3. The slope, Δs_r, of this straight line can be determined as the difference of drawdowns per log cycle of r. With this information one can determine the slope from Equation 10 as

$$\Delta s_r = \frac{2.3Q}{2\pi T} \qquad (11)$$

which is useful in determining either Q or T, provided that one of them is known. However, in practice most often Q is measured in the field by one of methods mentioned in Chapter 7 and T is determined as

$$T = \frac{2.3Q}{2\pi \Delta s_r} \qquad (12)$$

Calculation of the slope in practice demands preferably at least three observation wells so that drawdowns and corresponding distances are measured in the field without well loss effect.

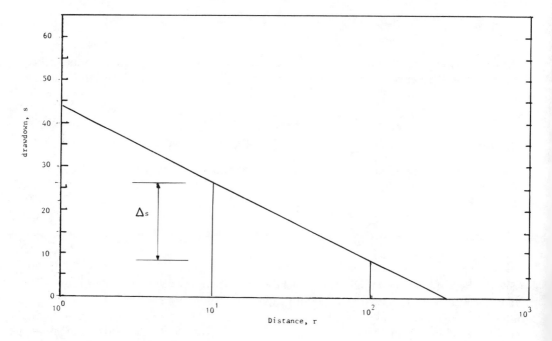

Figure 8 Distance-drawdown plot.

STEADY-STATE FLOW AQUIFER TESTS

Table 1 Confined Aquifer Quasi-Steady-State Data

Well number	MW	O1	O2	O3	O4
Distance (m)	1.1	25	65	105	180
Drawdown (m)	1.36	0.68	0.52	0.31	0.22

These measurements provide three points on a semilogarithmic paper. The best suitable straight line to the scatter of points gives the value of slope and accordingly the aquifer transmissivity is calculated from Equation 12. For accurate results, the drawdown measurements must be taken after pumping for a long time to assure steady or at least quasi-steady-state flow. An increase in the number of observation wells will increase the reliability in transmissivity calculations. This point is elaborated in Chapter 9.

4. The intercept of straight line for zero drawdown gives the radius of influence. Its value has a significant physical meaning in well location plannings as well as the interference studies in groundwater management and well spacing. It helps to define the boundary of depression cone.
5. The Thiem method, though, yields reliable estimates of aquifer transmissivity on the basis of the field data that it does not give any clue about the storage coefficient. However, Şen (1987) suggested a method based on the Thiem equation for calculating the storage coefficient from the steady-state flow drawdown measurements at a few observation wells. The basis of this method is the definition of the storage coefficient as in Equation 11 in Chapter 4. For confined aquifers after some algebraic calculations it is possible to arrive at

$$S = \frac{Qt - \pi r_w^2 s_w(t)}{\dfrac{r_w^2}{4T}[e^{\frac{4\pi T}{Q}s_w(t)} - 1] - \pi r_w^2 s_w(t)} \tag{13}$$

This expression is valid under the Thiem assumptions and its application requires first the calculation of transmissivity from Equation 12. The calculation of S by this method is very useful when there are late time or preferable quasi-steady-state flow measurements. It provides a supplementary technique to other methods of determining S values.

Example 1—Figure 9 shows the general location of a fully penetrating main (MW) and four observation wells (O1, O2, O3, O4) in a confined aquifer. The distances of each observation well to the MW of 2.2 m in diameter are given in Table 1. The quasi-steady-state drawdowns after a continuous water abstraction for 900 min are also presented. The constant discharge is 0.4 m³/min. The given data can be used to have many practically important conclusions and interpretations.

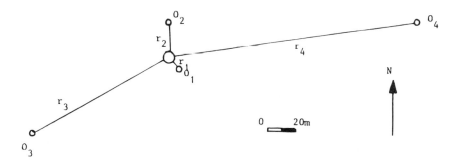

Figure 9 Well location map.

Table 2 Transmissivity (m²/d) Estimation Matrix

$$\begin{matrix} & O1 & O2 & O3 & O4 \\ MW & 412 & 465 & 398 & 410 \\ O2 & & 547 & 356 & 393 \\ O3 & & & 210 & 311 \\ O4 & & & & 549 \end{matrix}$$

Numerical method—The transmissivity estimation can be calculated from Equation 9 as

$$T = \frac{2.3Q}{2\pi} \frac{\log\left(\frac{R}{r}\right)}{s(r) - s(R)}$$

This expression is applicable between any pair of main and observation wells with the substitution of given data values on the r.h.s. The results are presented in matrix form in Table 2. There are ten different pairwise combinations of wells. Some examples for the transmissivity calculations, e.g., for MW-O3 and O2-O4 combinations, are as follows

$$T = \frac{2.3 \times 576}{2 \times 3.14} \frac{\log\left(\frac{105}{1.1}\right)}{(1.36 - 0.31)} = 398 \text{ m}^2/\text{d}$$

and

$$T = \frac{2.3 \times 576}{2 \times 3.14} \frac{\log\left(\frac{180}{65}\right)}{(0.52 - 0.22)} = 311 \text{ m}^2/\text{d}$$

The following conclusions and interpretations can be made from the transmissivity matrix:

1. Significant differences in transmissivity values indicate the heterogeneity of the aquifer material. The least transmissivity value results from calculation between wells O2 and O3, whereas the maximum transmissivity is between O3 and O4. The relative error between these two extreme values is about 61%. In practice any aquifers can be regarded as homogeneous for relative errors less than 5%.
2. A difference with bearing of each point and accordingly in the transmissivity values implies anisotropy of the aquifer.
3. The average transmissivity, $\overline{T}$, of the aquifer is 405 m²/d.
4. According to Table 5 in Chapter 4 on the basis of this average transmissivity value the aquifer has moderate potentiality.
5. The radius of influence can be calculated from Equation 10 as

$$R_0 = re^{\frac{2\pi\overline{T}}{Q}s(r)}$$

Substitution of relevant values on the r.h.s. for different observation well distances yields results as in Table 3. The average radius of influence is about 510 m.

Table 3 Radius of Influence

Well number	O1	O2	O3	O4
R_0 (m)	504	645	413	475

STEADY-STATE FLOW AQUIFER TESTS

Graphical method—The data in Table 1 can be treated graphically on semilogarithmic paper through the following procedure.

1. Plot the drawdowns on the vertical axis with an arithmetic scale against the corresponding distances on the logarithmic horizontal axis scale (see Figure 10).
2. Draw the best-fitting straight line through the scatter of data points. Fitting straight line is equivalent to assuming that the aquifer is homogeneous and isotropic. Therefore, a single transmissivity, storativity, and radius of influence values are expected from the graphical method.
3. Determine the slope, Δs_r, of this line, i.e., the drawdown difference corresponding to any log cycle of distance on the horizontal axis as shown in Figure 10. The slope value for the data at hand is read as $\Delta s_r = 0.47$ m.
4. Calculate the transmissivity value by substituting necessary quantities into Equation 12.

$$T = \frac{2.3 \times 575}{2 \times 3.14 \times 0.47} = 449 \text{ m}^2/\text{d}$$

This value is within the range of transmissivities obtained from the numerical method, (see Table 2). In fact, it is very close to the transmissivity value calculated for data from MW-02 combination. The relative error between the average transmissivity of numerical method and the graphical method transmissivity is about 10%. However, the graphical transmissivity value also shows that the aquifer has moderate potential as was the conclusion from the numerical method.

5. Find the intercept of the fitted straight line with the horizontal axis, i.e., the axis of zero drawdown. This intercept value is the graphical solution for the radius of influence which appears as 610 m.

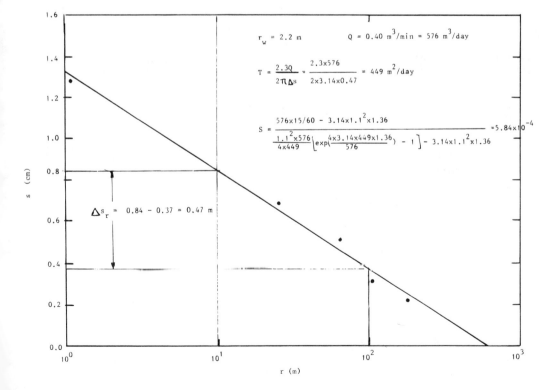

Figure 10 Distance-drawdown plot.

Storage coefficient determination—Application of Equation 13 with the graphical method values yields a storage coefficient estimate as

$$S = \frac{576 \times 15 \times 60 - 3.14 \times (1.1)^2 \times 1.36}{\frac{(1.1)^2 \times 576}{4 \times 449} \times (e^{\frac{4 \times 3.14 \times 449}{576} \times 1.36} - 1) - 3.14 \times (1.1)^2 \times 1.36} = 5.86 \times 10^{-4}$$

This storage coefficient value is smaller than 10^{-1} and hence the aquifer is confined.

B. UNCONFINED AQUIFER

Analytical treatment of flow in an unconfined aquifer is more complex because of the saturation thickness reduction and accordingly the transmissivity decreases as the groundwater flow approaches the well. In addition, the flow lines are not parallel to each other and the existence of seepage surface makes complex boundary conditions (Figure 11).

Under these circumstances the groundwater problems in unconfined aquifers are at least theoretically indeterminate. However, consideration of Dupuit-Forchheimer assumptions provide an opportunity to derive analytical expressions within practically acceptable error limits as good approximations to actual situations. The aquifer discharge at radial distance, r, is defined as

$$Q(r) = 2\pi r h(r) q(r) \qquad (14)$$

where $h(r)$ is the saturation thickness at distance r. Its substitution into Equation 6 yields after some mathematical manipulation

$$h^2(R) - h^2(r) = \frac{Q}{\pi K} \ln\left(\frac{R}{r}\right) \qquad (R > r) \qquad (15)$$

or in terms of drawdowns $s(r) = H - h(r)$, where H is the saturation thickness before the water abstraction (Equation 15) becomes

$$s^2(r) - s^2(R) = \frac{Q}{\pi K} \ln\left(\frac{R}{r}\right) \qquad (16)$$

This expression is known as Dupuit-Forchheimer equation. It is obvious that the relationship between the distance and drawdown is not linear as it was for the confined aquifer. Equation 16 does not give satisfactory results for drawdown measurements at distances when $r < 1.5H$.

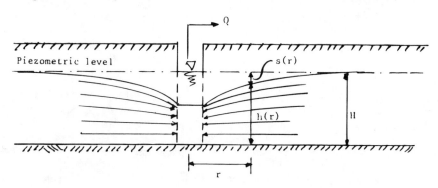

Figure 11 Seepage surface.

STEADY-STATE FLOW AQUIFER TESTS

Consideration of the following points is necessary for successful application of the Dupuit-Forchheimer expression:

1. The drawdown measurements must be taken at large distances (more than $1.5H$) and late times (quasi-steady state).
2. The Dupuit-Forchheimer equation can be rewritten mathematically with no loss of generality as

$$\left[s(r) - \frac{s^2(r)}{2H}\right] - \left[s(r) - \frac{s^2(R)}{2H}\right] = \frac{Q}{2\pi T} \ln\left(\frac{R}{r}\right) \quad (17)$$

or succinctly as

$$s'(r) - s'(R) = \frac{Q}{2\pi T} \ln\left(\frac{R}{r}\right) \quad (18)$$

which is equivalent to the Thiem equation for confined aquifers. Herein, $s'(r)$ and $s'(R)$ are referred to as the "corrected drawdowns" at distances r and R, respectively, that would occur in an equivalent confined aquifer. However, even to this correction there is a limitation in the sense that corrected drawdown cannot be less than 10% of the drawdowns observed in the field.

3. If the square of the drawdowns are plotted on a semilogarithmic paper vs. the distance on the logarithmic axis, the result should appear as a straight line according to Equation 16. This statement is correct when coupled with the condition in (1). The drawdown $s(r)$ close to a well in an unconfined aquifer can be estimated for $0.3H < r < 1.5H$ as

$$s(r) = \frac{Q\left[0.13 \ln\left(\frac{R_0}{r}\right)\right] \ln\left[10 \ln\left(\frac{R_0}{H}\right)\right]}{KH} \quad (19)$$

and for $r < 0.3H$ as

$$s(r) = \frac{Q\left[0.13 \ln\left(\frac{R_0}{r}\right) - 0.0123 \ln^2\left(\frac{R_0}{10r}\right)\right] \ln\left[10\left(\frac{R_0}{H}\right)\right]}{KH} \quad (20)$$

4. In thick unconfined aquifers with small drawdowns $s_w(t) < 0.10H$, the Dupuit-Forchheimer equation is approximated by

$$s(r) = \frac{Q}{\pi K(H + h)} \ln\left(\frac{R_0}{r}\right) \quad (21)$$

by defining the average transmissivity approximately as, $\overline{T} = K(H + h)/2 \cong KH$, this expression becomes identical to the Thiem equation given in Equation 10.

Example 2—An unconfined aquifer is tapped with a constant discharge, $Q = 0.32$ m³/min through a fully penetrating large-diameter well of diameter 3 m. The well location map is given in Figure 12 and quasi-steady-state drawdown after 2 h of abstraction including four observation wells (O1, O2, O3, O4) are presented in Table 4. The saturation thickness, H, prior to water abstraction is 20 m.

A first glance to this table shows that the drawdown in the main well is comparatively larger than the others. This is because of a well loss effect that will be explained in Chapter 11.

Numerical method—The hydraulic conductivity of an unconfined aquifer can be calculated from Equation 15 as

$$K = \frac{2.3Q}{\pi} \frac{\log\left(\frac{R}{r}\right)}{h^2(R) - h^2(r)}$$

The discharge is $0.32 \times 60 \times 24 = 460.8$ m³/d. The substitution of data for pairwise combination of wells lead to hydraulic conductivity estimation matrix as in Table 5. Some

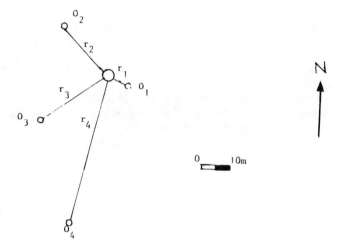

Figure 12 Well location map.

Table 4 Unconfined Aquifer Quasi-Steady-State Data

Well number		Observation wells			
	MW	O1	O2	O3	O4
Distance, r (m)	1.5	18	32	60	110
Drawdown, $s(r)$ (m)	14.4	4.5	2.7	1.6	0.7
Hydraulic head, $h(r)$ (m)	5.5	15.5	16.3	18.4	19.3
$h^2(r)$ (m²)	30.25	240.25	265.69	338.56	372.49
$H^2 - h^2(r)$ (m²)	369.75	159.75	134.31	614.40	375.40

Table 5 Hydraulic Conductivity (m/d) Matrix

	O1	O2	O3	O4
MW	1.73	1.90	1.75	1.84
O1		3.37	1.80	2
O2			1.26	1.69
O3				2.61

examples are presented for well combinations O1–O4 and O2–O3 as

$$K = \frac{2.3 \times 460.8}{3.14} \frac{\log\left(\frac{110}{18}\right)}{372.40 - 240.25} = 2.00 \text{ m/d}$$

and

$$K = \frac{2.3 \times 460.8}{3.14} \frac{\log\left(\frac{60}{32}\right)}{338.56 - 240.25} = 1.26 \text{ m/d}$$

The heterogeneity and anisotropy of the aquifer is obvious from this hydraulic conductivity matrix. The average hydraulic conductivity, $\overline{K}$ is about 2.00 m/d which gives $2 \times 20 = 40$ m²/d transmissivity when the whole saturation thickness is considered. This transmissivity value indicates that the unconfined aquifer has low potential. The greatest hydraulic conductivity value appears between wells O1 and O2 along a SE direction whereas the least value is along a NE direction and between wells O2 and O3.

The radius of influence, R_0 can be found from Equation 15 by considering average hydraulic conductivity value, $\overline{K}$ and that $h(R_0) = H$ as

$$R_0 = re^{\frac{\pi \overline{K}}{Q}[H^2 - h^2(r)]}$$

The results for each pair of wells are presented in Table 6. The average radius of influence is 175 m.

Graphical method—If the plot of drawdown vs. distance does not yield a straight line, the aquifer is unconfined. The Dupuit-Forchheimer equation represents a straight line between the logarithms of distances and corresponding values of $H^2 - h^2(r)$. Hence, in calculating unconfined aquifer hydraulic conductivity graphically the following procedure must be applied.

1. Plot distances on the logarithmically scaled horizontal axis vs. corresponding $H^2 - h^2(r)$ values on the vertical linearly scaled axis as shown in Figure 13.
2. Draw the best-fitting straight line through the field data points. This line corresponds to the graphical representation of the Dupuit-Forchheimer formula for the homogeneous and isotropic unconfined aquifer.
3. Find the slope, ΔS_r, of the line which can be read as equal to 171 m². Theoretically, one full cycle of the distance axis means that $\log(R/r) = 1$. Therefore, from Equation 15 one can estimate the hydraulic conductivity as

$$K = \frac{2.3Q}{\pi \Delta s_r}$$

4. The substitution of necessary values into this expression gives the hydraulic conductivity estimate as 1.97 m/d which has only 1.5% relative difference from the numerical result.

Table 6 Radius of Influence for an Unconfined Aquifer

Well number	O1	O2	O3	O4
R_0 (m)	159	200	139	205

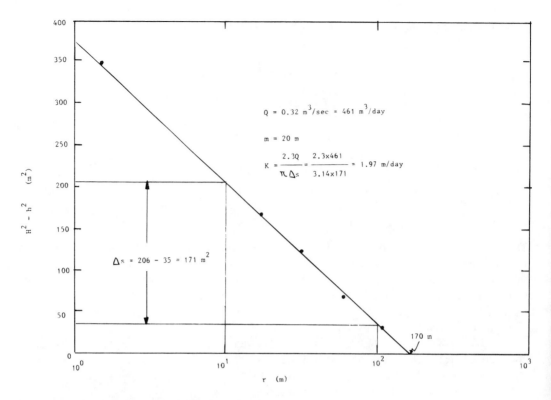

Figure 13 Semilog plot in unconfined aquifers.

Finally, a graphical value of the radius of influence is obtained by intersection of a straight line on the distance axis representing zero drawdown. Figure 13 yields $R_0 = 170$ m which is practically equal to the average radius of influence value found from the numerical method.

C. LEAKY AQUIFER

Confined and unconfined aquifers exist in nature as single layers of water reservoirs with no water exchange in the subsurface with other layers. In multiple layering, different combinations of hydrolithologies such as aquifer, aquiclude, aquitard, and aquifuge appear in sequence. If an aquifer is over- and/or underlain by any aquiclude and/or aquitard, the resulting two- or three-layer unit constitutes a leaky aquifer. For steady and unsteady flows in confined or unconfined aquifers the only source of abstracted water is the aquifer storage, whereas in the leaky aquifer an additional source is the leakage from adjacent aquitards. Let the thicknesses and the hydraulic conductivities of the aquifer and semipervious layers be m, K, m_L, m_U, K_L, and K_U, respectively. The first solutions for steady-state flow toward wells in leaky aquifers are due to De Glee (1930) and Steggewentz and Van Ness (1939).

A general solution of the problem is not possible if a set of simplifying assumptions is not considered. In addition to assumptions in Section II, the following special assumptions for leaky aquifers are necessary for the analytical solutions.

1. Practically applicable solutions are obtained when $K/K_U > 50$ ($K/K_L > 50$) and $m/m_U > 3$ ($m/m_L > 3$). Under these conditions, the flow in the aquifer is two dimensional since the vertical distance is deflected at the lower surface of the semipervious layer.
2. After a certain period of pumping the flow reaches a steady-state situation and the pump discharge is sustained entirely by the leakage.

3. Even after abstraction of water the water table in the upper unconfined aquifer maintains its constant head, h_0.

Physically, the total leakage into the aquifer through a concentric annulus of thickness, dr, on the lower boundary of the semipervious layer is expressible as

$$q_L(r) = -2\pi r dr K_U \frac{s(r)}{M_U} - 2\pi r dr K_L \frac{s(r)}{M_L} \qquad (22)$$

This is equal to the discharge rate with distance and hence,

$$\frac{dQ(r)}{dr} = -2\pi r dr \left(\frac{K_U}{M_U} + \frac{K_L}{M_L}\right) s(r) \qquad (23)$$

Consideration of $Q(r) = 2\pi m K r q_L(r)$ and elimination of $q_L(r)$ gives finally

$$\frac{d^2 s(r)}{dr^2} + \frac{1}{r}\frac{ds(r)}{dr} + \frac{s}{L^2} = 0 \qquad (24)$$

in which L is referred to as the leakage factor and explicitly

$$L = \left(\frac{MK}{\frac{K_U}{M_U} + \frac{K_L}{M_L}}\right)^{1/2} \qquad (25)$$

The leakage factor has the dimension of $[L]$ but it is difficult to give a physical meaning except that increase in L corresponds to decrease in total leakage. When the leakage occurs only through the upper or lower semipervious layer the leakage factor takes the forms of

$$L_U = \left(\frac{m m_U K}{K_U}\right)^{1/2} \qquad (26)$$

and

$$L_L = \left(\frac{m m_L K}{K_L}\right)^{1/2} \qquad (27)$$

respectively. According to the classification in Table 6 of Chapter 4, when the leakage factor is more than 10,000 m, aquifer can be considered as a confined unit (negligible leakage) and the procedures in Section A become applicable.

The solution of Equation 24 with the boundary conditions similar to Equations 1 and 4 gives

$$s(r) = \frac{Q}{2\pi T} \frac{K_0\left(\frac{R}{L}\right)}{(r_w/L) K_1(r_w/L)} \qquad (28)$$

in which $K_0(r/L)$ and $K_1(r_w/L)$ are the modified Bessel functions of the second kind of order zero and the second kind of first order, respectively. A practical point to be noticed at this stage is that r_w/L is very small and therefore $K_1(r_w/L) \equiv 1$ with an error of less than 1% for $r_w/L < 0.02$. Hence, Equation 28 may be approximated by

$$s(r) = \frac{Q}{2\pi T} K_0\left(\frac{R}{L}\right) \tag{29}$$

In dimensionless form this equation becomes

$$W\left(\frac{R}{L}\right) = K_0\left(\frac{R}{L}\right) \tag{30}$$

where

$$W\left(\frac{R}{L}\right) = \frac{2\pi T}{Q} s(r) \tag{31}$$

is referred to, herein, as the leaky aquifer well function for steady-state flow. Figure 14 gives a schematic representation of Equation 30 as $W(r/L)$ vs. r/L on double-logarithmic paper. Hantush (1956, 1964) has noticed that for $r/L < 0.05$ (in the vicinity of the pumping well), Equation 29 can be approximated for practical purposes by

$$s(r) = \frac{2.3Q}{2\pi T} \log\left(1.12 \frac{L}{r}\right) \tag{32}$$

Furthermore, Equation 29 can be used with the observation well records for the determination of the aquifer parameters, namely, L and T. For this purpose the ratio of drawdowns at two distances from Equation 29 leads to

$$\frac{s(r_1)}{s(r_2)} = \frac{K_0\left(\frac{r_1}{L}\right)}{K_0\left(\frac{r_2}{L}\right)} \tag{33}$$

Introduction of dimensionless drawdown and distance ratio as

$$\alpha = \frac{s(r_1)}{s(r_2)} \tag{34}$$

and

$$\beta = \frac{r_1}{r_2} \tag{35}$$

respectively, renders Equation 33 into the following useful form

$$\alpha = \frac{K_0(x_1)}{K_0\left(\frac{x_1}{\beta}\right)} \tag{36}$$

where $x_1 = r_1/L$. Because of the inverse relationship between the distance and drawdown the

STEADY-STATE FLOW AQUIFER TESTS

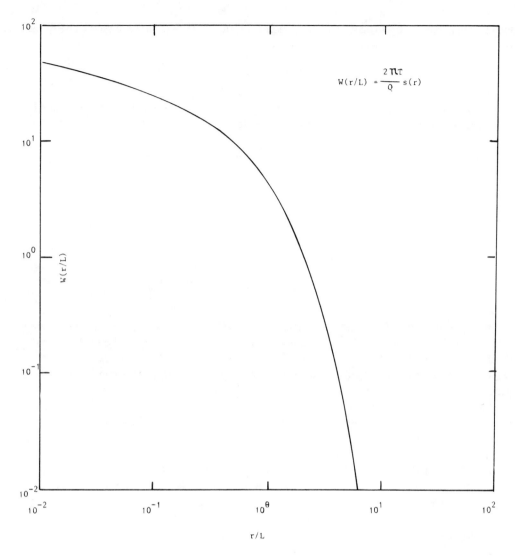

Figure 14 Leaky aquifer dimensionless curve.

domain of variation for α and β are $1 < \alpha < +\infty$ and $0 < \beta < 1$. The graphical representation of Equation 36 yields a variation of α with $\beta\gamma$ for a given set of x_1 values (see Figure 14). In practice, α and $\beta\gamma$ values are estimated from the field measurements and the corresponding x_1 value is read from the graph in Figure 14. The leakage factor is evaluated as $L = r_1/x_1$. Once L is known the aquifer transmissivity is found from the drawdown difference which leads to

$$T = \frac{Q}{2\pi[s(r_1) - s(r_2)]} \left[K_0\left(\frac{r_1}{L}\right) - K_0\left(\frac{r_2}{L}\right) \right] \tag{37}$$

This expression gives an opportunity to calculate the transmissivity between any two pairs of observation wells. In fact, even the main well can be used as one of the observation wells. Another parameter in characterizing the hydraulic conductivity of the semipervious layer is the leakage coefficient, L_c, and it is equal to the reciprocal of hydraulic resistance, R_h, which was already defined by Equation 21 in Chapter 4. Therefore, with the notations of this section

$$L_c = \frac{1}{R_h} = \frac{K_u}{m_u} = \frac{T}{L^2} \tag{38}$$

The dimension of L_c is [1/T] and it is defined physically as the leakage rate passing through unit area of the lower surface of semipervious layer on the condition that the head drop is equal to unity.

The discharge, $Q(r)$, flowing through the aquifer at a radial distance r is expressed as

$$Q(r) = Q \frac{r}{L} K_1\left(\frac{r}{L}\right) \tag{39}$$

The percentage, p, of the prevailing part of the flow rate is then

$$p = 100 \frac{Q - Q(r)}{Q} = 100\left[1 - \frac{r}{L} K_1\left(\frac{r}{L}\right)\right] \tag{40}$$

Figure 15 represents the graphical change of percentage with r/L. The leakage percentage increases with the value of r/L. For instance, when $r = 3.2L$, $p = 90\%$ then 90% of pumping discharge enters the cylinder of radius $3.2L$ through the semipervious layer. This radius is referred to as the recharge radius. Theoretically, the radius of influence has infinite value but practically it is assumed to correspond to $p = 90\%$ which yields from Figure 15 $r = 0.035L$.

Example 3—In a 30-m thick leaky aquifer water is abstracted constantly at a rate 0.8 m³/min. The thickness of overlying semipervious layer (aquitard) is 10 m and the steady-state drawdowns are reached in five observation wells after 4 h. The well location map is given in Figure 16 and the relevant field data in Table 7.

For the steady-state drawdown in a leaky aquifer De Glee (1930) derived the expression in Equation 29. In this equation there are two unknowns, T and L, and a mathematical function $K_0(r/L)$. Contrary to confined and unconfined aquifers the numerical method cannot be used. Therefore, a graphical procedure with the following steps leads to the leaky aquifer parameter estimations.

1. Plot the drawdowns vs. distances on a sheet of double-logarithmic paper of the same size as the standard curve in Figure 14.

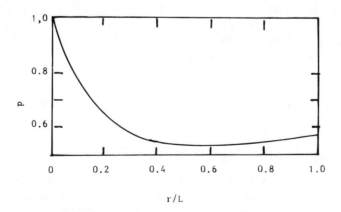

Figure 15 Percentage flow rate.

STEADY-STATE FLOW AQUIFER TESTS 183

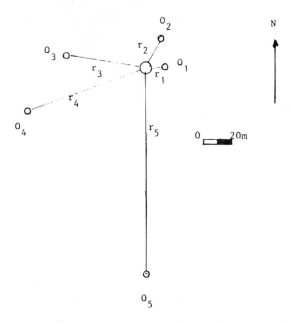

Figure 16 Well location map.

Table 7 Leaky Aquifer Steady-State Data

	Observation wells				
Well number	O1	O2	O3	O4	O5
Distance, r (m)	10	20	60	90	150
Drawdown, s (m)	0.66	0.55	0.35	0.25	0.15

2. Match the data plot with the standard curve by moving field data sheet such that the distance axis on the field sheet remains parallel to the r/L axis on the standard curve sheet. This procedure is presented in Figure 17.
3. Select an arbitrary point, M, on the overlapping portion of the sheets and note the coordinates of this point on the standard curve sheet, $(r/L)_p = 10^{-1}$, $[K_0(r/L)]_p = 10^1$, and coordinates on the field sheet $r_p = 1.73 \times 10^{-1}$ m and $s_p = 2.3 \times 10^{-1}$ m.
4. Calculate T from the De Glee equation (Equation 29) by substituting the known value of discharge and other values as

$$T = \frac{0.8 \times 60 \times 24}{2 \times 3.14 \times 2.3 \times 10^{-1}} \times 10^0 = 7971 \text{ m}^2/\text{d}$$

which implies a highly potential aquifer.
5. Calculate the leakage factor from information at point M as

$$L = \frac{r_p}{10^{-1}}, \text{ i.e., } L = \frac{1.73 \times 10^1}{10^{-1}} = 173 \text{ m}$$

6. The hydraulic resistance of the aquitard can be calculated from Equation 38 as

$$R_h = \frac{L^2}{T} = \frac{(173)^2}{7971} = 3.75 \text{ d}$$

Straight line method—Equation 32 appears as a straight line on a semilogarithmic paper with the horizontal logarithmic axis representing the distances vs. drawdown on the vertical

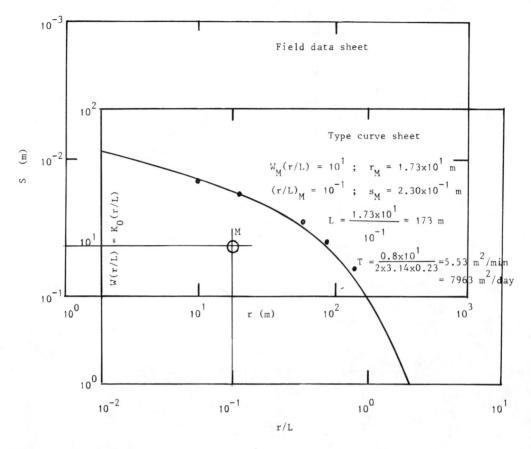

Figure 17 Curve matching.

axis. The slope, ΔS_r, of this line is related to the aquifer transmissivity as in the confined aquifer given by Equation 12. For the problem at hand, the semilogarithmic plot of data and the fitted straight line are presented in Figure 18 from where the slope value is read as 0.43 m. Substitution of this value into Equation 12 with other necessary values yield the aquifer transmissivity $T = 980$ m²/d.

The point representing the radius of influence corresponding to the intercept on the horizontal axis yields $R_0 = 325$ m. At the intercept point the drawdown is equal to zero and the implication of this condition into Equation 32 leads to

$$1.12 \frac{L}{R_0} = 1.12 \frac{\sqrt{TR_h}}{R_0} = 1 \text{ from which } L = \frac{R_0}{1.12} = \frac{325}{1.12} = 290 \text{ m}$$

According to Table 6 of Chapter 4 this implies potential leakage. On the other hand, the hydraulic resistance is

$$R_h = \frac{(R_0/1.12)^2}{T} = \frac{(325/1.12)^2}{980} = 85.9 \text{ d}$$

V. PARTIALLY PENETRATING WELLS

The equations for determination of aquifer parameters outlined in the preceding sections were based on the assumption that the main well penetrates the entire saturation thickness of the aquifer so that the radial flow toward the well is horizontal.

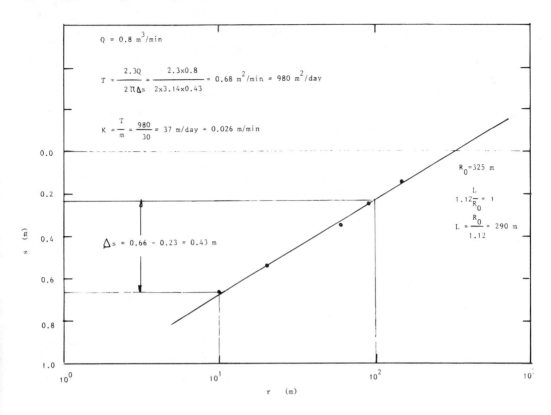

Figure 18 Leaky aquifer distance-drawdown plot.

Most often in practical studies the wells are partially penetrating due to mainly economic reasons for eliminating the unnecessary cost or reducing the quantity of water required to accomplish the desired result. In addition, partial penetration can occur unintentionally from the lack of knowledge about the true saturated thickness of aquifer where the Thiem equation is not applicable. Direct theoretical approach to the groundwater flow problems toward partially penetrating wells is not possible because of the complex flow net leading to three-dimensional flow as in Figure 19. At the sharp corners especially, the groundwater velocity is comparatively higher than any other part of the entrance section. The velocity distribution is nonuniform, violating the validity of Dupuit-Forchheimer assumptions. However, physically at large distances from the pumping well the velocity is again uniformly distributed and the streamlines are parallel to each other. Therefore, one can conclude that the Thiem equation is applicable at large distances but for small distances it is necessary to develop a new formulation. Practical

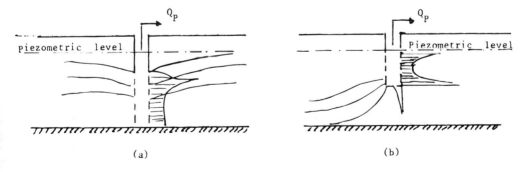

Figure 19 Streamlines in partial penetrations.

experiences have shown as a rule of thumb that beyond a radial distance equal to twice of the saturation thickness, the drawdown is approximately the same with fully penetration case. In partially penetrating wells the streamlines are longer than the fully penetrating wells and therefore the energy losses are comparatively greater (greater drawdowns). For a given drawdown, the discharge of a partially penetrating well is less than that a fully penetrating well because of increases in head loss. To minimize the deviations from a fully penetrating well the following points must be considered.

1. If horizontal bedding is strong in the flow domain and the observation wells are fairly close to the main well the aquifer is sometimes assumed to end at the bottom of the main well and hence fully penetrating well formulations are used directly on the basis of well penetration length.
2. Observations may be taken at such radial distances that the effects of partial penetration becomes negligible and the streamlines are substantially the same as if the well were fully penetrating. This distance can be calculated in terms of the aquifer thickness, m, horizontal K_h, and vertical K_v hydraulic conductivities as $2m\sqrt{K_h K_v}$
3. Jacob (1945) suggested the use of corrected drawdown, s_c, by measuring the drawdown at top and bottom of the aquifer separately at a radial distance using a pair of observation wells. The corrected drawdown is calculated as the arithmetic average of top and bottom drawdowns.

In certain cases the observed drawdown in a partially penetrating well may be adjusted for partial penetration according to theory and empirical formulations. To this end various investigators developed methods to correct for partial penetration in order to apply the formulations of full penetration.

A. UNCONFINED AQUIFERS

Forchheimer (1898) gave the discharge, Q_p, from a partially penetrating well similar to the Thiem equation as

$$Q_P = \pi K \frac{h^2(R) - [h(r) - t]^2}{ln\left(\frac{R}{r}\right)} \alpha \tag{41}$$

in which all of the variables are self explanatory as in Figure 20. The coefficient α is referred

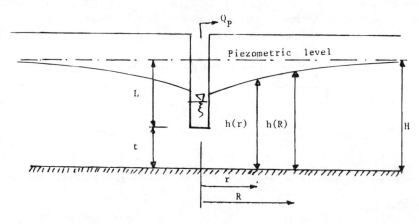

Figure 20 Partial penetration in unconfined aquifer.

to as the "reduction factor" which is expressed as

$$\alpha = \left(\frac{L}{H}\right)^{1/2}\left(2 - \frac{L}{H}\right)^{1/4} \tag{42}$$

in which H is the saturation thickness. For full penetration $L = H$ and Equation 42 becomes equivalent to Equation 15.

Kozeny (1933) gave an approximate empirical formula for discharge calculation in a partially penetrating well as

$$Q = \frac{2\pi T s_w \alpha \left[1 + 7\left(\frac{r_{ws}}{2am}\right)^{1/2} \cos\left(\frac{\pi\alpha}{2}\right)\right]}{\ln\left(\frac{R_0}{r_w}\right)} \tag{43}$$

where s_w is the steady-state drawdown in the main well; α the fractional part of the full thickness, m, tapped by main well; r_w is the well radius; R_0 is the radius of influence. The apparent, T', and true, T, transmissivities are related as follows

$$\frac{T}{T'} = \frac{c}{\alpha} \tag{44}$$

where c is discharge correction factor given as

$$c = \left[1 + 7\left(\frac{r_w}{2am}\right)^{1/2} \cos\left(\frac{\pi\alpha}{2}\right)\right]^{-1} \tag{45}$$

Muskat (1937) suggested that the flow of water toward a partially penetrating well in an isotropic aquifer becomes almost radial at a distance equal to twice the aquifer thickness. However, in anisotropic aquifers Jacob (1963) suggested the distance, r, for radial flow to be calculated as $r = 2m\sqrt{K_h/K_v}$. Jacob also gave a method for correction of δ, steady-state drawdown in partially penetrating well screening the top or bottom of a confined aquifer as

$$\delta = \frac{s_p}{Q/2\pi T} = \frac{\left(\frac{2}{\pi\alpha}\right)\sum_{n=1}^{\infty}\left[(-1)^n K_0\left(\frac{n\pi r}{b}\right)\sin(n\pi\alpha)\right]}{n} \tag{46}$$

in which s_p is the drawdown correction as the difference between the observed drawdown and the drawdown that would have occurred for full penetration. The plus sign is for the drawdown distribution along the top of the aquifer (Figure 21a) and minus sign is for the bottom case (Figure 21b). The graphical presentations of Equation for top and bottom are given in the same figure. The drawdown correction is given

$$s_p = \delta\left(\frac{Q}{2\pi T}\right) \tag{47}$$

This correction method is not applicable if the well taps the aquifer somewhere in between the top and bottom. The application procedure involves the following steps:

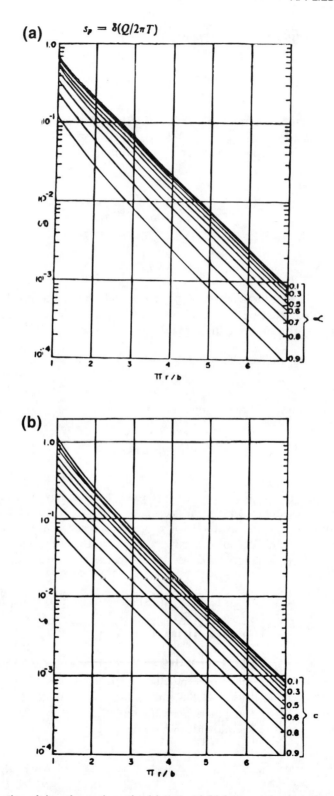

Figure 21 Correction of drawdown along the (a) top and (b) bottom of an aquifer.

1. Plot steady-state drawdowns in different observation wells vs. radial distances from the main well. First, preliminary T value is determined either from the Thiem or distance-drawdown Jacob method (Chapter 9).
2. Calculate values of $\pi r/m$, α, and then δ read for each observation well from Figure 21.
3. Calculate the corrected drawdown from Equation 47.
4. Recalculate the value of T by any method utilizing the corrected drawdown values.
5. Compare the transmissivities in steps (1) and (4): if they are different by 5% relative error then repeat the whole procedure starting with the recalculated transmissivity value.

Later, by the use of potential theory Nahrgang (1954) related the discharges of fully and partially penetrating wells as

$$Q_p = \alpha Q \qquad (48)$$

in which α is a factor depending on the two ratios, L/H and h/H, where L is the penetration length. Figure 22 represents the graphical relationship between h/H and α for a set of given L/H values.

Another alternative is due to purely theoretical consideration of Szechy (1955), who divided the whole flow domain into two parts:

1. If the flow domain along the penetration length is considered as full penetration, the discharge, Q, is expressed as

$$Q_1 = \pi K_h \frac{H^2 - h^2(r)}{\ln\left(\dfrac{R_0}{r}\right)} \qquad (49)$$

He also suggested that in this formula horizontal hydraulic conductivity, K_h, should be substituted.
2. The flow domain beneath the penetration thickness which is assumed to have the vertical

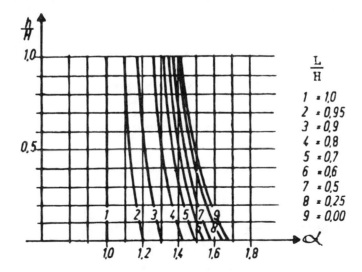

Figure 22 Dimensionless relationship in partially penetrating well.

hydraulic conductivity, K_v, and the discharge, Q_2, contribution is

$$Q_2 = \frac{2t}{w+l} \pi K_v \frac{H^2 - h^2(r)}{\ln\left(\frac{R_0}{r}\right)} \tag{50}$$

in which w is a factor assuming values in the range 5 to 10. The total discharge is then

$$Q_p = Q_1 + Q_2 \tag{51}$$

Finally, Hantush (1964) recommended the use of Equation 15 provided that the pumping time to reach the steady state is short or that the aquifer is relatively thick and the observed drawdowns, s_0, are corrected, s_c, according to

$$s_c = s_0 - \frac{s_0^2}{2L} \tag{52}$$

Example 4—The first numerical example for the partially penetrating wells is adopted from Jacob (1963). Assume that a well having an effective radius of 0.3 m and screened in the top 12 m of a 30-m thick confined aquifer is pumped at a constant rate of 3800 m³/d for a period long enough to establish quasi-steady-state condition. The drawdown in the main well is 6.2 m with negligible screen loss. Hence, estimate, (1) the hypothetical drawdown that would occur in a well fully screened in the aquifer, (2) the ratio of true to apparent transmissivity, and (3) the hypothetical yield.

First, the penetration fraction $\alpha = 12/30 = 0.40$ and from Kozeny's formula yields

$$\frac{1}{c} = 1 + \frac{7 \cos(3.14 \times 0.40/2)}{\sqrt{2 \times 0.40 \times 30/0.3}} = 1.633$$

Solving Equation 43 for hypothetical drawdown, $(2.3\, Q/2\pi T)\log R_0/r_w$ leads to

$$\frac{s_w \alpha}{c} = 6.2 \times 0.40 \times 1.6333 = 4.048 \text{ m}$$

Ratio of true, T, to apparent, T', transmissivity is

$$\frac{T}{T'} = \frac{c}{\alpha} = \frac{0.61}{0.40} = 1.5$$

Hypothetical yield, Q_0, from a fully penetration well becomes

$$Q_0 = \frac{Qc}{\alpha} = 3800 \times 15 = 5700 \text{ m}^3/\text{d}$$

B. CONFINED AQUIFER

If a partially penetrating well in a confined aquifer is pumped, the drawdown observed must be collected to obviate partial penetration effects. First studies concerning partial penetration in confined aquifers are due to De Glee (1930), who derived the additional steady-state drawdown, Δs_w, in the well as (see Figure 23)

$$\Delta s_w = \frac{2.3 Q_p}{2\pi Km} \frac{1-p}{p} \log \frac{(1.2-p)L_s}{\alpha r_w} \qquad (53)$$

in which flow entrance percentage, $p = L_s/m$, and α depends on the top or bottom screen ($\alpha = 1$) and middle screen ($\alpha = 2$). Equation 53 is valid only for screen length, L_s, between $10r_w$ and 0.8 m. The drawdown s_{wp} of partial penetration is the sum of this additional drawdown and drawdown s_w due to full penetration.

$$s_{wp} = \Delta s_w + s_w \qquad (54)$$

The simplest case with screen start from the confining layer as shown in Figure 24 was studied by Muskat (1937), who determined the potential distribution around the well. The lower well parts receive more water entrance because the velocity is maximum near the corners. Muskat assumed a uniform specific discharge distribution plus some correction factor

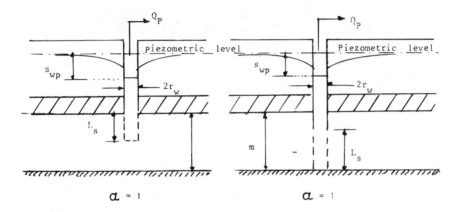

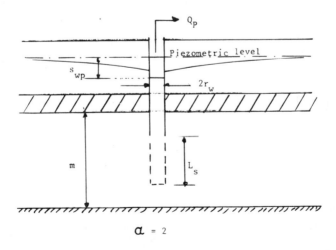

Figure 23 DeGlee representation.

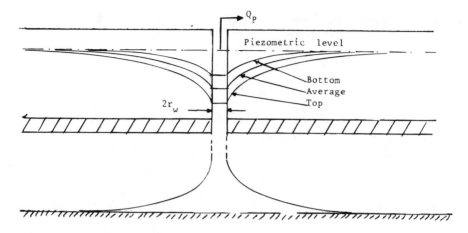

Figure 24 Partial penetration in the middle of aquifer.

for the flow at the corners and hence he obtained the drawdown in the well as

$$s_w = \frac{Q_W}{4\pi T} \frac{m}{L_S}\left[4.6 \log\left(4\frac{m}{r_w}\right) - G\left(\frac{L_s}{m}\right) - 4.6 \frac{L_s}{m} \log\left(\frac{4m}{R}\right)\right] \tag{55}$$

in which

$$G\left(\frac{L_s}{m}\right) = \frac{\Gamma\left(0.875 \frac{L_s}{m}\right)\Gamma\left(0.125 \frac{L_s}{m}\right)}{\Gamma\left(1 - 0.875 \frac{L_s}{m}\right)\Gamma\left(1 - 0.125 \frac{L_s}{m}\right)} \tag{56}$$

Kozeny (1933) summarized Muskat's analysis and developed a dimensionless formula which relates the discharges Q_p and Q of partially and fully penetrating wells respectively as

$$\frac{Q_p}{Q} = p\left[1 + 7\beta^{1/2} \cos\left(\frac{\pi}{2} p\right)\right] \tag{57}$$

in which β is referred to as the well slimness and defined as a ratio, $r_w/2L_s$. In short, this formula indicates the relationship between the discharge ratio, $C = Q_p/Q$, well slimness, β, and penetration ratio, p. Graphical representation of Equation 57 is shown in Figure 25 for quick calculation of the discharge ratios.

Kozeny's approach can be applied in unconfined aquifers when the drawdown is no more than 20 to 30% of the original saturation thickness.

Considering an electrical analog model Li (1954) developed simple dimensionless formulations for the ratios of drawdowns and discharges of the partial penetration to the fully penetration cases as

$$\frac{s_{w_p}}{s_w} = 1 + \frac{\log\left(\frac{m}{r_w}\right)}{\log\left(\frac{R}{r_w}\right)} \left[\left(\frac{m}{L_s}\right)^{3.4} - 1\right] \tag{58}$$

STEADY-STATE FLOW AQUIFER TESTS

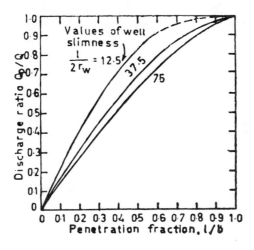

Figure 25 Discharge vs. penetration ratio.

and

$$\frac{Q_{wp}}{Q_w} = \left\{1 + \frac{\log\left(\frac{m}{r_w}\right)}{\log\left(\frac{R}{r_w}\right)}\left[\left(\frac{m}{L_S}\right)^{4/3} - 1\right]\right\}^{-1} \tag{59}$$

Based on the work of Muskat, Kozeny, and Jacob these formulas are applicable for $L_s/m > 0.1$ only.

Huisman (1972) developed an equation relating the additional drawdown due to the partially penetrating well to the penetration factor and the well radius as

$$\Delta s_w = \frac{2.3Q}{2\pi Km}\frac{1-p}{p}\log\frac{\alpha h}{r_0} \tag{60}$$

in which α is the function of p and the amount of eccentricity which is defined as a dimensionless ratio, $e = L_s/m$. This function is given in Table 8.

Example 5—A confined aquifer of 80-m thickness and with transmissivity 951 m²/d is tapped through a fully penetrating well at a constant rate 0.05 m³/s. The well has a diameter of 1.5 m and the resulting drawdown after a long abstraction period is 14.7 m. Assuming no well loss find what will be the drawdown value if the well is screened between 19 to 35 m below the top of the aquifer.

Table 8 α Values

	e = 0	0.05	0.10	0.15	0.20	0.25	0.30	0.35	0.40	0.45
p = 0.1	α = 0.54	0.54	0.55	0.55	0.56	0.57	0.59	0.61	0.67	1.09
0.2	0.44	0.44	0.45	0.46	0.47	0.49	0.52	0.59	0.89	
0.3	0.37	0.37	0.38	0.39	0.40	0.43	0.50	0.74		
0.4	0.31	0.31	0.32	0.34	0.36	0.42	0.62			
0.5	0.25	0.26	0.27	0.29	0.34	0.51				
0.6	0.21	0.21	0.23	0.27	0.41					
0.7	0.16	0.17	0.20	0.32						
0.8	0.11	0.13	0.13	0.22						
0.9	0.08	0.12								

The penetration percentage, p, and the eccentricity, e, become

$$p = \frac{35 - 19}{80} = 0.20, \quad e = \frac{40 - \frac{16}{2}}{80} = 0.4$$

and hence from Table 8 $\alpha = 0.89$. According to Equation 60 the additional drawdown due to partial penetration is

$$\Delta s_w = \frac{2.3 \times 0.5}{2 \times 3.14 \dfrac{95}{24 \times 60 \times 60}} \frac{1 - 0.2}{0.2} \log \frac{0.89 \times 16}{0.75} = 8.4 \text{ m}$$

This gives a total drawdown at the well face $s_{wp} = 14.7 + 8.4 = 23.1$ m.

VI. CONFINED-UNCONFINED AQUIFER

Overabstraction in a confined aquifer may result in unconfined conditions around the pumping well as a result of which significant subsidence occurs and may damage the well stability with irrecoverable consequences. Therefore, two important questions arise. First, if ever an unconfined situation occurs, what is its distance of influence, a, radially around the well? Second, what is the reduction in pumping discharge in such an occurrence? Figure 26 shows the transition from a confined into an unconfined aquifer situation.

The steady-state equations for confined and unconfined aquifer portions are expressible with convenient notations from Equations 9 and 16, respectively, as

$$Q = 2\pi Km \frac{h(R) - m}{Ln(R/a)} \tag{61}$$

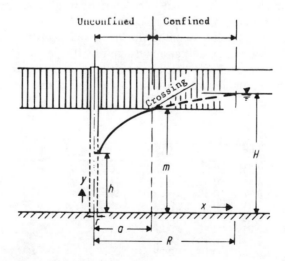

Figure 26 Confined-unconfined aquifer transition.

and

$$Q = \pi K \frac{m^2 - h^2(r_w)}{\ln\left(\dfrac{a}{r_w}\right)} \tag{62}$$

equating these two expressions to each other gives

$$a = \exp\left\{\frac{[m^2 - h^2(r_w)]\ln R + 2m[h(R) - m]\ln r_w}{2m[h(R) - m] + [m^2 - h^2(r_w)]}\right\} \tag{63}$$

which is the answer to the first question. The answer to the second question is found simply by substitution of Equation 63 into Equation 61 leading to

$$Q = \pi K \frac{h^2(R) - h^2(r_w)}{\ln\left(\dfrac{R}{r_w}\right)} - \pi K \frac{[h(R) - m]^2}{\ln\left(\dfrac{R}{r_w}\right)} \tag{64}$$

The second term on the r.h.s. of this expression actually shows the reduction in the discharge.

VII. PATCHY AQUIFERS

If the hydraulic conductivity in an unconfined aquifer changes sharply as in Figure 27, it is called patchy aquifer. The hydraulic conductivity, K, near the well is greater than hydraulic conductivity next to it.

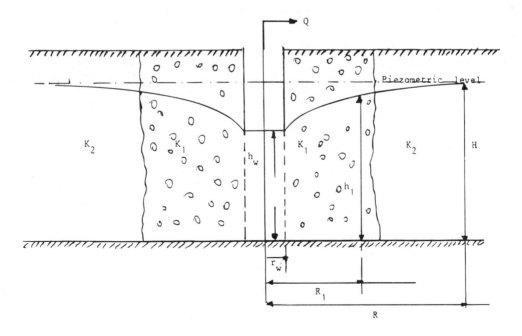

Figure 27 Steady-state flow in patchy aquifer.

For the first region the discharge formula from Equation 16 is

$$Q = \pi K_1 \frac{h_1^2 - h_w^2}{\ln\left(\dfrac{R_1}{r_w}\right)} \tag{65}$$

and for the second region it is

$$Q = \pi K_2 \frac{H^2 - h_1^2}{\ln\left(\dfrac{R}{R_1}\right)} \tag{66}$$

Addition of $h_1^2 - hw^2$ and $H^2 - h_1^2$ from these two equations leads to

$$Q = \frac{\pi}{2}(K_1 + K_2) \frac{H^2 - h_w^2}{\ln\left(\dfrac{R}{R_1}\right) + \ln\left(\dfrac{R_1}{r_w}\right)} \tag{67}$$

In terms of average hydraulic conductivity, $\overline{K} = (K_1 + K_2)/2$, it can be rewritten finally as

$$Q = \pi \overline{K} \frac{H^2 - h_w^2}{\ln\left(\dfrac{R}{R_1}\right)\left[1 + \dfrac{\ln(R_1/r_w)}{\ln(R/R_1)}\right]} \tag{68}$$

VIII. MULTIPLE WELL FIELDS

Another useful application of steady-state flow situations is the arrangement of multiple wells in such a way that there appears no interference between their depression cones. However, in the case of interference, the drawdown at any point is the sum of drawdowns due to each pumping well for which the distance of point from each well and their discharges should be known. Still a third possibility may occur in the case of already-existing wells as to manage their discharges in such a way that no interference is allowed. Last but not least, many dewatering problems consist of a number of wells that are pumped to provide the required relief of substratum pressure or reduction of piezometric levels. The common point in solutions of all these problems is to determine the number of spacing or individual well discharges required to give the desired shapes to the piezometric surface. Therefore, it is necessary to determine the relationship between the drawdown caused by a multiple well field and the flow from such arrangements. Considering the principle of drawdown superposition Dahler (1924) obtained the compound drawdown, s_i, at any desired ith point within the well field in confined aquifer from Equation 9 as

$$s_i = \frac{1}{2\pi T} \sum_{j=1}^{n} Q_j \ln \frac{R_j}{r_{ij}} \tag{69}$$

in which n is the number of wells. Q_j and R_j are the discharge and radius of influence of jth

STEADY-STATE FLOW AQUIFER TESTS

well; r_{ij} is the distance between the considered point, i, and jth well. The well field with seven wells is shown in Figure 28.

For well fields in unconfined aquifers the compound drawdown cannot be found directly but rather the difference between squares of static and dynamic water tables can be calculated by superposition principle as

$$H^2 - h_i^2 = \frac{1}{\pi K} \sum_{j=1}^{n} Q_j \ln \frac{R_j}{r_{ij}} \qquad (70)$$

Finally, for the leaky aquifer Huisman (1972) provided the compound drawdown within the well field as

$$s_i = \frac{1}{2\pi T} \sum_{j=1}^{n} Q_j K_0\left(\frac{r_{ij}}{L}\right) \qquad (71)$$

The aforementioned formulations of well fields are valid for irregular well positions.

For an array of n fully penetrating wells regularly spaced at distance D apart, each discharging at the same rate, Q parallel to a line source such as a river reach at a distance L as shown in Figure 29 the drawdown at any point x and y is given by Forchheimer (1898) in confined and unconfined aquifers as

$$s = \frac{2.3Q}{2\pi T} \log \frac{\cosh \frac{2\pi}{D}(x+L) - \cos \frac{2\pi y}{D}}{\cosh \frac{2\pi}{D}(x-L) - \cos \frac{2\pi y}{D}} \qquad (72)$$

and

$$s^2 - h^2 = \frac{2.3Q}{\pi T} \log \frac{\cosh \frac{2\pi}{D}(x+L) - \cos \frac{2\pi y}{D}}{\cosh \frac{2\pi}{D}(x-L) - \cos \frac{2\pi y}{D}} \qquad (73)$$

On the other hand, Muskat (1937) derived solutions for wells discharging by considering the following abstraction well configuration with the same radius of influence, R_0.

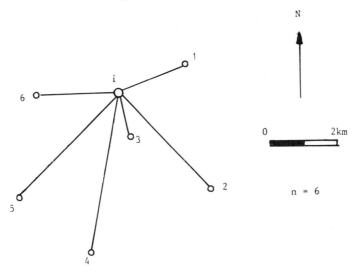

Figure 28 Well field plan view.

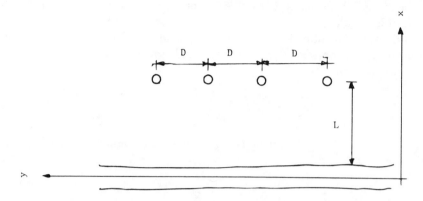

Figure 29 Wells parallel to a line.

1. For two wells spaced at a distance, D, each discharging Q, the valid formulation is

$$Q = 2\pi T \frac{s_w}{\ln(R_0^2/r_w D)} \tag{74}$$

2. For three wells discharging the same discharge

$$Q = 2\pi T \frac{s_w}{\ln(R_0^3/r_w D^2)} \tag{75}$$

3. For four wells forming a square of side D

$$Q = 2\pi T \frac{s_w}{\ln \frac{R_0^4}{\sqrt{2 r_w D^3}}} \tag{76}$$

IX. MULTIPLE AQUIFERS

In practice, wells may intersect two or more geological formations each with different geometric and hydraulic characteristics as in Figure 30. Such situations are inevitable consequences especially in the sedimentary basins. In fact, groundwater flow to a well penetrating more than one aquifer occurs rather frequently in nature than a well penetrating a single aquifer. Therefore, quantitative approximation to such flow situations is of highly practical value.

Sokol (1963) derived a simple steady-state equation relating natural groundwater fluctuations due to exchange between n aquifers as

$$H_w = \frac{\sum_{i=1}^{n} T_i H_i}{\sum_{i=1}^{n} T_i} \tag{77}$$

in which H_w is the composite piezometric level; H_is are individual piezometers of such aquifers

STEADY-STATE FLOW AQUIFER TESTS

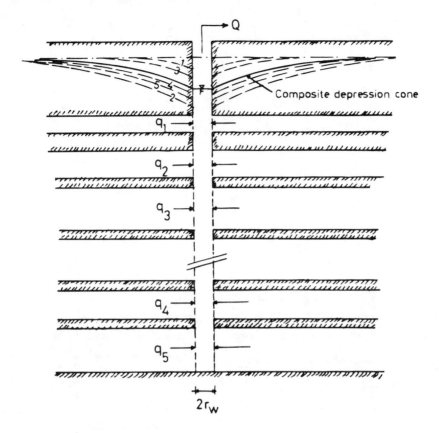

Figure 30 Multiple aquifers.

with respective transmissivity T_i. Prior to drilling each aquifer has its own piezometric level but their hydraulic connection through a well starts the interchange between the well and aquifers. Some aquifers release whereas others take water from the well and this exchange continues until each aquifer adjusts its piezometric level to a common static level H_w given in Equation 77. However, with the start of pumping each aquifer develops its own depression cone that converges to H_w at large distances (see Figure 30).

By assuming that each depression cone coincides with the groundwater level in the well, Sokol showed that the ratio of the water level fluctuation, h_w, in a well to the piezometric surface fluctuation h_i in an aquifer is equal to the ratio of transmissivity of the affected aquifer to the total transmissivities of all aquifers penetrated by the given well as

$$\frac{h_w}{h_i} = \frac{T_i}{T_1 + T_2 + \cdots + T_n} \tag{78}$$

However, Şen (1987) assumed on the contrary that not the individual depression cones but the composite depression cone coincides with the groundwater level in the well and hence Equation 9 is applicable as

$$Q = \frac{2\pi \sum_{i=1}^{n} T_i}{\ln\left(\dfrac{r}{r_w}\right)} [h(r) - h(h_w)] \tag{79}$$

in which

$$h(r) = \frac{\sum_{i=1}^{n} T_i h_i(r)}{\sum_{i=1}^{n} T_i} \quad (80)$$

and

$$h_w = \frac{\sum_{i=1}^{n} T_i h_{wi}}{\sum_{i=1}^{n} T_i} \quad (81)$$

X. RADIUS OF INFLUENCE

The radius of influence is the distance between the pumping well center and the point in the aquifer where the drawdown is equal to zero. The radius of influence should be interpreted

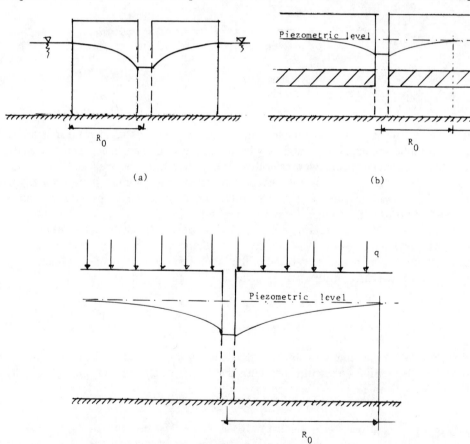

Figure 31 Radius of influence.

as a distance beyond which the drawdown is negligible. In nature exact zero points occur only in the cases of recharges to the aquifers as such finite aquifers as an island in an ocean, a leaky aquifer, or uniform direct infiltration to an unconfined aquifer shown in Figure 31.

In the quasi-steady-state flow case as it is clear from Equations 9 and 16, the radius of influence is a function of the drawdown in the well as well as the aquifer hydraulic conductivity. In groundwater investigations a priori knowledge about the radius of influence helps solve many problems quickly and cheaply. Unfortunately, it is not known directly and therefore, in general, a good estimate has to be made either from the past experiences or from the empirical formulas. Fortunately, it appears in the aforementioned formulations always in the form of $\log R_0$ so that even a large error in its estimation does not affect significantly the final results. Experiences in the field indicate that it varies between 50 and 500 m for unconfined aquifers whereas the range in confined aquifers is larger, from 300 m to 30 km. The most commonly used empirical formulation is due to Sichard (1927) as

$$R_0 = 3000 s_w \sqrt{K} \tag{82}$$

where the basic units should be in meters and in seconds. The empirical estimations should be supported by topography and surface, subsurface, and structural geology. The most reliable method of estimating R_0 is by the use of the Jacob method of pumping test data, as will be evaluated in the next chapter.

REFERENCES

Cooper, H. H. and Jacob, C. E., 1946 A generalized graphical method for evaluating formation constants and summarizing well field history, *Am. Geophys. Union Trans.*, 27, 526.

Dahler, R., 1924 *Grundwasserströmung*, Verlag Julius Springer, Vienna, 92.

De Glee, G. J., 1930 Over grondwaterstromingen bij teronttrekking door middel van putten, Thesis, J. Waltman, Delft, The Netherlands.

Dupuit, J., 1863. Etude theoretique et pratique sur le movement des eaux dans les canaux decouverts et a travers les terrains permeables, Dunod, Paris.

Ferris, J. G., 1948 Ground Water Hydraulics as a Geophysical Aid, Tech. Report No. 1, Michigan Department of Conservation.

Forchheimer, P. V., 1898 Grundwasserspiegel bei Brunnen Anlagen, *Z. Oster. Ing. Verein.*

Forchheimer, P. V., 1901 Wasserbewegung durch Boden, *Z. Ver. Deutsch. Ing.*, 1, 1736.

Forchheimer, P. V., 1930 *Grundwasserbewegung und Hydraulik*, Tubrier, Leipzig.

Hantush, M. S., 1956 Analysis of data from pumping test in leaky aquifers, *EOS Trans. AGU*, 37, 702.

Hantush, M. S., 1964 Hydraulics of wells, in *Advances of Hydrosciences*, Vol. 1, Chow, V. T., Ed., Academic Press, p. 281.

Huisman, L., 1972 *Ground-Water Recovery*, MacMillan, London, England.

Jacob, C. E., 1945 Partial Penetration of Wells, Adjustment for, Water Resources Bulletin, U.S. Geol. Survey, p. 175.

Jacob, C. E., 1963 Correction of drawdowns caused by a pumped well tapping less than the full thickness of aquifer in Methods of Determining Permeability, Transmissivity and Drawdown, Water Supply Paper 1536-I, U.S. Geol. Surv., p. 272.

Karanth, K. R., 1978 *Ground Water Assessment. Development and Management*, Tata McGraw-Hill, New Delhi. 720.

Kozeny, J., 1933 Theorie und Berehnung der Brunnen, *Wasserkraft Wasserwirtschaft*, 39, 101.

Kruseman, G. P. and de Ridder, N. A., 1990 Analysis and Evaluation of Pumping Test Data, Bulletin 11, Institute for Land Reclamation and Improvement, Wageningen, p. 41.

Li, W. H., 1954 A new formula for flow into particular penetrating wells in aquifers, *Proc. Am. Geomorph. Union*, 55(5), 805.

Muskat, M., 1937 *The Flow of Homogeneous Fluids through Porous Medium,* McGraw-Hill, New York.

Nahrgang, G., 1954 *Zur Theorie des Vollkommenen und Unvollkommenen Brunnen,* Springer-Verlag, Berlin.

Papadopulos, I. S. and Cooper, H. H., 1967 Drawdown in a well of large diameter well, *Water Resour. Res.,* 3, 241.

Sichard, V., 1927 Das Fassungsvermögen von Bohrbrunnen und Eine Bedeutung für die Grundwasserersenkung inbesondere für grossere Absentiefen, Dissertation, Tech. Hochschule, Berlin.

Sokol, D., 1963 Position of fluctuation of water level in wells penetrated in nature more than one aquifer, *J. Geophys. Res. Union,* 68, 1079.

Stallman, R. W., 1963 Type curves for solutions of single boundary problems, in Short Cuts and Special Problems in Aquifer Tests, U.S. Geol. Survey, Water Supply Paper, 1545-C, 45.

Streggewentz, J. H. and Van Ness, B. A., 1939 Calculating the yield of a well taking account of replenishment of the ground-water from above, *Water Water Eng.,* 41, 561.

Szechy, K., 1955 Beitrag zur Theorie der Grundwasserabsenkung. Aus Gedanken für Prof. Dr. Jaky Yekademini Kidao, Budapest.

Şen, Z., 1987 Storage coefficient determination from quasi-steady state flow, *Nordic Hydrol.,* 18, 101.

Şen, Z., 1991 Recovery type curves for large diameter wells, *Nordic Hydrol.,* 22, 253.

Thiem, G., 1906 *Hydrologische Methoden,* Gebhardt, Leipzig,

Walton, W. C., 1962 Selected Analytical Methods for Well and Aquifer Evaluation, Illinois State Water Survey, Bull. 49, Urbana, 81.

CHAPTER 9

Porous Medium Aquifer Tests

I. GENERAL

The objective of unsteady flow tests is to determine the aquifer parameter estimations from time-drawdown field measurements in the main and/or observation wells by processing them through physically based and simplified mathematical models. These models help to infer aquifer, well, or pump properties and finally to make the future predictions of discharges, drawdowns, or distances between wells for the groundwater resources planning, design, operation, and management. This chapter concentrates on the development of physically based models without entering the mathematical morass. The geological and physical procedures are presented as necessary tools for analyzing complex natural aquifers. For those readers interested in detailed mathematical involvements, consult reference books by Muskat (1937), Hantush (1964), Bear (1981), and Marino and Luthin (1982).

Hydraulic properties of an aquifer can be determined by an aquifer test that involves abstraction of water from a well at a constant rate and observing with respect to time the water level changes in the pumped and/or observation wells. Properly completed aquifer tests provide the following useful information about

1. The general aquifer parameters, such as transmissivity, storativity, leakage factor, drainage factor, and hydraulic resistance. It is also possible to determine hydraulic conductivity if the aquifer thickness is known.
2. The geometry of the aquifer including distance, direction, and nature of impermeable barriers and recharge boundaries.
3. The lateral variations in the well vicinity.

A complete picture of predictions is possible first by identifying the aquifer parameters such as the porosity, hydraulic conductivity, transmissivity, specific yield, storage coefficient, leakage factor, and delayed yield index and secondly their exploitation in a suitable model by considering the subsurface geological environment to have future realizations of different possible alternatives for the groundwater resources assessments. An essential step in such an approach is the field work of the study area from the subsurface geological composition point of view. Especially, time series records of drawdown during an unsteady-state flow will have imbedded reflections of various geological effects. For instance, an impervious layer will appear as a jump in the records. Successful application of the methods in this chapter depends greatly on the previous geological studies, if available, in the study area; otherwise, the view taken in this book is that at least a preliminary reconnaissance study should be carried out prior to the field applications of these tests to have reliable interpretations of test results. Concise hydrogeological field investigations may provide adequate information concerning subsurface geological facies changes in sedimentary basins on hydrostratigraphic sequences including water levels. Only then will the methods presented in this chapter become beneficial, the results correct, and future predictions reliable for any application in the study area.

II. PARAMETERS VS. VARIABLES

Interpretation of field time-drawdown measurements or aquifer parameters calculated thereof is appreciated with great confidence if the planner should have a sound knowledge about the mutual relationships between the aquifer parameters and basic test variables. As described in Chapter 4 aquifer parameters are reflections of the aquifer storage, transmission, and leakage properties. In general, there are four basic test variables: discharge, time, drawdown, and distance. In steady-state flow the number of variables is three with the exception of time. However, for prediction and management studies time plays a vital role by itself and also by its influence on the drawdown as well as on the distance in the form of an expanding radius of influence. In the groundwater studies it is almost impossible to consider the variation of all four basic variables simultaneously during one test. A common practice is to keep one of them constant depending on the purpose of the test. For instance, the discharge is kept constant in many situations because the mathematical treatment of the problem becomes easier.

Let us consider the qualitative relationships between aquifer parameters and the test variables by keeping all of the aquifer parameters and discharge constant. Figure 1 shows two confined aquifers with different transmissivities. In an aquifer with low transmissivity the depression cone is deep, with steep hydraulic gradients and high drawdowns while in a high-transmissivity aquifer the cone is wider with lower hydraulic gradients, smaller drawdowns, and a comparatively larger radius of influence.

If Darcy law is valid then the specific discharge is great in the well vicinity but becomes smaller within a short distance away from the well. There exists a direct relationship between the radius of influence and the transmissivity as mentioned in Chapter 8. On the other hand, in Figure 2 the two aquifers have different storage coefficients with all other parameters being constant. Under a given set of conditions the smaller the storativity the greater is the drawdown, and more extensive is the depression cone. These imply an inverse relationship between the

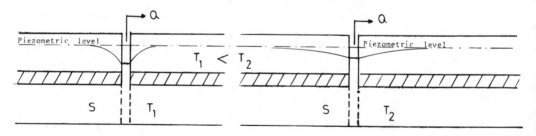

Figure 1 Different transmissivities.

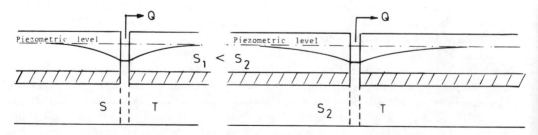

Figure 2 Different storage coefficients.

radius of influence and the storage coefficient. Figure 3 represents the effect of leakage factor on the development of the depression cone. Small leakage factors give rise to extensive depression cones.

It is clear from the foregoing discussions that whatever the relative values of aquifer parameters are there is always a depression cone development around the well because of water abstraction. The greater the discharge value the more extensive will be the depression cone for the same aquifer. The following qualitative relationships are invariably valid in any groundwater study:

1. The drawdowns are proportional with the time and the proportionality factor is a combined function of the aquifer parameters and discharges. Figure 4 shows different time drawdown plots on arithmetical graph paper. They all have similar concave shape with decreasing curvature as time increases.
2. The drawdowns are inversely proportional with the distance. The form of this decrease is of the negative exponential type and the maximum value corresponds to the drawdown at the well face within the aquifer. Figure 5 gives possible shapes for such a relationship.
3. There is always a direct proportional relationship between the drawdown and the discharge. Marsily (1986) refers to such an association as the "characteristic curve" of a well. The general form of this curve is the convex type and parabolic (see Figure 6a).

However, from the pumping machine point of view an inverse relationship is valid as in Figure 6b. Such a relationship is named the pump characteristic curve. This curve implies that as the drawdown increases the water pump is able to haul less discharge. The specific forms of all the aforementioned qualitative figures are dependent on the aquifer parameter and the initial as well as the boundary conditions of flow domain. Although it is possible to find the best matching curve to any field data statistically by use of the least-squares technique, the results are in terms of statistical parameters which are difficult to interpret and are not explicitly related to the aquifer physical properties. Therefore, it is essential to develop

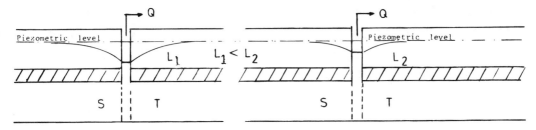

Figure 3 Different leakage factors.

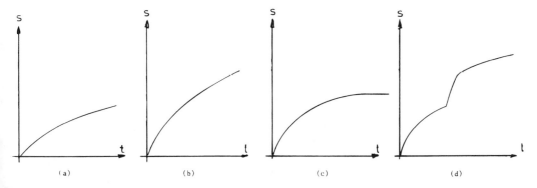

Figure 4 Possible time-drawdown relationships.

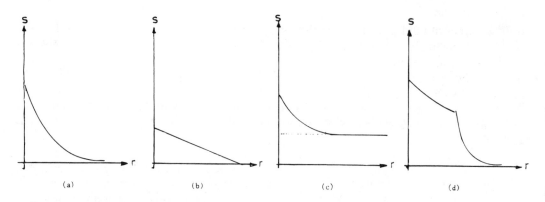

Figure 5 Possible distance-drawdown relationships.

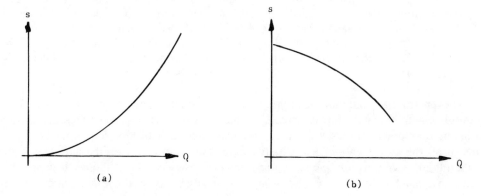

Figure 6 Characteristic curves.

physically based models of idealized aquifers and to match their curves to the field data to determine the parameter estimates.

III. FIELD TEST PROCEDURES

The purpose of any test is to determine the aquifer parameters or well characteristics. The performance of such tests depends on the availability of facilities in terms of time, capital, and personnel. Any field study may include one or more of the following points.

1. Determination of relevant aquifer parameters only. The basic parameters are the storativity, transmissivity, leakage factor, delayed yield, etc.
2. Determination of the well characteristics which are the well losses, safe yield, specific drawdown, and the specific capacity of the well.
3. Determination of the geological and/or hydrological boundaries, especially the hidden ones in the subsurface. The boundaries are either impervious barriers or geological facies changes or structures such as fractures, joints, in addition to water seepages and leakages.
4. Determination of geometrical quantities such as the cone of depression shape and radius of influence may lead to appropriate location, design, management, and construction of wells.
5. Determination of some proposed developments on the recharge, discharge, and salt-water intrusion.
6. Determination of dewatering area dimensions, especially depths in construction engineering.

Quantitative evaluation of these points requires data recorded in the field. There are different types of field tests through which the data on the variables discussed in the previous section are obtained. These data have similar importance as what blood pressure and body temperature data mean to a physician. An experienced hydrogeologist can interpret data even prior to any systematic processing to end up with significant conclusions concerning the aquifer and well characteristics.

In a pumping test the main well is equipped with any type of pump by which a certain quantity of water is withdrawn and time-drawdown measurements are recorded in the main and/or observation wells. There are two types of pumping tests commonly employed in the groundwater studies. The first one is referred to as the aquifer test, the data of which yield aquifer properties only, such as transmissivity, storativity, leakage factor, etc. The aquifer test is a part of the geological investigation and should last long enough so that the test operator has adequate boring information including water levels and laboratory analysis. Carefully conducted aquifer tests may provide data that serve as a basis for solutions of many local and regional groundwater flow problems. Most sections in this chapter will present different models and procedures for the identification of the aquifer parameters from given field data. The second type of pumping test is known as the "well test" which provides data for the estimates of the well characteristics only. It is the main topic of this chapter.

On the other hand, the nonpumping tests are performed in any well with or without a pump. Their application is comparatively cheaper and takes short time periods. Among these tests the recovery type is always coupled with the aquifer test and it supplies supplementary data.

In practice all the tests have a common point in providing time-drawdown, time-distance, discharge-time-drawdown, or time-distance-drawdown data. For their presentation on any graph paper one can simply plot invariably drawdown vs. any other variable. These graphs are referred to, herein, as the field data sheet. Contrary to the unified field plotting each field data sheet necessitates different theoretical models for the parameter identifications. These models depend on different combinations of aquifer, well, flow, material, penetration, and boundary types. Theoretically, the number of models required to cover all the solutions is equal to the number of these combinations. However, in practice not all these solutions are applicable but the most commonly encountered combinations will be treated in this chapter. After reading this chapter the reader will appreciate that there are still many practical combinations awaiting practical solutions.

IV. AQUIFER MODEL

The aquifer parameters are hidden in the field test data and their identifications are possible only after a physically plausible model is developed for the prevailing field circumstances. The word "model" herein implies a theoretical form of relationships between variables and parameters. The development of any aquifer model requires five stages as shown in Figure 7.

The pure mathematical derivation stage in model development is skipped throughout this book by giving relevant references only. Readers interested in the mathematical background

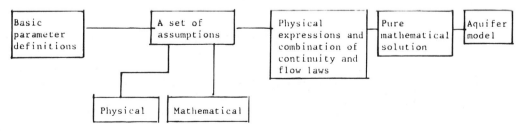

Figure 7 Flow chart for modeling.

of the problems can have detailed information from these references. The purpose of this chapter is to prepare the reader in the first three stages up to the domain of mathematics and finally to equip him with the mechanical use of the aquifer models in addition to physical interpretations of the conclusions. The author's experience shows that avoiding mathematical derivations presents no problem but saves time and affords without loss of basic appreciation. Furthermore, any reader uses this pattern of reasoning may well be successful in tackling some other groundwater problems.

The last stage in Figure 7 may have three different methods to arrive at a convenient tool in the forms of type curves, equations, slopes, or adaptive procedures as presented in Figure 8.

It is not necessary that all the mathematical models in this figure are applicable for every aquifer. The selection of the suitable model depends on the aquifer flow and well types as well as their interactions.

V. CONFINED AQUIFER MODELS

In light of the above-mentioned discussions it is rather clear that the models depend on the type of aquifer. Researchers have tackled different aquifers and come out with specific solutions that are valid only for the stated assumptions and initial and boundary conditions. These specific models are referred to with the names of their pioneers. Figure 9 gives different methods that can be applied to the field data from confined aquifers.

In the following sequel each one of these models is presented up to the boundary of mathematics and then the necessary end products are presented either in terms of equations or curves or tables as basic tools for application to field data.

In general, the assumptions stated in Section II of Chapter 8 for the steady-state flow remain the same with some additions or deletions as the unsteady-state model may require. General additional assumptions for the unsteady-state flow toward the well are

1. The groundwater is compressible and aquifer material has an elastic behavior.
2. Aquifer response to abstraction is instantaneous.

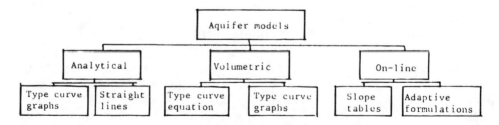

Figure 8 Different aquifer models.

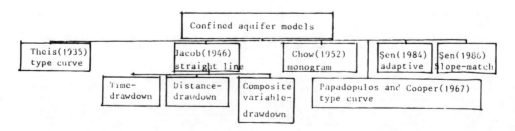

Figure 9 Confined aquifer models.

A. THEIS MODEL

The first study in the unsteady-state domain of Darcian (linear) groundwater flow is credited by Theis (1935), who developed a useful solution for the confined aquifer parameter identification with fully penetrating wells. He assumed an infinitesimally small well diameter as a mathematical convenience. Hence, a pumping well becomes equivalent to a point sink. Figure 10 shows a confined aquifer with thickness, m, storage coefficient, S, and hydraulic conductivity, K.

The well fully penetrates the aquifer and is pumped at a constant discharge rate, Q. With the start of pump the static piezometric level takes the form of a depression cone. Since the flow is unsteady, equality of discharges at every imaginary concentric cylinder surfaces as in Equation 5 in Chapter 8 is no more valid.

The continuity equation in terms of volumes between two concentric cylinders of radius r and $r + \Delta r$ during time period Δt can be written as (see Figure 10)

$$Q(r,t)\Delta t - Q(r + \Delta r,t)\Delta t = S2\pi r\Delta r[h(r,t) - h(r,t + \Delta t)] \qquad (1)$$

in which $Q(r,t)$ and $h(r,t)$ are the discharge and piezometric head at time t and radial distance r from the main well center. The first and the second terms on the left-hand side (l.h.s.) are the total input and output from the annular portion of the aquifer, respectively. The right-hand side (r.h.s.), however, expresses the total change in the aquifer storage in this annular region. For large times $h(r,t + \Delta t) \cong h(r,t)$ and this equation becomes equivalent to the steady-state case given succinctly by Equation 5 in Chapter 8.

However, for an infinitesimally small time period ($\Delta t \to 0$) and annular thickness

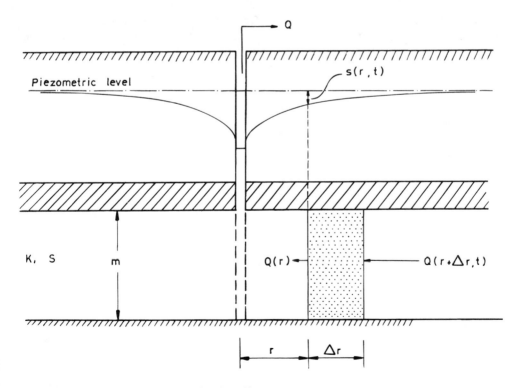

Figure 10 Unsteady-state flow in confined aquifer.

($\Delta r \to 0$), Equation 1 takes the form of partial differential as

$$\frac{\partial Q(r,t)}{\partial r} = 2\pi Sr \frac{\partial h(r,t)}{\partial t} \qquad (2)$$

In general, the discharge is defined for any concentric cylinder as

$$Q(r,t) = 2\pi mr q(r,t) \qquad (3)$$

where $q(r,t)$ is the specific discharge. Elimination of $Q(r,t)$ between Equations 3 and 2 leads to

$$\frac{\partial q(r,t)}{\partial r} + \frac{1}{r} q(r,t) = \frac{S}{m} \frac{\partial h(r,t)}{\partial t} \qquad (4)$$

This expression shows the general relationship between the piezometric level and the specific discharge around the well, irrespective of any condition. If the flow is Darcian then

$$q(r,t) = K \frac{\partial h(r,t)}{\partial r} \qquad (5)$$

the substitution of which into Equation 4 gives

$$\frac{\partial^2 h(r,t)}{\partial r^2} + \frac{1}{r} \frac{\partial h(r,t)}{\partial r} = \frac{S}{T} \frac{\partial h(r,t)}{\partial t} \qquad (6)$$

in which T is the aquifer transmissivity. This equation describes groundwater flow toward a fully penetrating well in a confined aquifer. Theis (1935) recognized that this equation represents the heat flow from a point source in a uniform thickness of layer with the solution given by Carslaw and Jeager (1959). Under the relevant assumptions the solution is

$$W(u) = \int_{-\infty}^{u} \frac{e^{-u}}{u} du \qquad (7)$$

Furthermore, the explicit form of this expression is given by an infinite series as

$$W(u) = -0.5572 - Lnu + u - \frac{u^2}{2.2} + \frac{u^3}{3.3} - \frac{u^4}{4.4} + \cdots \qquad (8)$$

where $W(u)$ and u are referred to as the well function and the dimensionless time factor, respectively. They are related to aquifer parameters and variables as

$$W(u) = \frac{4\pi T}{Q} s(r,t) \qquad (9)$$

and

$$u = \frac{r^2 S}{4tT} \qquad (10)$$

From a practical point of view, it is very important to notice at this stage that

POROUS MEDIUM AQUIFER TESTS

1. $W(u)$ and u are dimensionless variables.
2. $W(u)$ and u are composite variables of unknown aquifer parameters (S and T) and variables (Q, r, t and s) that can be measured in the field.
3. According to Equation 7 there is a relationship between u and $W(u)$, the graphical presentation of which is referred to as the type curve in groundwater literature. Corresponding values of u and $W(u)$ are given in Table 1.
4. S and T are determined from Equations 9 and 10, provided that field data on Q, r, t, and s are known in addition to theoretically defined variables $W(u)$ and u.

Type curve shows the variation of $W(u)$ against u or the reverse dimensionless time factor, $1/u$, on a double-logarithmic paper. However, it is frequently more convenient to use the latter version in practical application. The standard Theis-type curve is given in Figure 11. This

Table 1 Theis-Type Curve Values

u	1.0	2.0	3.0	4.0	5.0	6.0	7.0	8.0	9.0
× 1	0.219	0.049	0.013	0.0038	0.0011	0.00036	0.00012	0.000038	0.000012
× 10^{-1}	1.82	1.22	0.91	0.70	0.56	0.45	0.37	0.31	0.26
× 10^{-2}	4.04	3.35	2.96	2.68	2.47	2.30	2.15	2.03	1.92
× 10^{-3}	6.33	5.64	5.23	4.95	4.73	4.54	4.39	4.26	4.14
× 10^{-4}	8.63	7.94	7.53	7.25	7.02	6.84	6.69	6.55	6.44
× 10^{-5}	10.94	10.24	9.84	9.55	9.33	9.14	8.99	8.86	8.74
× 10^{-6}	13.24	12.55	12.14	11.85	11.63	11.45	11.29	11.16	11.04
× 10^{-7}	15.54	14.85	14.44	14.15	13.93	13.75	13.60	13.46	13.34
× 10^{-8}	17.84	17.15	16.74	16.46	16.23	16.05	15.90	15.76	15.65
× 10^{-9}	20.15	19.45	19.05	18.76	18.54	18.35	18.20	18.07	17.95
× 10^{-10}	22.45	21.76	21.35	21.06	20.84	20.66	20.50	20.37	20.25
× 10^{-11}	24.75	24.06	23.65	23.36	23.14	22.96	22.81	22.67	22.55
× 10^{-12}	27.05	26.36	25.96	25.67	25.44	25.26	25.11	24.97	24.86
× 10^{-13}	29.36	28.66	28.26	27.97	27.75	27.56	27.41	27.28	27.16
× 10^{-14}	31.66	30.97	30.56	30.27	30.05	29.87	29.71	29.58	29.46
× 10^{-15}	33.96	33.27	32.86	32.58	32.35	32.17	32.02	31.88	31.76

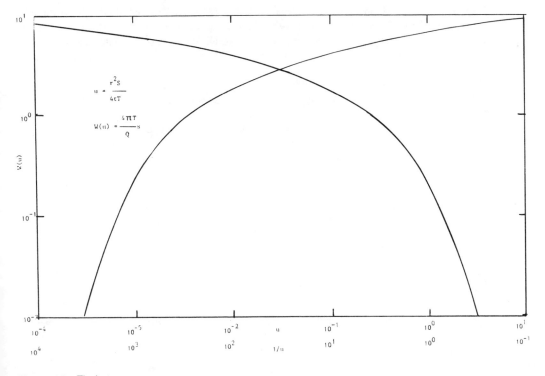

Figure 11 Theis-type curve.

curve is similar to the depression cone shape, in that initially it has a large slope but it decreases steadily with increasing reverse dimensionless time factor. In fact, it is equivalent to the dimensionless form of the depression cone.

Equation 6 in its present form is more attractive to a mathematician than to a hydrogeologist. However, each one of the terms has physically valuable interpretations in the groundwater domain. For instance, the derivative in the r.h.s. is, in fact, equal to the slope of time-drawdown curve (Figure 12a), whereas the second derivative on the l.h.s is equivalent to the distance drawdown curve slope which is named as the hydraulic gradient (Figure 12b). Finally, the first derivative on the l.h.s. is equal to the slope of hydraulic gradient-distance curve (Figure 12c).

It is, therefore, possible to conclude that the general groundwater movement equation is composed of piezometric level variations with respect to time and distance. In practice, the Theis method finds application for different purposes ideally such as for

1. Fully penetrating well in a confined aquifer.
2. Early and late time responses of fractured aquifer with a double-porosity model (Chapter 10).
3. Early responses of unconfined and leaky aquifers with fully penetrating well.
4. Assessment of time drawdown (or distance drawdown data) deviations from the confined aquifer.

1. Type Curve-Matching Procedure

This is a mechanical procedure through which the most suitable type curve position is sought for the field data at hand. As a general rule, any type curve leads to aquifer parameter values only after the proper application of the matching procedure. In any matching procedure the following sequence of steps is necessary in arriving at the final target.

1. The well function variation with the dimensionless time factor is drawn on a double-logarithmic paper. Such a paper is referred to as the type curve sheet. In practice, it is preferable to draw these curves on a transparent paper and to write explicitly the well function as well as the dimensionless time factor expressions in terms of aquifer parameters and field measurements, (see Figure 11).
2. The observed values of drawdown are plotted against time (or distance if many observation wells are used) on the same scale logarithmic paper as was for type curve sheet. This plot has been referred to as the field data sheet in Chapter 7.
3. The type curve sheet is superimposed on the field data sheet and translated horizontally and/or vertically, keeping the corresponding coordinate axes (well function corresponds to drawdown) of the two sheets parallel, until a position of the best fit between the data plots

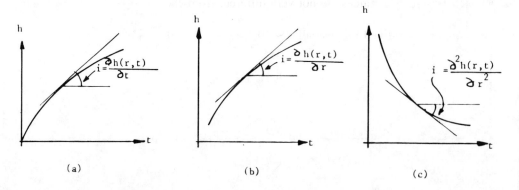

Figure 12 Graphical representation of terms.

and the type curve is reached. At this position the two sheets are kept without movement, preferably, by using a sticker.
4. An arbitrary point is chosen on the overlapping area between the two sheets. This point is called the "match point".
5. The coordinates of match point are read from the two sheets. As a result, four readings, namely two abcissca, time or distance and the dimensionless time factor, and two ordinates, drawdown and well function values, are available for further use. At this step, also, if there is any label as a factor attached to the type curve its value is also recorded. Hence, the matching procedure ends and the readings are kept for their substitution in the necessary well function and dimensionless time factor expressions in order to calculate the numerical values of aquifer parameters.

In general, field data may deviate from the Theis-type curve at small time instances even though the aquifer is confined. These deviations may be due to some of the following assumptions:

1. In theory the well diameter is equal to zero. However, it is never the case in practice and therefore a time lag may occur between the piezometric level decline and the water release into the well from the aquifer. Such deviations are very systematic for a large diameter as will be discussed in Section D.
2. Again in theory the pump discharge is constant but there is always a warm-up period for any pump and therefore initially the discharge may not be constant. During this period the pump adjusts itself to the piezometric level changes.
3. The aquifer is homogeneous. Most often during deep well drilling the material in the vicinity of the well is disturbed and, accordingly, the aquifer may have two different regions as will be discussed in Chapter 10 for the fractured medium.
4. The flow abides with the linear flow (Darcy law). However, due to high gradients especially near the well, there may appear turbulent nonlinear (non-Darcian) flow (Chapter 12). In practice all these theoretical assumptions become satisfactory at large radial distances or late times.

Example 1—The time-drawdown data from three observation wells at 25, 60, and 150-m radial distance from the main well center are given in Table 2 with the general configuration of the wells.

A first glance of this table indicates a direct relationship between the time and drawdown, whereas the distance from the main well is inversely related to the drawdown. Field data and Theis-type curve sheets are matched individually for each well in light of the above-mentioned matching procedure steps in Figures 13 through 15 and the results are given in Table 3.

According to the criterion given in Table 5 in Chapter 4 the aquifer is highly potential since $T > 500$ m²/day. T values are not very different from each other and hence the aquifer is isotropic and homogeneous. The S values have the same order of magnitude and confirm that the aquifer is confined because all $S < 10^{-1}$.

B. JACOB MODEL

Cooper and Jacob (1946) have noticed for small u values less than 0.01 that the Theis well function given in Equation 8 has negligible contributions from terms beyond $Ln(u)$. Hence, they have expressed the well function as

$$W(u) = -0.5772 - Lnu \qquad (11)$$

The error involved in adopting Equation 11 instead of the Theis well function is less than

Table 2 Field Data

Time (min) t	Drawdown (cm) s			1/t (1/min)
	W1 r = 25 m	W2 r = 60 m	W3 r = 150 m	
1	40.2	10	0.24	1×10^0
2	60.4	23.2	2.44	0.5
3	73.8	32.2	5.48	0.33
4	83.0	40.8	9.72	0.25
5	97.0	52.9	16.40	0.20
8	106.8	60.0	22.60	1.25×10^{-1}
10	113.4	68.5	28.00	1.00
15	127.0	80.0	36.00	6.67×10^{-2}
20	134.0	87.5	43.60	5.00
25	144.0	96.0	53.00	4.00
30	152.0	102.6	57.92	3.33
40	161.6	114.8	68.20	2.50
50	169.6	121.0	75.00	2.00
60	175.8	128.6	80.50	1.67
80	185.4	132.6	91.00	1.25
100	193.0	145.2	98.80	1.00
130	200.0	156.0	103.60	7.69×10^{-3}
160	208.4	160.0	111.60	6.25
190	225.0	166.0	118.40	5.26
290	220.0	171.4	124.00	3.45
400	222.0	175.6	130.50	2.50

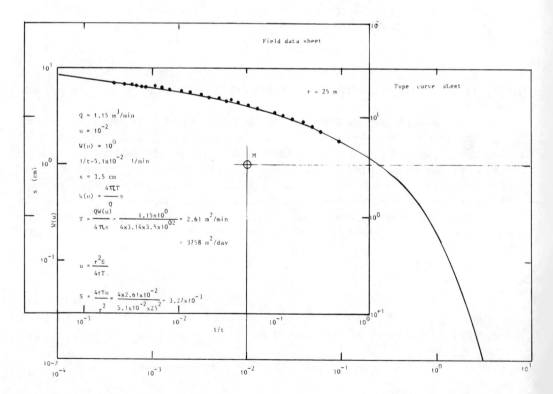

Figure 13 Field data plot ($r = 25$ m).

POROUS MEDIUM AQUIFER TESTS

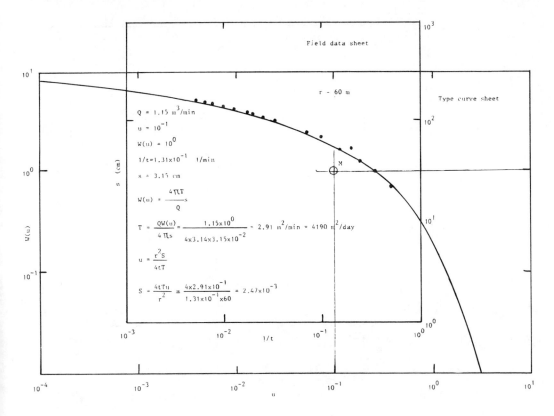

Figure 14 Field data plot ($r = 60$ m).

1% and it yields a straight line on a semilogarithmic paper as shown in Figure 16. Substitution of Equations 9 and 10 into Equation 11 after some algebra leads to the following form

$$s(r,t) = \frac{2.3Q}{4\pi T} \log \frac{2.25Tt}{r^2 S} \qquad (12)$$

This equation represents the drawdown variations with time and distance and is valid for the following error levels compared to the Theis curve.

Error (%)	Validity
0.25	$u < 0.01$
2.00	$u < 0.05$
5.00	$u < 0.10$
10.00	$u < 0.15$

Usually in hydrogeological studies 5% error is acceptable and hence the Jacob method for any specific aquifer is valid when $t > (10r^2 S/4T)$. For instance, with the data in Table 2 the Jacob method is applicable for wells W1, W2, and W3 when $t > 2$, 8, and 31 min, respectively. Equation 12 gives rise to three different relationships between the dependent (s) and independent, (r and t) variables. Accordingly, three different versions of the Jacob method emerge: time-drawdown, distance-drawdown, and time-distance-drawdown (composite variable t/r^2) methods. All the Jacob methods are basically concerned with the last portion of the Theis-type curve and are not as a consequence of physical reality but rather a mathematical convenience for

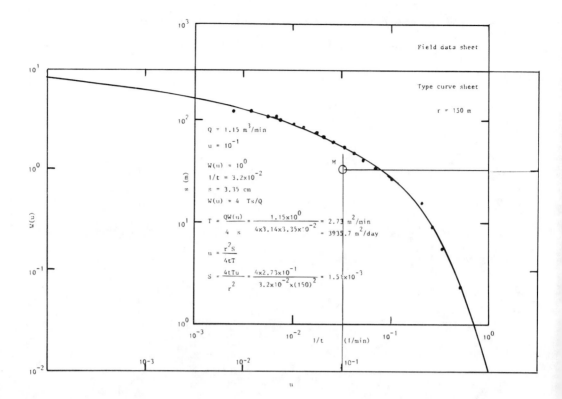

Figure 15 Field data plot ($r = 150$ m).

Table 3 Theis-Type Curve Calculations

| | Radial well distance r (m) | Match point coordinates | | | | Parameter estimates | |
| | | Type curve | | Data sheet | | Equation 9 | Equation 10 |
		$W_M(u)$	u_M	s_M(m)	$(l/t)_M$	T (m²/d)	S
W1	25	10^0	10^{-2}	3.50	5.1×10^{-2}	3758	3.27×10^{-3}
W2	60	10^0	10^{-1}	3.20	1.3×10^{-1}	4190	2.47×10^{-3}
W3	150	10^0	10^{-1}	3.40	3.2×10^{-2}	3936	1.51×10^{-3}
					Averages	3957	2.42×10^{-3}

Note: Discharge $Q = 1.15$ m³/min.

$u < 0.01$. To use a proper field test data for the application of the Jacob method either the time must be large or the distance small or the ratio t/r^2 must be large.

1. Time-Drawdown Model

If the drawdowns are recorded in individual observation wells then the distance r has a constant value that is measured in the field. Accordingly, Equation 12 will have two variables, time and drawdown, and their plot on a semilogarithmic appears as a straight line.

In general, any straight line gives two useful facts: its slope, Δs_t, and intercept, t_0, corresponding to zero drawdown. The slope on a semilogarithmic paper is calculated as the difference in the drawdown values at the end and the beginning of a complete log cycle. The steeper the slope the greater is the drawdown change implying physically that the specific discharge is rather small and correspondingly the hydraulic conductivity has small values. There is an inverse relationship between the slope and the hydraulic conductivity. On the other hand, the

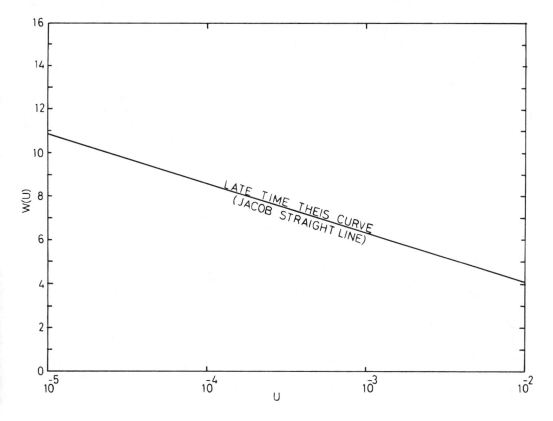

Figure 16 Jacob straight line.

intercept does not have any physical meaning but provides a mathematical convenience. The analytical expression for the slope and the intercept can be found from Equation 12 leading to aquifer parameter identification by the following equations:

$$T = \frac{2.3Q}{4\pi \Delta s_t} \tag{13}$$

and

$$S = \frac{2.25 t_0 T}{r^2} \tag{14}$$

In these equations, Q and r are directly available from the field measurements but Δs_t and t_0 are calculated in the office from the field time-drawdown data.

Table 4 Time Drawdown Model Calculation

Well no.	Distance r (m)	Straight-line readings		Parameter calculations	
		Slope Δs (m)	Intercept t_0 (min)	Equation 13 T (m²/d)	Equation 14 S
W1	25	0.77	1.05×10^0	788	1.4×10^{-3}
W2	60	0.77	1.30×10^0	788	4.4×10^{-4}
W3	150	0.72	4.70×10^0	421	1.4×10^{-4}
			Averages	666	6.6×10^{-4}

Note: Discharge $Q = 1.15$ m³/min.

Example 2—For the application of this method the drawdown data in Table 2 are plotted against the corresponding time on semilogarithmic paper as shown in Figures 17 through 19. The slopes of these straight lines are measured on the vertical axis per log cycle of time. The intercept with the logarithmic axis is also read from these figures. The results of calculations are shown in Table 4.

Numerical values of T and S indicate that the aquifer is highly potential and confined. However, the values of both parameters are comparatively smaller than the Theis method values presented in Table 3. The u values for the validity of the Jacob method with parameter estimates are 0.38, 0.56, and 0.06, respectively.

2. Distance-Drawdown Model

We have discussed in Chapter 3 that the depression cone has temporal as well as spatial variations. To depict the spatial variation drawdown measurements should be taken at different observation wells for any given time or preferably in a steady- or quasi-steady-state situation. If the time is regarded as fixed then Equation 12 provides a relationship between the distance and drawdown which appears as a straight line on a semilogarithmic paper.

The physical interpretation of the slope, Δs_r, is the same as was for the time-drawdown model. However, herein the intercept, r_0, has a meaningful physical interpretation corresponding to the radius of influence. What is the relationship of distance-drawdown plot to the Thiem method discussed in Chapter 8? In fact, they have the same concept and lead to the same result. If Equation 12 is written for two different distances, r_1 and r_2, and subtracted from each other, then the result is equivalent to the Thiem formula given in Equation 19 in Chapter 8.

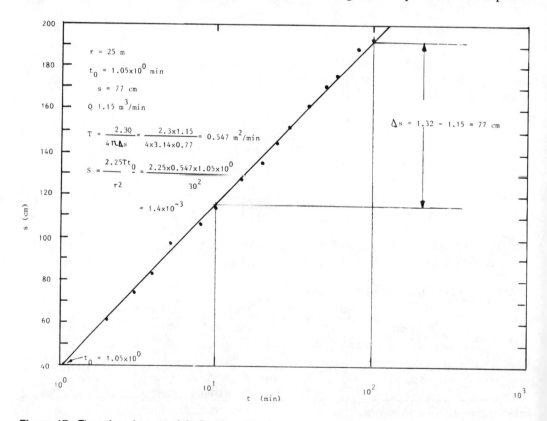

Figure 17 Time-drawdown straight line ($r = 25$ m).

POROUS MEDIUM AQUIFER TESTS

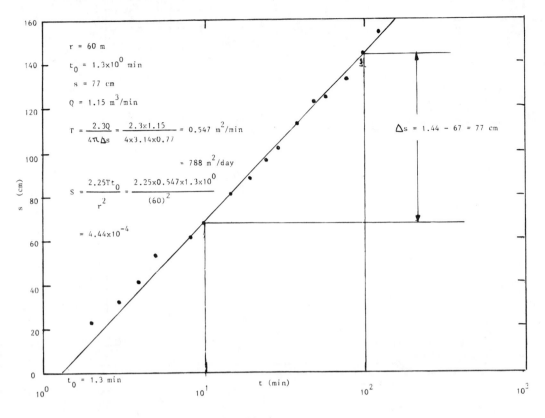

Figure 18 Time-drawdown straight line ($r = 60$ m).

In practice, the distance-drawdown method requires readings of at least at three observation wells. The slope and intercept can be expressed analytically from Equation 12 in terms of aquifer parameters which lead after some simple manipulations to

$$T = \frac{2.3Q}{2\pi\Delta s_r} \quad (15)$$

and

$$S = \frac{2.25 t_0 T}{r_0^2} \quad (16)$$

Comparison of Equations 13 and 15 indicates that for the same aquifer and pumping test the time-drawdown slope is equal to the half of the distance-drawdown slope. This is useful in determining the radius of influence from the time-drawdown measurements only. The maximum drawdown is plotted vs. the distance on a semilogarithmic paper and hence a point is obtained on the drawdown-distance straight line. On the other hand, by considering twice the slope of time-drawdown line, the slope of distance-drawdown line can be constructed. The intercept of this line gives an estimate of the radius of influence.

Example 3—The aquifer test data in Table 2, plotted on a semilogarithmic paper as drawdown vs. logarithms of distances for late but different times, are shown in Figure 20. The slope and intercept values together with parameter estimations are presented in Table 5.

Values of S and T indicate that the aquifer is next to moderately potential and confined. Transmissivity values compare quite well with each other and therefore aquifer is almost homogeneous and isotropic.

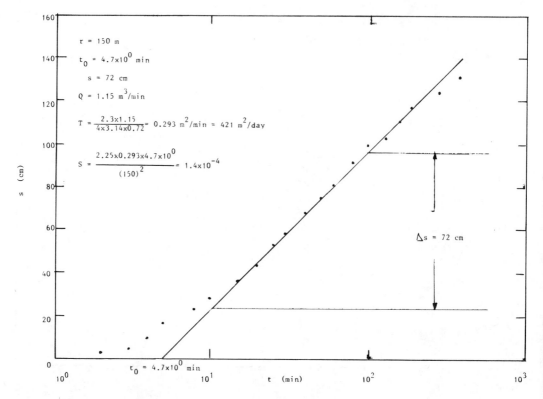

Figure 19 Time-drawdown straight line (r = 150 m)

3. Composite Variable Model

The spatial and temporal variability in the depression cone can be expressed by a composite variable which appears as t/r^2 in Equation 12. This variable is a mathematical combination of the previous time- and distance-drawdown models. Obviously, the change of drawdown with corresponding composite variable will appear as a straight line on a semilogarithmic paper.

Similar to time-drawdown model the intercept $(t/r^2)_0$ does not give any physical clue but it is a mere mathematical convenience. The analytical expressions for the slope and intercept from Equation 12 lead to aquifer parameter estimations as

$$T = \frac{2.3Q}{4\pi \Delta s_{r^2/t}} \tag{17}$$

and

$$S = 2.25T \left(\frac{t}{r^2}\right)_0 \tag{18}$$

The composite variable-drawdown model slope is exactly equal to the slope of time-drawdown model and the following chain of equations is valid for slopes of various methods:

$$\Delta s_t = \frac{\Delta s_r}{2} = \Delta s_{r^2/t} \tag{19}$$

Example 4—The application of this method is given by plotting the data in Table 2 on a

POROUS MEDIUM AQUIFER TESTS

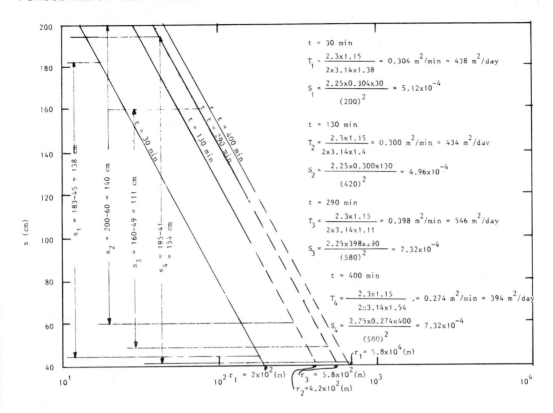

Figure 20 Distance-drawdown straight lines.

Table 5 Distance-Drawdown Model Calculations

	Straight-line readings		Parameter calculations	
Time (min)	Slope, Δs (m)	Intercept, r_0 (m)	Equation 15 T (m²/d)	Equation 16 ($\times 10^{-4}$)
30	1.38	200	438	5.1
130	1.40	420	434	5.0
290	1.11	580	546	7.3
400	1.54	680	394	7.3
		Averages	454	6.2

Note: Discharge $Q = 1.15$ m³/s.

semilogarithmic paper as shown in Figure 21. The slope and the intercept of the most suitable straight line are $\Delta s_{r/t}^2 = 0.80$ m and $(t/r^2) = 4.1 \times 10^{-4}$ min/m². Equations 17 and 18 yield the parameter estimations as $T = 379$ m²/day $S = 3.3 \times 10^{-4}$. The aquifer is moderately potential and of confined type.

Comparison of Jacob methods—An interesting question is whether the applications of different Jacob methods yield the same aquifer parameters for the same aquifer test. Theoretically, they should yield the same results if the aquifer and well configuration satisfy the assumptions listed in Chapter 8. However, in real field situations they give different values and the difference depends on the subsurface geological compositions. The following points are worth considering:

1. The time-drawdown model is the most economical version because it requires a single observation well. The parameter estimations depend on the drawdown changes at the observa-

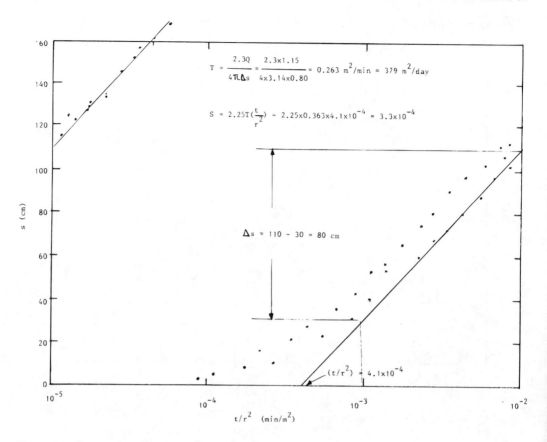

Figure 21 Composite variable method straight line.

tion well and are regarded as locally best estimations. This is the most widely used version of the Jacob method.

2. Both the distance-drawdown and composite variable methods require the same number of observation well records. The most reliable way of estimating the radius of influence is the distant-drawdown method. In the absence of an aquifer test, it is necessary to make rough estimates of r_0 from topography and areal geology. However, if the aquifer test data are available without recharge, then r_0 is a function of the aquifer parameters and the duration of the test only. This last statement is obvious from Equation 16, which leads to

$$r_0 = 1.5\left(\frac{T}{S} t_0\right)^{1/2} \qquad (20)$$

The difficulty in the application of this formula is the duration that depends on cost considerations as well as the aquifer type. However, this distance, which is sometimes called as "fictional radius of influence", is purely theoretical, especially as the Jacob straight line is grossly inaccurate for early time values. On the other hand, the distance-drawdown model provides reliably regional aquifer parameters within the radius of influence.

3. The composite variable model provides temporally and spatially averaged aquifer estimates as an overall average response of the aquifer. According to author's experience this version of the Jacob method is suitable for regional groundwater storage and flow evaluations.

4. Dimensionless Straight Line Method

In practice, the Jacob straight line method is taken for granted by groundwater hydrologist, hydrogeologists, and engineers and applied almost universally. Not every straight line on semilogarithmic paper, however, warrants the application of the Jacob method. The Jacob straight line method has physical restrictions in addition to the mathematical requirements as mentioned earlier in this chapter, which are often overlooked in practice. Therefore, the determination of storage coefficient and transmissivity from the semilogarithmic plot of data yields either over- or underestimations. The condition $u < 0.01$ does not guarantee the validity of all Theis assumptions which must also be valid simultaneously. The slope of the Jacob straight line on semilogarithmic paper is equal to 2.3 (see Figure 16). These assumptions imply physically that the medium is porous (not fractured or karstic), homogeneous, and isotropic and that the flow lines are parallel and the flow itself is Darcian, i.e., laminar. If any one of these assumptions is violated, is it still possible to obtain a straight line on the semilogarithmic plot and to apply the Jacob method? For example, if the transmissivity is high, the hydraulic gradient may become very steep near the well, and consequently the flow of groundwater may become turbulent. Is the Jacob method still applicable in this situation?

Because of turbulence, drawdown is always greater than the Darcy (laminar) flow case. Furthermore, the mathematical requirement of r being small for the Jacob method application might be contradictory with the basic assumptions. The smaller the radial distance, the steeper the hydraulic gradient in the well vicinity. As the well is approached the hydraulic gradient becomes steeper, and the flow per unit distance becomes greater, violating the Jacob method assumptions.

On the other hand, for leaky aquifers the final portion of the type curves is expected to be horizontal. If the slope of the straight line is less than the Jacob line, then either leakage or gravity drainage in unconfined aquifers or recharge or a combination of these may occur and Equation 12 yields higher transmissivity but lower storage coefficient values. If the slope is greater than the Jacob line, either a nonlinear flow takes place and/or there is a local decrease in permeability or limited water supply in storage. The steeper slope implies lower transmissivity but higher storage coefficient than usual (see Figures 1 and 2). The following procedure is proposed for the verification of the Jacob method.

1. Find the aquifer parameters by using the conventional Jacob method.
2. Calculate the dimensionless field drawdown, w_f, and time, u_f, values by using Equations 9 and 10 as

$$u_f = \frac{r^2 S}{4 t_f T} \quad (21)$$

and

$$w_f = \frac{4\pi T}{Q} s_f \quad (22)$$

where t_f and s_f are the time and corresponding drawdown measurements in the field.

Example 5—The dimensionless values for the original field data in Table 2 are calculated with transmissivity and storativity estimations given in Table 4; the results are presented in Table 6.

The plots of dimensionless values are given in Figures 22 through 24 together with the Jacob straight line. The Jacob straight line method is applicable only when the dimensionless line has a slope equal to 2.3. Late-time dimensionless field data on semilogarithmic paper can provide guidance in three directions.

1. If the plot of dimensionless field data has a slope of 2.3, the Jacob method is applicable reliably.

Table 6 Dimensionless Field Data

Time (min)	r = 25 m T = 788 m²/d S = 1.4 × 10⁻³		r = 60 m T = 788 m²/d S = 4.44 × 10⁻³		r = 150 m T = 421 m²/d S = 1.4 × 10⁻³	
	u_f	w_f	u_f	w_f	u_f	w_f
1	3.99×10^{-1}	240.3	7.30×10^{-1}	59.8	2.69×10^{-0}	0.76
2	1.99	361.0	3.65	138.7	1.35	7.78
3	1.33	441.1	2.43	192.5	8.99×10^{-1}	17.48
4	9.994×10^{-2}	496.1	1.83	243.8	6.74	31.01
5	7.99	579.9	1.46	316.2	5.39	52.33
8	4.99	638.3	9.13×10^{-2}	358.6	3.37	72.11
10	3.99	677.8	7.30	409.4	2.69	89.34
15	2.66	759.1	4.87	478.2	1.79	114.86
20	1.99	800.9	3.65	523.0	1.35	138.12
25	1.59	860.7	2.92	573.8	1.08	169.11
30	1.33	908.5	2.43	613.3	8.99×10^{-2}	184.80
40	9.99×10^{-3}	965.9	1.83	686.2	6.74	217.60
50	7.99	1013.7	1.46	723.2	5.39	239.30
60	6.66	1050.8	1.22	768.7	4.49	256.85
80	4.99	1108.2	9.13×10^{-3}	792.6	3.37	290.36
100	3.99	1153.6	7.30	867.5	2.69	315.25
130	3.08	1195.4	5.62	932.4	2.07	330.56
160	2.49	1245.6	4.57	956.3	1.68	356.10
190	2.10	1344.8	3.85	992.2	1.42	377.80
290	1.37	1314.9	2.52	1024.5	9.30×10^{-3}	395.60
400	9.99×10^{-3}	1326.9	1.83	1049.6	6.74	416.40

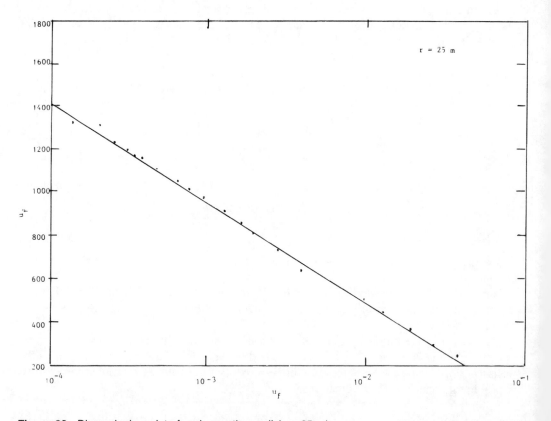

Figure 22 Dimensionless data for observation well ($r = 25$ m).

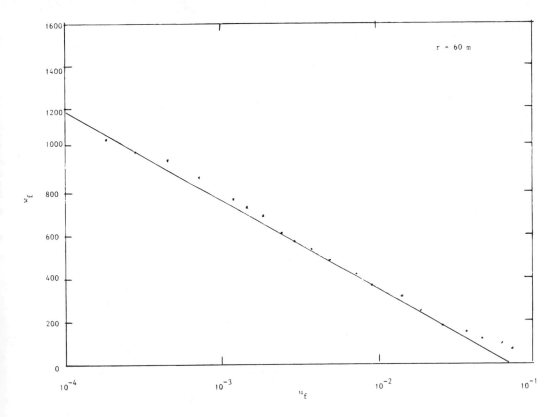

Figure 23 Dimensionless data for observation well ($r = 60$ m).

2. If the dimensionless plot of field data slope is greater that 2.3, either the non-Darcian flow exists or the medium is coarse grained or fractured or may have less-permeable boundaries.
3. If the slope of the dimensionless plot of the data is less than 2.3, there is a recharge or leakage to the aquifer.

C. CHOW NOMOGRAM MODEL

This model is proposed by Chow (1952) and avoids two practically undesirable points: the matching of type curve and deletion of early data values in Jacob methods. The basic assumptions are the same as for the Theis method. Chow introduced a third dimensionless function, $F(u)$, in terms of $W(u)$ and u as

$$F(u) = \frac{W(u)e^u}{2.3} \qquad (23)$$

$F(u)$ is referred to as the Chow's function in this book. He proposed the graphical representation of this relationship on a double-logarithmic paper by making use of the Theis relationship in Equation 8. Chow presented a nomogram (Figure 25) showing the relationship between $W(u)$, $F(u)$ and u.

This nomogram helps to know the values of any two variables, provided that the third one is known. Only $F(u)$ is dependent directly on the field data and obtained from the plot of all the data on a semilogarithmic paper as time-drawdown graph. The points are connected by a smooth curve on which an arbitrary point is chosen. The slope, Δs_c, of the tangent to the curve at this point and the corresponding drawdown, s_c, are read from the graph. Then the value of the Chow's function for this point is calculated as

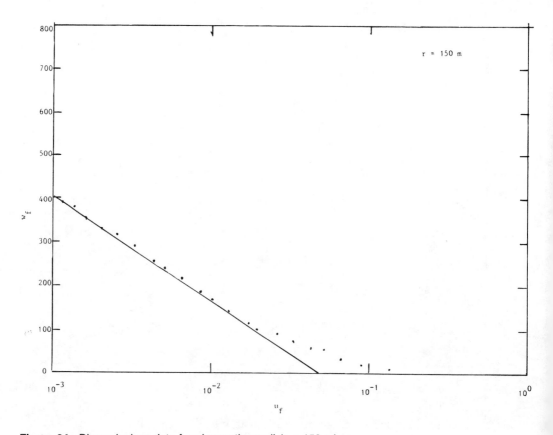

Figure 24 Dimensionless data for observation well ($r = 150$ m).

$$F(u) = \frac{s_c}{\Delta s_c} \qquad (24)$$

If the nomogram in Figure 25 is entered with $F(u)$ one can determine u and $W(u)$ values. The aquifer parameters are then calculated similar to the Theis model from Equations 9 and 10.

Example 6—The numerical application is worked out with data in Table 2 for $r = 150$ m. First the data are plotted on semilogarithmic paper as in Figure 26. A smooth curve is drawn through the field data. Two arbitrary points, A1 and A2, are chosen on this curve and the corresponding tangents are drawn. For each point two readings are taken from the figure as the drawdown and the slope of tangent. A summary of the results is given in Table 7.

It is obvious from this table that the points chosen at early times yield overestimates.

D. PAPADOPULOS AND COOPER MODEL

One of the frequently encountered problems in practice is large well diameter contrary to the above-mentioned models. The significance of large-diameter wells has been discussed in Chapter 6. Papadopulos and Cooper (1967) have provided analytical solutions for such situations. All the assumptions are the same as were set for the Theis method except that the well diameter is large and not infinitesimally small (Figure 27). Well storage has a distinctive effect on drawdown readings at initial time instances during an aquifer test. The water in the well has a dampening effect on the aquifer contribution to the pumping discharge. Furthermore, well losses are comparatively smaller than the small well diameters. The same groundwater

POROUS MEDIUM AQUIFER TESTS

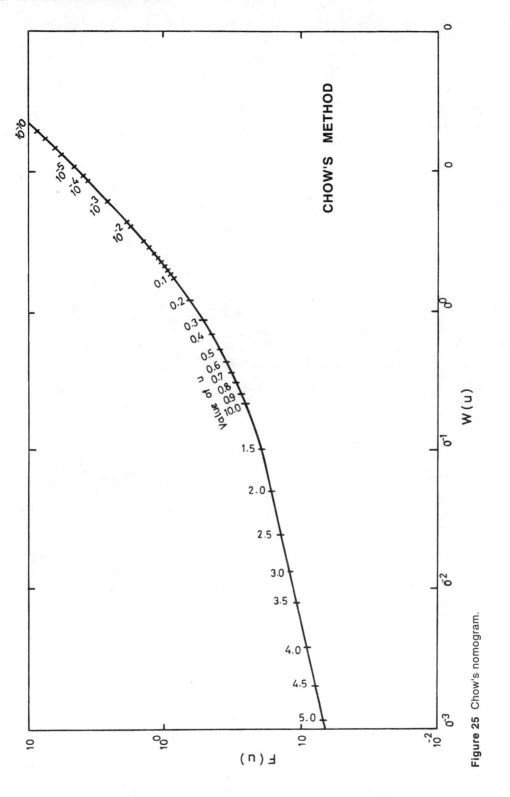

Figure 25 Chow's nomogram.

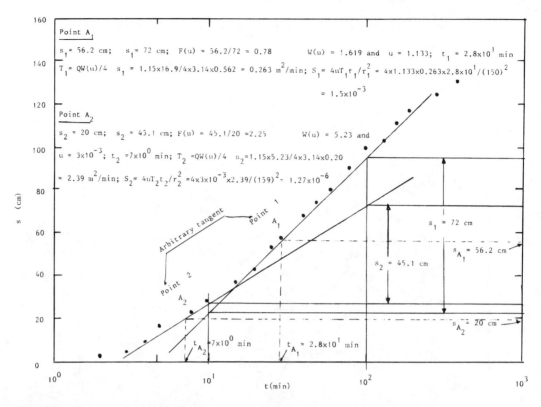

Figure 26 Chow method application.

Table 7 Chow Method Calculations

	Arbitrary readings from figure F(u)						Parameter values	
Point	Drawdown (m)	Time (min)	Slope (m)	Equation (23)	Chow u	Monog. $W(u)$	Equation 15 T (m²/d)	Equation 16 S (×10⁻³)
A1	0.56	28	0.72	0.78	2.1	1.62	379	1.5
A2	0.20	7 × 10⁻⁶	0.45	2.25	3 × 10⁻³	5.23	3442	1.3

Note: Discharge $Q = 1.15$ m³/min.

equation (Equation 6) and boundary conditions are valid except that the pumping discharge has the condition given in Equation 2 in Chapter 8. The solutions for the large-diameter well itself are given in terms of three dimensionless factors, two of which are similar to the Theis approach as

$$W(u_w, \beta) = \frac{4\pi T}{Q} s_w \quad (25)$$

$$u_w = \frac{r_w^2 S}{4tT} \quad (26)$$

and

$$\beta = \left(\frac{r_w^2}{r_c}\right)^2 \quad (27)$$

in which $W(u_w, \beta)$ and u_w are large-diameter well function and corresponding dimensionless

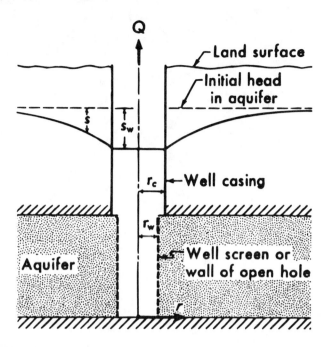

Figure 27 Large-diameter wells.

time factor, whereas β is a dimensionless constant and subscripts w and c stand for the pumped well and the casing radius. In shallow wells, there is no casing; therefore, $r_w = r_c$ and hence $\beta = S$. The necessary type curves are given on double-logarithmic paper in Figure 28 for a set of β values. The following points are the common characteristics of the type curves:

1. For small u_w values the type curve portions are straight lines with 45° of slope. This is only possible when all of the pumping discharge comes from the well storage. For such a case,

$$Q = \pi r_w^2 \frac{ds_w(t)}{dt} \qquad (28)$$

the integration of which leads to

$$Qt = \pi r_w^2 s_w(t) + C \qquad (29)$$

Since there is no aquifer contribution to the pumping discharge the straight line portion does not give any useful information about the aquifer parameters and hence it is superfluous for parameter determination. To determine aquifer parameters from a large-diameter well time-drawdown data, Şen (1983) suggested test durations as shown in Table 8.
2. The final portion of each type curves merges with the Theis curve, indicating that there remains no well storage effect. This justifies the validity of the Jacob method even for the large-diameter wells.
3. The curvature of middle portions are very close to each other and therefore determination of S value by this model has questionable reliability. In practice, observed points most often equally fit several of the type curves, corresponding to different values of β (or S). However, the values of T calculated from these curves differ slightly, but never to too great extent. The best method thus consists of first selecting a type curve where the label corresponds to the most probable storage coefficient values based on the geological and hydrogeological conditions and then calculating T.

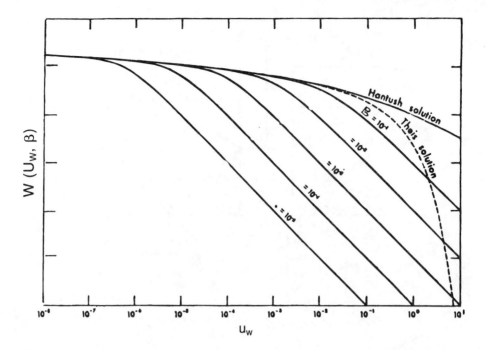

Figure 28 Large-diameter well-type curves.

Table 8 Straight Line Portion

S	$\dfrac{t}{x(r_w^2/T)}$
10^{-1}	0.25×10^3
10^{-2}	0.25×10^4
10^{-3}	0.25×10^5
10^{-4}	0.25×10^6
10^{-5}	0.25×10^7

4. Because the drawdowns measured in the pumping well are used, they include well losses. It is necessary to install piezometers near the pumping well and to check whether the well losses are negligible. On the other hand, Papadopulos (1967) presented type curves for observation wells located in the vicinity of a large-diameter well. These type curves are presented in Figure 29. Any specific type curve depends on the distance ratio, r/r_w^2. As the ratio becomes closer to 1 the type curves approach to the Papadopulos and Cooper curves, otherwise, for large r/r_w^2 values they converge to the Theis-type curve.

Example 7—The aquifer test data from Quaternary deposits in wadi Uoranah within the eastern provinces of the Kingdom of Saudi Arabia are given in Table 9.

The measurements are performed in a large-diameter well of diameter 3.0 m and the well fully penetrates the aquifer. However, the fine- and coarse-grained aquifer is overlain by a thin layer of silt which makes the aquifer a confined type. The pumping discharge is 0.48 m^3/min.

The time drawdown data are plotted on a double-logarithmic field data paper and the conventional-type curve matching procedure is carried out similar to the Theis method. Figure 30 gives the most suitable matching and the results are summarized in Table 10. The storativity can also be calculated from Equation 13 in Chapter 8 with the data at hand which yields $S = 1.7 \times 10^{-4}$.

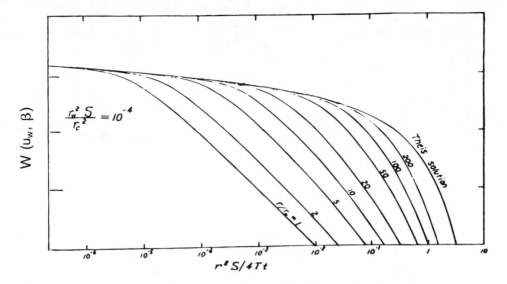

Figure 29 Observation well-type curves ($S = 10^{-4}$).

Table 9 Large-Diameter Well Field Data

Time (min)	Drawdown (cm)	Time (min)	Drawdown (cm)
0	0	30	1.27
1	0.06	35	1.34
2	0.16	40	1,41
3	0.24	45	1.48
4	0.30	50	1.54
5	0.39	55	1.60
6	0.47	60	1.66
7	0.54	65	1.68
8	0.61	75	1,71
9	0.66	80	1.73
10	0.71	90	1.75
12	0.81	100	1.76
14	0.89	110	1.765
16	0.96	120	1.775
18	1.02	135	1.790
20	1.07	150	1.805
22	1.12	165	1.815
24	1.16	180	1.820
26	1.20	210	1.821
28	1.23	217	1.8215

E. ADAPTIVE MODELS

Previously explained conventional models for parameter estimation from aquifer test analysis do not account for any possible uncertainty or provide any measure of reliability in parameter estimates. The field data do not necessarily satisfy the theoretically derived flow equations due to the deviations from basic assumptions that give rise to uncertainties in the system and noise model of the measurements. Furthermore, conventional aquifer parameter estimations are dependent on the analyst's personal judgment and experience. To avoid these subjectivities and assess existing uncertainties in the aquifer system as well as the errors in measurements, an adaptive parameter estimation procedure has been developed. The basis of the adaptive models is the Kalman (1962) filter technique. Figure 31 shows schematically the sequences in the aquifer model.

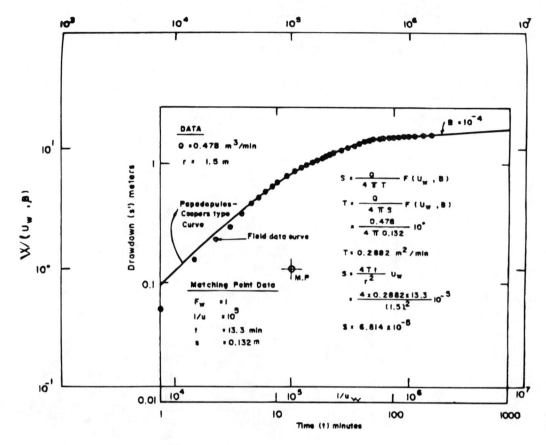

Figure 30 Large-diameter well-type curve matching.

Table 10 Parameter Estimation

	Match point coordinates				Parameter estimates	
	Field data sheet		Type curve sheet		Equation 25	Equation 26
	s_M (m)	t_M (min)	u_M	$W_M(u_w,\beta)$	T (m²/min)	S
MW	0.13	13.3	10^{-5}	1×10^0	415	6.8×10^{-5}

Note: Discharge $Q = 0.48$ m³/min; well diameter $r_w^2 = 3$ m.

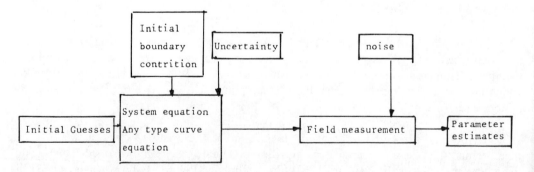

Figure 31 Adaptive model block diagram.

Detailed explanation of this model is given by Chander et al. (1981a) and Şen (1984). The parameter identification by this model can be achieved once the proper system and measurement equations are given. In general, the system equation depends on the type of aquifer. In fact, it is a version of the type curve equations. For instance, in the case of confined aquifers the system equation can be written from Equations 9 and 10 as

$$\log(s) = -\log T + \log W(u) + \log\left(\frac{Q}{4}\right) + u_1 \quad (30)$$

and

$$\log(t) = -\log T - \log(u) + \log S + \log\left(\frac{r^2}{4}\right) + u_2 \quad (31)$$

where $\log s$ and $\log t$ are the measurement variables; $\log u$ and $\log W(u)$ theoretical part of aquifer system variables; $\log S$ and $\log T$ are the aquifer system parameters; Q and r are system constants; and u_1 and u_2 are additional terms that account for the uncertainty.

In an aquifer test analysis the measurement equation is completely independent of the aquifer type as is the field test procedures for measuring the drawdown values at the present time. The measurement equation has invariably a linear form as

$$s_k = z_k + n_k \quad (32)$$

where s_k is the field measurement value of drawdown, z_k is the error intact drawdown, and n_k is an error term which is assumed to have a Gaussian distribution. The following adaptive formulations have been proposed by Şen (1984) for the confined aquifer parameter identification:

$$(\log T)_{k/k} = (\log T)_{k-1/k-1} + (k_{11})_k \log \frac{4\pi(T)_{k-1/k-1}}{QW(u_{k-1})} s_k + (k_{12})_k \log \frac{4u_{k-1}(T)_{k-1/k-1}}{1440 r^2 (S)_{k-1/k-1}} t_k \quad (33)$$

and

$$(\log S)_{k/k} = (\log S)_{k-1/k-1} + (k_{21})_k \log \frac{4\pi(T)_{k-1/k-1}}{QW(u_{k-1})} s_k + (k_{22})_k \log \frac{4u_{k-1}(T)_{k-1/k-1}}{1440 r^2 (S)_{k-1/k-1}} t_k \quad (34)$$

where k_{11}, k_{12}, k_{21}, and k_{22} are the Kalman gain elements defined as

$$K_k = P_{k/k-1} H_k^T [H_k P_{k/k-1} H_k^T + R_k]^{-1} \quad (35)$$

Example 8—Confined aquifer tests from "Qude Korendijk" at an observation well 30 m from the main well have been analyzed by Kruseman and de Ridder (1979) using conventional methods. The same data have been treated by the Kalman filter procedure by Şen (1984). For successive application of Equations 33 to 35, in a recursive manner on a programmable hand calculator or personal computer, initial values $T_{0/0}$, $S_{0/0}$, $P_{0/0}$, and R_0 are needed. Initial estimates of $T_{0/0}$ and $S_{0/0}$ can be found by standard methods using the first and second data points. In the case of vague initial state-variable information, the Kalman procedure is initiated with large diagonal elements in the $P_{0/0}$ matrix. Using the Jacob (1944) approximation for the well function, T, can be found, approximately from the first two values as

$$T_{0/0} = \frac{QLn(t_2/t_1)}{4\pi(s_2 - s_1)} \tag{36}$$

and approximation for storage coefficient can be evaluated by finding the value of well function, $W(u_w) = 4\pi T_{0/0} s_1/Q$, for the first drawdown measurement. Correspondingly, u_w is found either from available tables or from Equation 8 by dropping all the terms after $u^4/(4.4!)$. Subsequently the initial estimate of S is obtained as

$$S_{0/0} = \frac{4t_1 T_{0/0}}{r^2} \tag{37}$$

The approximation used in Equation 36 is valid only for $u_w < 0.01$, which is seldomly satisfied in the initial stage of an aquifer test. Therefore, the initial estimates found in this way are rather small T values, and convergence of the Kalman filter will require long time periods. The initial T and S estimates are obtained from Equations 36 and 37, leading to approximate values of 25 m²/day and 1.5×10^{-5}, respectively. Another way of determining the initial parameter estimates is to select them from the geological considerations of the pumping site and the type of aquifer. Although 25 m²/day is rather low, it is employed herein to show that even in such unfeasible cases the Kalman algorithm converges to final parameter estimation although convergence is slow. In addition, two other values as 200 and 1100 m²/day are adopted arbitrarily to show the applicability of the Kalman procedure. On the other hand, $(p_{11})_{0/0}$ and $(p_{12})_{0/0}$ are the variances of the transmissivity and storativity and, because there is little or no initial information about their variability, rather large values are needed. As the adoptive filter digests new information, they decrease continuously, so that the larger they are the longer is the convergence period. Herein, $(p_{11})_{0/0} = (p_{12})_{0/0} = 10^5$ which implies that all errors have equal variances. The value $(p_{12})_{0/0}$ indicates the covariance between the initial values of the transmissivity and storativity. Since it is not known a priori, it has been taken as a large value equal to 10^3. By similar reasoning, the diagonal elements of the system noise covariance matrix are set at 10^3, and off-diagonal elements are taken as zero, which implies that the system variable measurements are uncorrelated, i.e., it is a diagonal matrix. However, measurement noise is also taken as zero which means that the drawdown measurements are very accurate.

Sequential execution of Equations 33 through 35 with these initial values leads to parameter estimations that are updated by each new drawdown measurement. The final parameter estimates are given in Table 11 along with the results obtained from other methods (Kruseman and de Ridder, 1979). The three Kalman estimates, with different initial values of $T_{0/0}$, are in good agreement with other solutions. If the Theis equation is assumed as the proper choice for the data, the three Kalman estimates yield relative errors of 22, 4.5, and 0.5%, respectively.

Error variance for storativity and transmissivity is shown in Figures 32 and 33, respectively. To get an idea about the convergence of procedure, successive estimates of T and S are presented in Figures 34 through 36.

Table 11 Aquifer Parameter Estimates by Different Methods

Method	Transmissivity (m²/d)	Storativity ($\times 10^{-4}$)
Theim	343	—
Theis	418	1.7
Chow	375	2.2
Jacob	401	1.7
Kalman ($T_{0/0}$ = 25 m²/d)	342	2
Kalman ($T_{0/0}$ = 200 m²/d)	400	1.6
Kalman ($T_{0/0}$ = 1100 m²/d)	420	1.6

POROUS MEDIUM AQUIFER TESTS

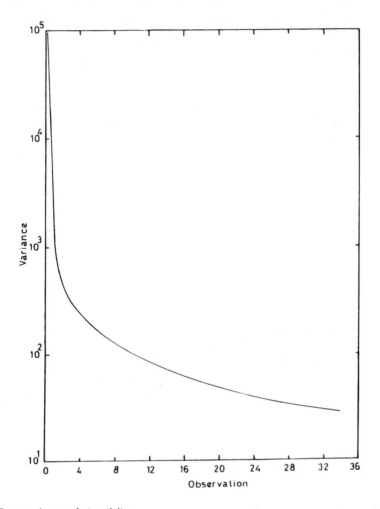

Figure 32 Error variance of storativity.

The initial value of 25 m²/day gives successive increasing T and S estimates, without consideration of asymptotic values, with the available number of observations (see Figure 36). However, this figure gives the impression that a few more observations will lead to better estimates, hence reduction in the error percentage. On the other hand, initial values of 200 and 1100 m²/day give asymptotic values of S and T even before all the available data are processed by the adaptive method and almost half the observations suffice to yield asymptotic value. On the other hand, convergence is more rapid with initial $T_{0/0} = 1100$ m²/day value and four or five observations are enough to provide the desired asymptotic values.

F. SLOPE MATCHING METHOD

Sen (1986a) devised a method that is applicable to any type curve equation and avoids the curve-matching technique in aquifer parameter determination from given field data. The elegance of the method is the aquifer parameter estimations right after the second time-drawdown measurement and subsequently, as the new record is measured, new estimates of the concerned parameters are obtained. The basic concept of the model is very simple and it derives from the fact that matching of two curves is equivalent to saying that they have the same slopes at corresponding points. The aquifer parameters are dependent on the geological

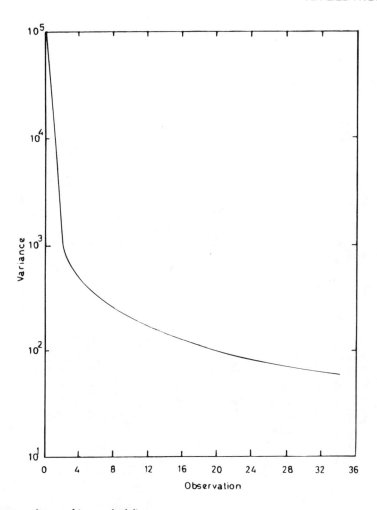

Figure 33 Error variance of transmissivity.

characteristics that are controlled by the evolution of rocks (sedimentation, volcanic eruptions, etc.) and by subsequent secondary geological events such as folding, faulting, fracturing, and fissuring in addition to the chemical changes, especially in limestones. Although the conventional aquifer tests tend to average these conditions in the aquifer response to pumping, the field test data will still have some local deviations from any analytically derived type curves; therefore, the slopes of type and field curves are not equal over the entire time domain. In the conventional-type curve-matching procedures the aquifer is assumed to represent an equivalent homogeneous aquifer and consequently the parameters are expected to be temporally and spatially invariable constants. However, due to the aforementioned deviations, one should physically expect some variation in these parameters. The main objective of slope model is to identify the likely variations in aquifer parameters by matching slopes of the type and field curves. Şen (1986a) gave the analytical slope, α, an expression for the Theis curve on a double-logarithmic paper as

$$\alpha = -\frac{e^{-u}}{W(u)} \tag{38}$$

This expression helps to convert the Theis-type curve table into Theis slope values as presented

POROUS MEDIUM AQUIFER TESTS

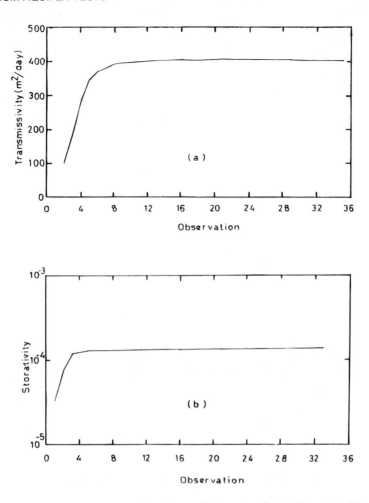

Figure 34 Recursive estimations with initial estimate of $T_{0/0} = 25$ m²/d. (a) Transmissivity; (b) storativity.

in Table 12. Considering this table, the processing of the aquifer test data can be achieved through the following procedure without type curve matching.

1. After the second time-drawdown measurement, calculate the slope between the two successive points in the double logarithmic scale as $\alpha_i = Ln(s_i/s_{i-1})/Ln(t_{i-1}/t_i)$ where $i = 2,3,\ldots, n$ and n is the number of drawdown records.

Table 12 Slope Method Calculations for a Confined Aquifer

	1.0	2.0	3.0	4.0	5.0	6.0	7.0	8.0	9.0
$\times 10^0$	−1.6798	−2.7619	−3.8298	−4.8199	−6.1254	−6.8854	−7.5990	−8.8279	−10.2841
$\times 10^{-1}$	−0.4971	0.6711	0.8141	0.9576	−1.0831	−1.2196	−1.3421	−1.4494	−1.5637
$\times 10^{-2}$	−0.2451	−0.2926	−0.3278	−0.3585	−0.3851	−0.4095	−0.4337	−0.4547	−0.4760
$\times 10^{-3}$	−0.1578	−0.1769	−0.1906	−0.2014	−0.2104	−0.2189	−0.2262	−0.2329	−0.2394
$\times 10^{-4}$	−0.1159	−0.1259	−0.1322	−0.1379	−0.1424	−0.1461	−0.1494	−0.1525	−0.1551
$\times 10^{-5}$	−0.0914	−0.0976	−0.1016	−0.1047	−0.1072	−0.1094	−0.1112	−0.1128	−0.1144
$\times 10^{-6}$	−0.0755	−0.0797	−0.0824	−0.0844	−0.0859	−0.0873	−0.0886	−0.0896	−0.0906
$\times 10^{-7}$	−0.0643	−0.0673	−0.0692	−0.0707	−0.0728	−0.0727	−0.0735	−0.0743	−0.0745
$\times 10^{-8}$	−0.0560	−0.0583	−0.0597	−0.0607	−0.0616	−0.0623	−0.0629	−0.0634	−0.0640
$\times 10^{-9}$	−0.0496	−0.0514	−0.0524	−0.0533	−0.0539	−0.0545	−0.0549	−0.0553	−0.0557
$\times 10^{-10}$	−0.0445	−0.0459	−0.0468	−0.0475	−0.0480	−0.0484	−0.0488	−0.0491	−0.0494

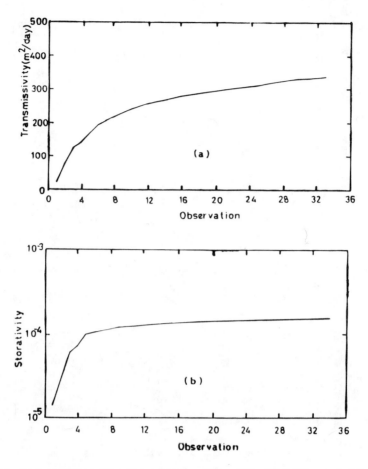

Figure 35 Recursive estimations with initial estimates of $T_{0/0} = 200$ m²/d. (a) Transmissivity; (b) storativity.

2. Find the u_i value corresponding to this slope from Table 12 (if needed, interpolate).
3. Knowing the slope and the u value, find the well function value from Equation 38 as

$$W_i(u) = \frac{e^{-u_i}}{\alpha_i} \tag{39}$$

4. Calculate local T and S values from Equations 9 and 10, respectively.
5. Repeat the previous steps with the next time-drawdown record. Finally, sequences of estimation are obtained for each aquifer parameters.

Example 9—Two sets of aquifer test data were selected for confined aquifer parameter estimation by the slope method. The first data set was analyzed by Shultz (1973) using the Theis method. The results of the parameter estimation calculations for Shultz's data are presented in the first two columns of Table 13. Shultz found that $T = 1240$ m²/day and $S = 1.9 \times 10^{-4}$ by type curve matching procedure. Transmissivity and storativity values calculated by the slope matching method are shown in the last two columns of the same table. The average values of the parameter estimation sequence in Table 13 using the slope method give 1233 m²/day and $S = 1.98 \times 10^{-4}$ with the relative differences 5 and 4%, respectively. It is interesting to compare any value in the parameter estimation sequence with the Theis method results. Each parameter pair (S and T) compares reasonably well with the Theis result. This

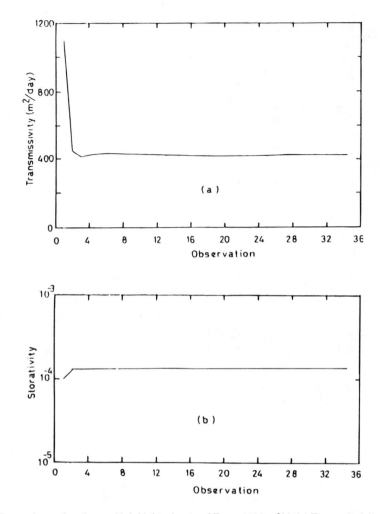

Figure 36 Recursive estimations with initial estimate of $T_{0/0} = 1100$ m²/d. (a) Transmissivity; (b) storativity.

is especially true for S values because Q, s, and T are not related to natural S, but rather to $\log(S)$, so that any error in Q, s, and T results in a relatively small error in S. Even the first estimates of S and T are rather close to the final averages.

Another advantage of the slope method in the case of confined aquifers is that, if the aquifer test is stopped due to any undesirable reason after a short time, one can still obtain reliable estimates of the parameters. For instance, if the aquifer test data in Table 13 were not available after 5.57×10^{-3} days, the average estimates of parameters resulting from this incomplete test would turn out to be $T = 1290$ m²/day and $S = 2.06 \times 10^{-4}$, the relative differences with respect to the Theis solution being 4 and 8%. These differences are within practically acceptable limits. This example shows the effectiveness of the slope method in determining the aquifer parameters from a short-term aquifer test. Such situations are encountered very frequently, especially in arid regions of the world, where limited groundwater storage does not allow aquifer tests for long durations.

On the other hand, the slope method exposes the inherent erratic variations in S and T and it provides a basis for the statistical analysis of these parameters in terms of frequency functions and confidence limits. Assuming that a Gaussian distribution is valid for the parameter estimates, the confidence limits can be found by calculating the mean and the standard deviation

Table 13 Slope-Matching Values

Time (day)	Drawdown (m)	Slope	u	$W(u)$	Transmissivity (m^2/d)	Storativity
6.96×10^{-4}	0.201	−0.691	1.6×10^{-1}	1.23×10^0	1262	1.558×10^{-4}
1.02×10^{-3}	0.266	−0.441	7.2×10^{-2}	2.11×10^0	1636	1.363×10^{-4}
1.39×10^{-3}	0.302	−0.517	1.2×10^{-1}	1.72×10^0	1174	2.174×10^{-4}
1.74×10^{-3}	0.339	−0.465	8.0×10^{-2}	1.96×10^0	1192	1.840×10^{-4}
2.09×10^{-3}	0.369	−0.408	5.7×10^{-2}	2.31×10^0	1291	4.529×10^{-4}
2.78×10^{-3}	0.415	−0.412	6.1×10^{-2}	2.28×10^0	1133	2.133×10^{-4}
3.48×10^{-3}	0.455	−0.350	3.9×10^{-2}	2.75×10^0	1247	1.876×10^{-4}
4.17×10^{-3}	0.485	−0.334	3.2×10^{-2}	3.00×10^0	1276	1.890×10^{-4}
5.57×10^{-3}	0.534	−0.269	1.4×10^{-2}	3.66×10^0	1414	1.222×10^{-4}
6.96×10^{-3}	0.567	−0.319	2.9×10^{-2}	3.05×10^0	1109	2.481×10^{-4}
8.33×10^{-3}	0.601	−0.357	4.0×10^{-2}	2.69×10^0	923	3.428×10^{-4}
9.72×10^{-3}	0.635	−0.225	7.0×10^{-3}	4.41×10^0	1432	1.083×10^{-4}
1.25×10^{-2}	0.672	−0.240	9.0×10^{-3}	4.13×10^0	1268	1.585×10^{-4}
1.67×10^{-2}	0.720	−0.242	9.2×10^{-3}	4.09×10^0	1172	1.997×10^{-4}
2.09×10^{-2}	0.760	−0.221	6.1×10^{-3}	4.50×10^0	1221	1.724×10^{-4}
2.78×10^{-2}	0.810	−0.216	5.8×10^{-3}	4.60×10^0	1171	2.096×10^{-4}
3.48×10^{-2}	0.850	−0.190	3.0×10^{-3}	5.25×10^0	1274	1.474×10^{-4}
4.17×10^{-2}	0.880	−0.181	2.2×10^{-3}	5.51×10^0	1291	1.414×10^{-4}
5.57×10^{-2}	0.927	−0.180	2.2×10^{-3}	5.54×10^0	1233	1.674×10^{-4}
6.96×10^{-2}	0.965	−0.195	3.4×10^{-3}	5.11×10^0	1092	2.949×10^{-4}
8.33×10^{-2}	1.000	−0.176	2.0×10^{-3}	5.67×10^0	1170	2.166×10^{-4}
1.02×10^{-1}	1.040	−0.156	9.2×10^{-4}	6.40×10^0	1104	1.175×10^{-4}
1.25×10^{-1}	1.070	−0.179	2.1×10^{-3}	5.57×10^0	1073	3.129×10^{-4}
1.46×10^{-1}	1.100	−0.135	3.5×10^{-4}	7.41×10^0	1455	0.800×10^{-4}

of the parameter estimation sequences. The average parameter estimates are shown in Table 14 along with the corresponding upper and lower limits for the 95% confidence interval. It is obvious from this table that the upper and lower confidence limits include most of the parameter estimates obtained by the slope method. Adaptive and slope matching methods help to explore any heterogeneity involved in the aquifer material around the well vicinity.

VI. UNCONFINED AQUIFER MODELS

Groundwater movement in unconfined aquifers are geologically, hydrologically, geometrically, and physically different from the confined aquifer cases. Unconfined aquifers present more complicated mathematical treatments as a result of these differences. Therefore, the literature concerning unsteady-state flow in unconfined aquifers relates to relatively recent times and rather few studies have been carried out.

As explained in Chapter 8, the nonexistence of an upper confining layer makes the groundwater flow in unconfined aquifers three dimensional especially near the water table. For the same reason the saturation thickness does not remain constant in the vicinity of the pumping well. The change in saturation thickness by definition is reflected in the aquifer parameters, i.e., the transmissivity and the storage coefficient estimations. Therefore, in the unconfined aquifers, the determination of the hydraulic conductivity and the specific yield

Table 14 Confidence Limits of Parameters from Slope Matching

Parameter	Lower confidence limit	Average value	Upper confidence limit	Standard deviation
Shultz data				
S	1.648×10^{-4}	1.985×10^{-4}	2.231×10^{-4}	8.407×10^{-4}
T (m^2/d)	1173	1235	1297	154
Oude Korendijk				
S	1.038×10^{-4}	1.800×10^{-4}	2.573×10^{-4}	2.300×10^{-4}
T (m^2/d)	438	489	540	153

values become significant instead of composite parameters. Furthermore, in addition to water release from the aquifer due to compaction corresponding to effective stress increase and water compaction as a result of piezometric pressure reduction, there exists a third physical cause for water movement that is the gravity drainage within the dewatered zone (depression cone). It depends on the depression cone creation and therefore is not appreciable at the beginning of the pumping but rather delayed, depending on the expansion of the depression cone. In a way this gravity drainage plays the role of groundwater recharge to the water table after some time since the pump start.

A significant physical point that should receive careful consideration in unconfined aquifer analysis is that, with the gravity drainage effect, as a noninstantaneous response, the storage coefficient reduces with time and ultimately becomes equal to the specific yield.

Nwankwor et al. (1985) suggest that the specific yield increases with time during an aquifer test, whereas Neuman (1987) proposes that the water balance is not applicable due to its implication that the specific yield grows with time as the pumping test progresses.

A. RESTRICTIVE MODELS

The basic idea of these models stems from the question of how one can apply confined aquifer models to the unconfined aquifers. For the application of restrictive models, an unconfined aquifer must have physical conditions at least similar to those of confined aquifers. The following points are worth noting:

1. In rather thick saturation thickness, m, the maximum drawdown, s_M, is relatively small. In practice, $(s_M/m) \times 100 < 10$ is a good criterion for the application of confined aquifer models to unconfined aquifers. Under these conditions, the three-dimensional flow can be approximated as radial flow and gravity drainage contribution is not significant.
2. Based on the theoretical studies of Boulton (1963), the time-drawdown record must be taken at observation wells at distances of 0.2 to 0.6 times the saturation thickness.
3. Boulton (1963) gave the minimum time interval after the start of pumping for confined aquifer models to be applicable as

$$t \geq \frac{5S_y m}{K} \qquad (40)$$

   in which S_y and K are the specific yield and hydraulic conductivity, respectively. With this condition, Jacob methods become applicable for unconfined aquifers at large times.
4. For thin saturation thickness aquifers, Jacob suggested correction of drawdowns prior to application of confined aquifer models. Hence, the equivalent confined aquifer drawdown, s_c, is expressed as

$$s_c = s_u - \frac{s_u^2}{2m} \qquad (41)$$

   where s_u is the observed unconfined aquifer drawdown and m is the initial saturation thickness in unconfined aquifer. He concluded that, if $s^2/2m < 3 \times 10^{-3}$ m, then the correction in Equation 41 not necessary because the condition in (1) is satisfied only in this case.

None of the above-mentioned conditions guarantee the full applicability of confined aquifer models. Most often in practice the physical plausibility of the final parameter estimates is questionable.

B. BOULTON MODEL

Boulton (1963) gave a specific treatment of groundwater flow problem toward wells in unconfined aquifers with specific initial and boundary conditions. The assumptions listed for the Theis model are the same except that the aquifer is unconfined and shows delayed yield phenomenon due to gravity drainage from a dewatered zone. There are three distinct portions in time-drawdown plots of unconfined aquifers:

1. The initial portion covers only a short period after the start of pumping. The response of the aquifer is the same as for the confined aquifer. There is no significant delayed yield in this portion because the depression cone has not yet developed to the extent that the gravity drainage can take place toward the water table (see Figure 37a).
2. After this first portion the depression cone expands rather rapidly and accordingly effective gravity drainage takes place toward the saturation zone. There is a decrease in the rate of drawdown increase relative to the confined aquifer. Figure 37b indicates the contribution of vertical delayed yield from the dewatered zone to the saturated zone. The amount of gravity drainage decreases with increasing distance from the well. Of course, the drawdown is theoretically equal to zero beyond the radius of influence.
3. At later stages the increase in the depression cone volume rate is reduced due to approach to the steady or quasi-steady state. Because this increment is not significant, the gravity drainage and contributing area will be negligible. As a result, the water will come from the storage available in the aquifer.

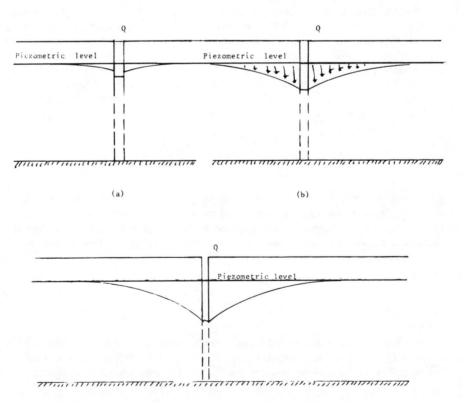

Figure 37 Gravity drainage stages.

The initial and final portions are expressible in terms of Theis-type curves; however, the intermediate portion needs special attention. Boulton (1963) produced a semiempirical mathematical solution that reproduces all three portions in an unconfined aquifer. A critical point in his derivation is that the delayed yield was not related clearly to any physical phenomenon. The Boulton well function for the initial portion is given as

$$W(u_E, r/\beta) = \frac{4\pi T_E}{Q} s_E \qquad (42)$$

where β is referred to as the drainage factor and defined as follows:

$$\beta^2 = \frac{T_E}{S_L} \qquad (43)$$

in which $1/\beta$ is called the Boulton delay index and it is an empirical constant. For this initial portion the dimensionless time factor is

$$u_E = \frac{r^2 S_E}{4 t_E T_E} \qquad (44)$$

S_E is early time storage coefficient when the unconfined aquifer reacts as a confined aquifer. The late portion of the time-drawdown plot is modeled through

$$W(u_L, r/\beta) = \frac{4\pi T_L}{Q} s_L \qquad (45)$$

with dimensionless time factor as

$$u_L = \frac{r^2 S_L}{4 t_L T_L} \qquad (46)$$

in which S_L is the specific yield of the unconfined aquifer. The importance of the vertical flow component is directly related to the magnitude of $\eta = S_L/S_E$. As the value of S_E approaches zero ($\eta \to \infty$) the duration of the first portion becomes infinitesimally small. The medium segment is horizontal in this situation. On the contrary, as S_E assumes large values ($\eta \to 0$), the time duration of the first portion increases so that the aquifer behaves as a confined aquifer with storativity S_L. The above-mentioned formula are valid theoretically for $\eta \to 0$, however, practically for $\eta > 100$. Otherwise, the intermediate section is not horizontal and the time-drawdown curve for this portion is given with an expression similar to the steady-state leaky aquifer formulation. The necessary tables for constructing Boulton-type curves are given in Table 15, whereas the type curves for a set of parameters are presented in Figure 38. The type curves are bounded by early and late Theis-type curves and their applications require two matching procedures, for early and late field data, by taking into consideration connection through the moderate data values. For the early field data matching points $(s_E)_M$, $(t_E)_M$, $(1/u_E)_M$, and $W_M(u_E, r/\beta)$ are read from the type curve and field data sheet. In addition, the most representative curve label, $(r/\beta)_M$, is also recorded. Substitutions of these values into Equations 42 and 44 with known Q yields T_E and S_E, respectively. Then the field data sheet is moved to the right and matched with late time-type curve preferably with the same $(r/\beta)_M$ label. The coordinates corresponding to the second match point are read as $(s_L)_M$, $(t_L)_M$, $(1/t_L)_M$, and $W_M(u_L, r/\beta)$. Rearrangements of Equations 45 and 46 and substitution of the match point coordinates lead to the determination of T_L and S_L, respectively. For the successful application, early and

Table 15 Values of Boulton-Type Curves

Note: $1/u_E, 1/u_L$: $N \times 10^n$

$r/\beta = 0.01$			$r/\beta = 0.1$			$r/\beta = 0.2$			$r/\beta = 0.316$		
$1/u_E$			$1/u_E$			$1/u_E$			$1/u_E$		
N	n	$W(u_E,r/\beta)$	N	n	$W(u_E,r/\beta)$	N	n	$W(u_E,r/\beta)$	N	n	$W(u_E,r/\beta)$
1	1	1.82	1	1	1.80	5	0	1.19	1	0	0.216
1	2	4.04	5	1	3.24	1	0	1.75	2	0	0.544
1	3	6.31	1	2	3.81	5	1	2.96	5	0	1.15
5	3	7.82	2	2	4.30	1	2	3.29	1	1	1.56
1	4	8.40	5	2	4.71	5	2	3.50	5	1	2.50
1	5	9.42	1	3	4.83	1	3	3.51	1	2	2.62
1	6	9.44	1	4	4.85				1	3	2.65

$r/\beta = 0.04$			$r/\beta = 0.6$			$r/\beta = 0.8$			$r/\beta = 1.0$		
$1/u_E$			$1/u_E$			$1/u_E$			$1/u_E$		
N	n	$W(u_E,r/\beta)$	N	n	$W(u_E,r/\beta)$	N	n	$W(u_E,r/\beta)$	N	n	$W(u_E,r/\beta)$
1	0	0.213	1	0	0.216	5	−1	0.046	5	−1	0.0444
2	0	0.534	2	1	0.504	1	0	0.197	1	0	0.185
5	0	1.11	5	0	0.996	2	0	0.466	1	0	0.421
1	1	1.56	1	1	1.31	5	0	0.857	5	0	0.715
5	1	2.18	2	1	1.49	1	1	1.05	1	1	0.819
1	2	2.22	5	1	1.55	2	1	1.12	2	1	0.841

$r/\beta = 0.04$			$r/\beta = 0.6$			$r/\beta = 0.8$			$r/\beta = 1.0$		
$1/u_L$			$1/u_L$			$1/u_L$			$1/u_L$		
N	n	$W(u_L,r/\beta)$	N	n	$W(u_L,r/\beta)$	N	n	$W(u_L,r/\beta)$	N	n	$W(u_L,r/\beta)$
1	−1	2.23	4.44	−1	1.59	2.5	−2	1.13	4	−2	0.844
1	0	2.26	2.22	0	1.71	2.5	−1	1.16	4	−1	0.901
5	0	2.40	4.44	0	1.85	1.25	0	1.20	4	0	1.36
1	1	2.55	1.67	1	2.45	2.5	0	1.39	4	1	3.14
3.75	1	3.20	4.44	1	3.26	9.37	0	1.94			
1	2	4.05				2.5	1	2.70			

$r/\beta = 1.5$			$r/\beta = 2.0$			$r/\beta = 2.5$			$r/\beta = 3.0$		
$1/u_L$			$1/u_L$			$1/u_L$			$1/u_L$		
N	n	$W(u_L,r/\beta)$	N	n	$W(u_L,r/\beta)$	N	n	$W(u_L,r/\beta)$	N	n	$W(u_L,r/\beta)$
7.11	−2	0.444	4	−2	0.239	2.56	−2	0.132	1.78	−2	0.0743
3.55	−1	0.509	2	−1	0.283	1.28	−1	0.162	8.89	−2	0.0939
7.11	−1	0.587	4	−1	0.337	2.56	−1	0.199	1.78	−1	0.119
2.67	0	0.963	1.5	0	0.614	9.6	−1	0.399	6.67	−1	0.262
7.11	0	1.57	4	0	1.11	2.56	0	0.798	1.78	0	0.577

$r/\beta = 1.5$			$r/\beta = 2.0$			$r/\beta = 2.5$			$r/\beta = 3.0$		
$1/u_E$			$1/u_E$			$1/u_E$			$1/u_E$		
N	n	$W(u_E,r/\beta)$	N	n	$W(u_E,r/\beta)$	N	n	$W(u_E,r/\beta)$	N	n	$W(u_E,r/\beta)$
5	−1	0.0394	3.33	−1	0.01	5	−1	0.0271	5	−1	0.0210
1	0	0.151	5	−1	0.0335	1	0	0.0803	1	0	0.0534
1.25	0	0.199	1	0	0.114	1.25	0	0.0961	1.25	0	0.0607
2	0	0.301	1.25	0	0.144	2	0	0.117	2	0	0.0681
5	0	0.413	2	0	0.194	5	0	0.125	5	0	0.0695
1	1	0.427	5	0	0.227	1	1	0.125	1	1	0.0695
2	1	0.428	1	1	0.228						

$r/\beta = 0.01$			$r/\beta = 0.1$			$r/\beta = 0.2$			$r/\beta = 0.316$		
$1/u_L$			$1/u_L$			$1/u_L$			$1/u_L$		
N	n	$W(u_L,r/\beta)$	N	n	$W(u_L,r/\beta)$	N	n	$W(u_L,r/\beta)$	N	n	$W(u_L,r/\beta)$
2	2	9.45	4	0	4.86	4	−1	3.52	4	−1	2.66
4	3	9.54	4	1	4.95	4	0	3.54	4	0	2.74
4	4	10.2	4	2	5.64	2	1	3.69	4	1	3.38
4	5	12.3	4	3	7.72	4	1	3.85	4	2	5.42
4	6	14.6	4	4	10.00	1.5	2	4.55	4	3	7.72

POROUS MEDIUM AQUIFER TESTS

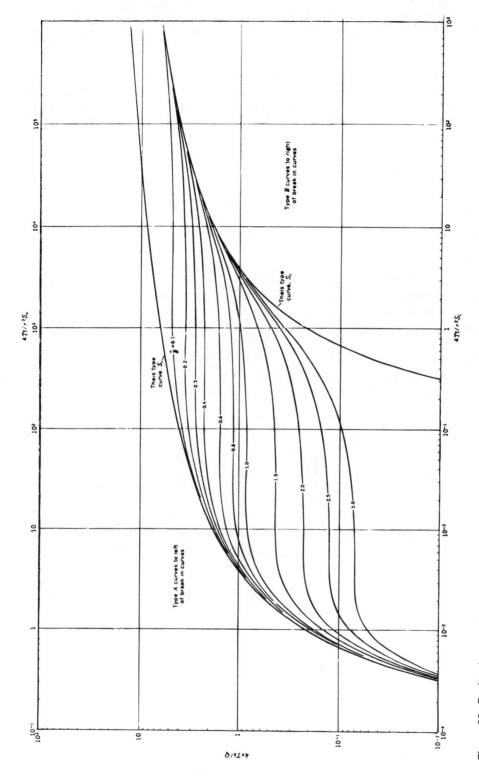

Figure 38 Boulton-type curves.

late time calculations should give almost the same value for the transmissivity, i.e., $T_E \cong T_L$. However, as discussed in Chapter 4 specific yield and storage coefficient are different such that always $S_L > S_E$.

Example 10—The numerical application of the method is presented through the following field data. A well, fully penetrating an unconfined aquifer, 25 m thick, is pumped at a constant discharge of 4.6 m³/sec. The drawdown registrations in an observation well at 73 m from the pumped well are given in Table 16. The type curve-matching procedure is given in Figure 39. The calculations are shown succinctly in Table 17.

C. NEUMAN MODEL

Neuman (1972,1973) also reproduced all three portions of the time-drawdown curve in an unconfined aquifer without any empirical constants definition. The procedure of Neuman's type curve method is very similar to that of Boulton's model described in the previous section. Although Boulton's method fits field data quite well, it nevertheless fails to provide insight into the physical nature of the delayed yield phenomenon.

Neuman's method differs from that of Boulton by considering well-defined physical parameters of the flow phenomenon and no longer involves semiempirical quantities such as Boulton's delay index. This method accounts for the entire delayed yield phenomenon by regarding the following physical points.

Table 16 Unconfined Aquifer Test Data

Time, t (min)	Drawdown, s (cm)	Time, t (min)	Drawdown, s (cm)
0.17	4	10	31
0.25	5.9	12	31.4
0.34	7.8	15	31.7
0.42	10	18	32
0.50	11.89	20	32.3
0.58	13.1	25	32.9
0.66	14.9	30	34.4
0.75	16.1	35	35
0.83	17.4	40	35.7
0.92	18.6	50	36.3
1	19.5	60	37.2
1.08	20.5	70	38.1
1.16	21.3	80	39
1.24	21.9	90	39.3
1.33	22.6	100	39.9
1.42	23.2	120	41.4
1.50	23.8	150	44.2
1.68	25	200	46.3
1.85	25.6	250	48.5
2	26.2	300	50.3
2.15	26.5	350	51.8
2.35	27.4	400	53.3
2.50	27.7	500	56.4
2.65	28	600	59.4
2.80	28.3	700	59.4
3	28.7	800	61.3
3.50	28.9	900	63.7
4	29.6	1000	67
4.50	29.7	1200	69.2
5	29.9	1500	71.6
6	30.2	2000	75.9
7	30.5	2500	78.8
8	30.5	3000	81.07
9	32		

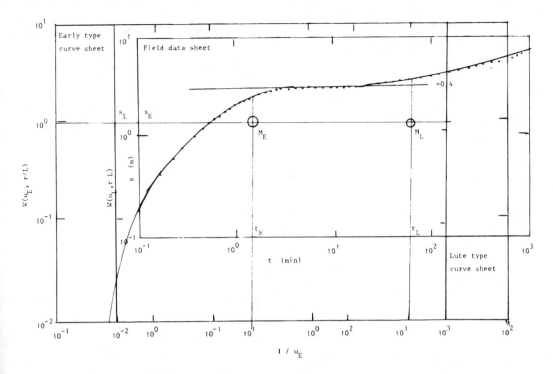

Figure 39 Boulton-type curve matching.

1. The water table moves as a material boundary or free surface.
2. The unsaturated flow above the water table has very small influence on the time-drawdown response of unconfined aquifer during gravity drainage. As a result the flow contribution to the water table from dewatered zone is neglected, i.e., unsaturated zone is not considered at all.
3. The elastic storage in the aquifer is assumed to play the main role in the groundwater flow toward the well due to the compaction of the aquifer and expansion of water.
4. The groundwater flow to well is axially symmetrical with three-dimensional flow patterns.

Besides, Neuman's method takes into account the aquifer's anisotropy and the partial penetration of well. His rather complex solution can be written simply for practical purposes, in general, as

$$W(u_E, u_L, \beta) = \frac{4\pi T_E}{Q} s_E \qquad (47)$$

in which $W(u_E, u_L, \beta)$ is the unconfined well function and $\beta = r^2/m^2$. The complete set of type

Table 17 Boulton Method Parameter Calculation

	Match points coordinates				Parameters estimate curve		
	Field data sheet		Type curve sheet		Equation 42, 45	Equation 44, 46	Label, β
Early times	s_E (m) 1.51	t_E (min) 4.6	$1/u_E$ 10	$W(u_E, r/\beta)$ 1	T_E (m²/day) 351	S_E 8.4×10^{-4}	0.4
Late times	s_L (m) 1.51	t_L (min) 59.0	$1/u_L$ 19	$W(u_L, r/\beta)$ 1	T_L (m²/d) 351	S_L 1.1×10^{-2}	0.4

Note: Discharge $Q = 4.6$ m³/s.

curves is given in Figure 40. However, for the early portion of type curves the well function $W(u_E,\beta)$ and the dimensionless time factor are

$$W(u_E,\beta) = \frac{4\pi T_E}{Q} s_E \qquad (48)$$

and

$$u_E = \frac{r^2 S_E}{4t_E T_E} \qquad (49)$$

where S_E is the elastic storativity responsible for the critical release of water to the well. Similarly, late time portion of type curves approaches asymptotically the Theis curve but with the new definition of parameters as

$$W(u_L,\beta) = \frac{4\pi T_L}{Q} s_L \qquad (50)$$

and

$$u_L = \frac{r^2 S_L}{4t_L T_L} \qquad (51)$$

in which S_L is the specific yield responsible for the delayed release of water to the well.

Example 11—Neuman-type curve matching is similar to the Boulton method. Therefore, the same data set in Table 16 is adopted for numerical application. The resulting match point coordinates from Figure 41 and the necessary calculations are provided in Table 18.

Independently from Neuman, Streltsova (1972,1973) also obtained type curves for drawdowns in unconfined aquifer. She had a different physical reasoning than Neuman as follows:

1. At any point in the aquifer two types of heads are considered: the free surface of water table and the average head along the depth. The free surface head is greater than the average head. This renders the three-dimensional flow into the two-dimensional axisymmetric flow that depends on the radial distance from the well and vertical coordinate.
2. The water table is considered as a free surface and there is no delayed yield in the unsaturated zone.
3. The rate of vertical gravity drainage is linearly proportional to the difference between the free surface and average heads.

Hence, Neuman and Streltsova had a similar physical look at the problem but the latter's mathematical treatment is different as a result of which estimated aquifer parameters by two methods differ slightly.

The methods proposed by Boulton, Neuman, and Streltsova yield practically identical values of transmissivity, storativity, and specific storage for known aquifer thickness but Boulton's method does not yield a value for vertical permeability, but instead yields only the lumped parameter of delayed yield index, β.

1. Neuman Straight Line Method

Neuman (1975) has shown that type curves plotted on a semilogarithmic in Figure 42 tend to fall on a straight line rather than that proposed by Jacob (1944) as

POROUS MEDIUM AQUIFER TESTS

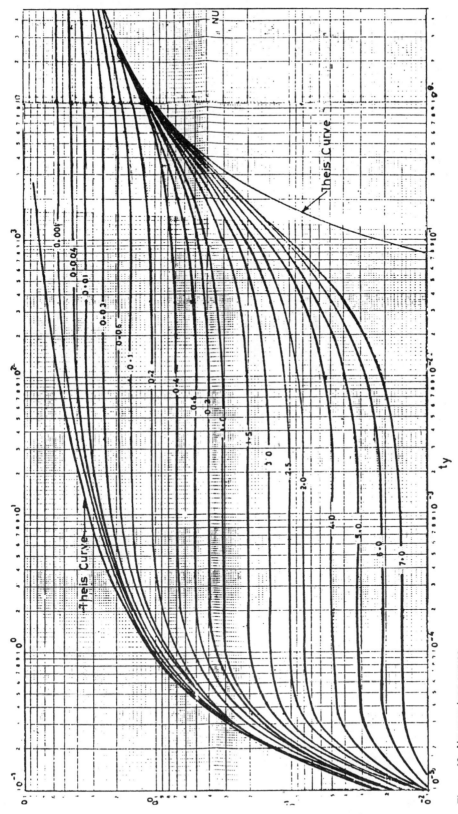

Figure 40 Neuman-type curves.

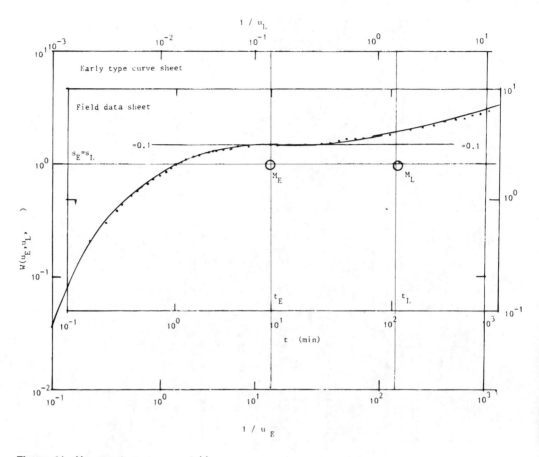

Figure 41 Neuman-type curve matching.

$$s_L = \frac{2.3Q}{4\pi T_L} \log\left(\frac{2.25 T_L t_L}{r^2 S_L}\right) \tag{52}$$

The intermediate data fall on a horizontal line, whereas some of the early data are near the straight line as

$$s_E = \frac{2.3Q}{4\pi T_E} \log\left(\frac{2.25 T_E t_E}{r^2 S_E}\right) \tag{53}$$

Let t_β be the value of t_L corresponding to the intersection of any horizontal line with the inclined line given by Equation 52. For example, Figure 42 shows that the value of t_β for $\beta = 0.03$ is equal to 5.2. When the values of $1/\beta$ are plotted vs. t_β on a semilogarithmic paper,

Table 18 Neuman Method Parameter Estimations

	Match point coordinates				Parameter Estimates Curve		
	Field data sheet		Type curve sheet		Equation 47, 50	Equation 48, 51	Label, β
Early times	s_E (m)	t_E (min)	$1/u_E$	$W(u_E, u_E, \beta)$	T (m²/day)	S	
	2.2	7.7	10	1	240	2.4×10^{-4}	0.1
Late times	s_L (m)	t_L (min)	$1/u_L$	$w(u_L, \beta)$			
	2.2	102	1		240	3.2×10^{-2}	0.1

Note: Discharge $Q = 4.6$ m²/sec.

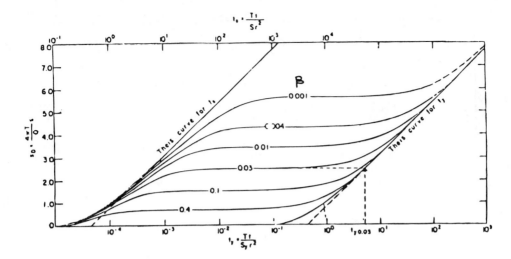

Figure 42 Neuman straight lines.

the result is a unique curve shown in Figure 43. A good approximation for the relationship between β and t_L within the range $4.0 < t_\beta < 100.0$ is given by the equation

$$\beta = \frac{0.195}{1.1053} t_\beta \tag{54}$$

Example 12—The procedure of aquifer parameter determination by this method from the field data in Table 16 consists of the following steps:

1. Plot the field data from Table 16 on a semilogarithmic paper as in Figure 44.
2. A straight line is fitted to the late portion of the time drawdown data. The intercept of this line with the time axis corresponds to $s = 0$ and is denoted by t_L. The slope of the same line is denoted by Δs_L. According to Equation 52 the transmissivity of the aquifer is determined as

$$L = \frac{2.3Q}{\Delta \pi \Delta s_L} \tag{55}$$

and the specific yield is

$$S_L = \frac{2.25 T_L t_L}{r^2} \tag{56}$$

3. A horizontal line is fitted to the intermediate portion of the time drawdown data in Figure 44. The value of time corresponding to the intersection of this horizontal line with the straight line in the previous step is denoted by t_β. Knowing T_L and S_L from the previous step it is possible to calculate the dimensionless time t_L by

$$t_\beta = \frac{T_L t_L}{S_L r^2} \tag{57}$$

The value of β can now be obtained directly from the curve in Figure 43.
4. A straight line is fitted to the portion of the early time drawdown data. If the slope of this line differs markedly from that of the late time straight line then this step must be skipped. In such a case the storativity must be determined by type curve-matching procedure. However, if the two lines are nearly parallel, then the intersection of the early line with the time axis

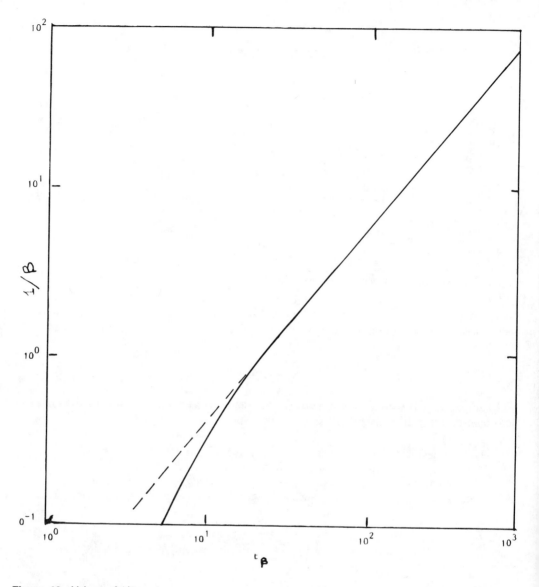

Figure 43 Values of $1/\beta$ vs. t_β.

is denoted by t_E. In addition, the slope of this line is Δs_E. The transmissivity can then be calculated as

$$E = \frac{2.3Q}{4\pi \Delta s_E} \tag{58}$$

The value of T should be approximately equal to that previously obtained from the late drawdown data. The storativity is obtained from

$$S_E = \frac{2.25 T_E t_E}{r^2} \tag{59}$$

The calculations are presented explicitly in Figure 44.

POROUS MEDIUM AQUIFER TESTS

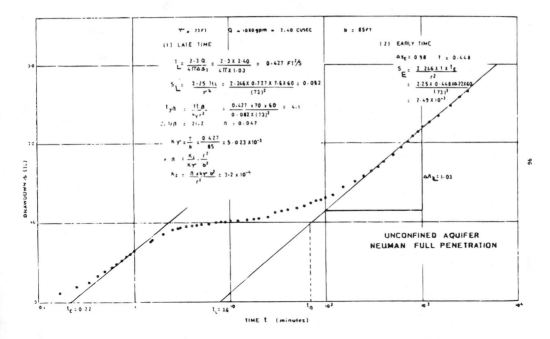

Figure 44 Field data plot.

VII. LEAKY AQUIFER MODEL

Aquifers that receive or deliver water through adjacent semipervious layers (aquitards) are defined as the "leaky aquifers" in Chapter 2. In nature, although perfectly confined aquifers are rarely encountered, leaky aquifers are common features of many alluvium deposits in sedimentary basins. The clay-marly deposits at shallow depths and shales at deeper horizons are formed due to the consolidation and cannot be considered either a completely impervious layer or a perfect aquifer. These layers present a certain permeability value that is rather small and in a regional study they allow a significant amount of water transmissions. For instance, compared to sandstones, shales have five to six times less hydraulic conductivity but act as important water transmitters. Owing to the hydraulic conductivity differences within a hydrostratigraphic sequence, vertical flow occurs between the aquifers and aquitards.

If a well in a leaky aquifer is pumped, the rate, Q, will have contributions from two sources: (1) the internal source because of aquifer compaction and water expansion as a result of hydraulic head reduction and (2) alternatively, from outside the aquifer as leakage through semipervious layers that are in contact with other potential aquifers. It is physically plausible that at initial periods of pumping the internal source contributes to Q more effectively and with the passage of time the leakage becomes dominant and finally sustains the whole contribution when the flow reaches steady state in the main aquifer. At the steady state the main aquifer only transmits the leakage water to the well playing a similar role as the semipervious layers.

Leaky aquifer models have restrictions as will be explained later in the following sections. The assumptions mentioned in Chapter 8 for the steady-state flow situation are the same with the exception that the discharge Q has two contributors, aquifer storage and leakage.

A. HANTUSH-JACOB MODEL

The theory of leaky aquifer for nonequilibrium was developed concisely first by Hantush and Jacob (1955) under the assumption of vertical leakage, i.e., the aquifer replenishment from adjacent aquifers varies linearly with the instantaneous difference in heads across the producing bed (Figure 45).

The notations for aquifer properties in this figure are self-explanatory. Prior to pumping all the aquifers in the system are assumed to have the same static piezometric level so that there is neither horizontal nor vertical groundwater movement. With the pump start the well abstracts water directly from the main aquifer and the piezometric surface of this aquifer will deviate from the static position, whereas the piezometric surfaces of upper and lower aquifers will remain the same throughout the pumping period. Because of leakage, there will be some deviations even in these two piezometric surfaces of not more than 5% (Neuman and Witherspoon, 1972). After some time upward and downward leakages will contribute to the main aquifer; therefore, the whole system will be leaky. Physically, right at the beginning, these leakages will not be prompt depending on the semipervious layer properties and the main aquifer will react similarly to confined aquifer. It is, therefore, expected to have time-drawdown variation at the beginning similar to the Theis-type curve.

The continuity equation given by Equation 1 for nonleaky aquifers must be modified by considering the leakage. Hence,

$$Q(r,t)\Delta t - Q(r + \Delta r,t)\Delta t + 2\pi r \Delta r [L_u(r,t) + L_l(r,t)] = S 2\pi r \Delta r [h_a(r,t) - h_a(r,t + \Delta t)] \quad (60)$$

in which $h_a(r,t)$, $L_u(r,t)$, and $L_l(r,t)$ are the main aquifer hydraulic head, and upper and lower leakage rates, respectively. For an infinitesimally small time period and distance ($\Delta t \to 0$ and $\Delta r \to 0$) Equation 60 becomes

$$\frac{\partial Q(r,t)}{\partial r} + 2\pi r [L_u(r,t) + L_l(r,t)] = 2\pi r S \frac{\partial h_a(r,t)}{\partial r} \quad (61)$$

The leakage rates can be written explicitly as

$$L_u(r,t) = \frac{K_a}{m_a} s(r,t) \quad (62)$$

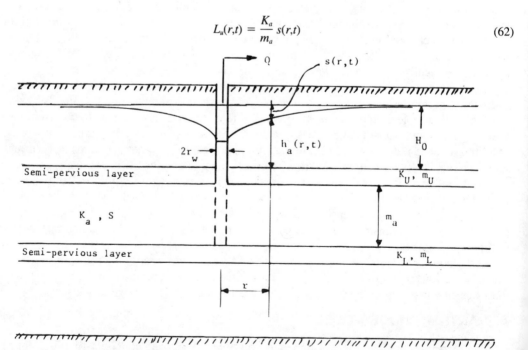

Figure 45 Leaky aquifer system.

and

$$L_l(r,t) = \frac{K_l}{m_l} s(r,t) \qquad (63)$$

where the drawdown is $s(r,t) = H_0 - H_a(r,t)$ with H_0 being the static hydraulic head. Considering Darcy law valid within the aquifer the aquifer discharge is

$$Q(r,t) = -2\pi m_a r K_a \frac{\partial s(r,t)}{\partial r} \qquad (64)$$

in which m_a and K_a are the main aquifer thickness and hydraulic conductivity, respectively. Substitution of Equations 62 through 64 into Equation 61 and completion of the necessary algebraic manipulations lead to

$$\frac{\partial^2 s(r,t)}{\partial r^2} + \frac{1}{r}\frac{\partial s(r,t)}{\partial r} + \frac{s(r,t)}{L^2} = \frac{S}{T}\frac{\partial s(r,t)}{\partial t} \qquad (65)$$

where L is the leakage factor defined by Equation 26 in Chapter 8. The solution of this partial differential equation has been given in a compact form by Hantush and Jacob (1955) and accordingly the result can be represented similar to Theis solution succinctly as

$$W\left(u, \frac{r}{L}\right) = \frac{4\pi T}{Q} s(r,t) \qquad (66)$$

and

$$u = \frac{r^2 S}{4tT} \qquad (67)$$

where $W(u,r/L)$ is referred to as the leaky aquifer well function which is given explicitly as

$$W\left(u, \frac{r}{L}\right) = \int_{-\infty}^{u} \frac{1}{x} \exp\left(x - \frac{r^2}{4L^2 x}\right) dx \qquad (68)$$

The evaluation of aquifer parameters, namely, S, T, and L is possible similar to the Theis-type curve-matching method, provided that leaky aquifer-type curves are drawn. In this case the type curves are defined as the plot of $W(u,r/L)$ vs. u for each r/L value. Table 19 shows the corresponding values of u and $W(u,r/L)$ for a given set of r/L values. The presentation of this table in the graphical form is available in Figure 46 as type curves first drawn by Walton (1970).

In general, each type curve in this figure has three distinctive portions. The first portion coincides with the Theis-type curve, implying that the aquifer behaves as a confined one at initial stages. Increase in the observation well distance, r, and/or decrease in the leakage factor reduces the duration of this portion. The second portion starts at the time of particular type curve deviation from the Theis curve and continues until it becomes horizontal. During this portion both the aquifer storage and the leakage are active in contributing to the pump discharge. However, the third part is completely horizontal physically, indicating a steady-state flow during which all pump discharge comes from the leakage water only.

Example 13—The time drawdown data for unsteady flow in four observation wells at different distances from the main well are given in Table 20. The main well is pumped at a constant discharge of 0.42 m³/min. Well locations are shown in Figure 47.

Table 19 Leaky Aquifer-Type Curves

u\r/L	0	0.002	0.004	0.005	0.007	0.01	0.02	0.04	0.06	0.08	0.10
0		12.6611	11.2748	10.8286	10.1557	9.4425	8.0569	6.6731	5.8456	5.2950	4.8541
1×10^{-6}	13.2383	12.4417	11.2711	10.8283	10.1557						
2×10^{-6}	12.5451	12.1013	11.2259	10.8174	10.1554						
5×10^{-6}	11.6289	11.4384	10.9642	10.6822	10.1290	9.4425					
8×10^{-6}	11.1589	11.0377	10.7151	10.5027	10.0602	9.4313					
1×10^{-5}	10.9357	10.8382	10.5725	10.3963	10.0034	9.4176	8.0569				
2×10^{-5}	10.2426	10.1932	10.0522	9.9530	9.7126	9.2961	8.0558				
5×10^{-5}	9.3263	9.3064	9.2480	9.2052	9.0957	8.8827	8.0080	6.6730			
7×10^{-5}	8.9899	8.9756	8.9336	8.9027	8.8224	8.6625	7.9456	6.6726			
1×10^{-4}	8.6332	8.6233	8.5937	8.5717	8.5145	8.3983	6.6828	6.2748	5.7658	5.2749	4.8510
2×10^{-4}	7.9402	7.9352	7.9203	7.8958	7.8800	7.8192	6.5508	6.1917	5.7274	5.2618	4.8478
5×10^{-4}	7.0242	7.0222	7.0163	7.0118	6.9999	6.9750	6.8346	6.3626	5.8011	5.2848	4.8530
7×10^{-4}	6.6879	6.6865	6.6823	6.6790	6.6706	6.6527	6.5508	6.1917	5.7274	5.2618	4.8478
1×10^{-3}	6.3315	6.3305	6.3276	6.3253	6.3194	6.3069	6.2347	5.9711	5.6058	5.2087	4.8292
2×10^{-3}	5.6394	5.6389	5.6374	5.6363	5.6334	5.6271	5.5907	5.4516	5.2411	4.9848	4.7079
5×10^{-3}	4.7261	4.7259	4.7253	4.7249	4.7237	4.7212	4.7068	4.6499	4.5590	4.4389	4.2990
7×10^{-3}	4.3916	4.3915	4.3910	4.3908	4.3899	4.3882	4.3779	4.3374	4.2719	4.1839	4.0771
0.01	4.0379	4.0378	4.0375	4.0373	4.0368	4.0351	4.0285	4.0003	3.9544	3.8920	3.8190
0.02	3.3547	3.3547	3.3545	3.3544	3.3542	3.3536	3.3502	3.3365	3.3141	3.2832	3.2442
0.05	2.4679	2.4679	2.4678	2.4678	2.4677	2.4675	2.4662	2.4613	2.4531	2.4416	2.4271
0.07	2.1508	2.1508	2.1508	2.1508	2.1507	2.1506	2.1497	2.1464	.1408	2.1331	2.1232
0.10	1.8229	1.8229	1.8229	1.8229	1.8228	1.8227	1.8222	1.8200	1.8164	1.8114	1.8050
0.20	1.2227	1.2226	1.2226	1.2226	1.2226	1.2226	1.2224	1.2215	1.2201	1.2181	1.2155
0.50	0.5598	0.5598	0.5598	0.5598	0.5598	0.5598	0.5597	0.5595	0.5592	0.5587	0.5581
0.70	0.3738	0.3738	0.3738	0.3738	0.3738	0.3738	0.3737	0.3736	0.3734	0.3732	0.3729
1.00	0.2194	0.2194	0.2194	0.2194	0.2194	0.2194	0.2194	0.2193	0.2192	0.2191	0.2190
2.00	0.0489	0.0489	0.0489	0.0489	0.0489	0.0489	0.0489	0.0489	0.0489	0.0489	0.0488
5.00	0.0011	0.0011	0.0011	0.0011	0.0011	0.0011	0.0011	0.0011	0.0011	0.0011	0.0011
7.00	0.0001	0.0001	0.0001	0.0001	0.0001	0.0001	0.0001	0.0001	0.0001	0.0001	0.0001
8.00	0.0000	0.0000	0.0000	0.0000	0.0000	0.0000	0.0000	0.0000	0.0000	0.0000	0.0000

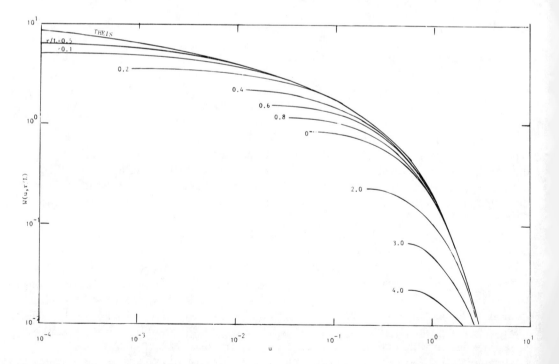

Figure 46 Leaky aquifer-type curves.

Table 20 Leaky Aquifer Field Data

Time, t (min)	(1/t) (1/min)	Drawdown, s (cm)			
		r = 20 m	r = 70 m	r = 90 m	r = 120 m
2	$5.00 \times 10^{10-1}$	35	25	28	15.5
3	3.33	54	34	22.5	18.0
4	2.50	66	39	26	19.0
5	2.00	78	43	29	20.0
10	1.00	120	59	37	24.0
15	6.67×10^{-2}	138	65	44	26.0
20	5.00	157	70	47	27.0
30	3.33	185	76	49	27.5
40	2.50	190	80	51	28.0
50	2.00	200	84	54	28.5
60	1.67	212	88	55	29.0
90	1.11	245	93	60	31.0
120	8.33×10^{-3}	260	99	61	31.5
180	5.56	270	101	62	32.0
240	4.17	284	110	65	33.0
300	3.33	293	115	66	33.5
360	2.78	298	118	67	34.0
420	2.38	300	121	68	34.0
480	2.08	304	123	70	34.0
570	1.75	307	124	71	34.0
650	1.53	309	124	71	34.0

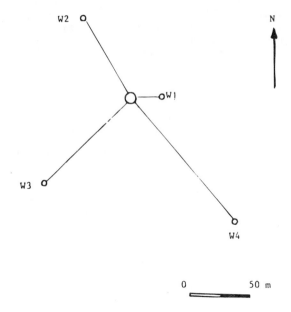

Figure 47 Well location map.

The thickness of the aquifer and the semipervious layer are 25 and 5 m, respectively. Four field data sheets, each for one observation well, are presented in Figures 48 to 51 with the best matching-type curve. The match point coordinates with corresponding parameter estimates are presented in Table 21.

It is obvious from this table that the aquifer is highly heterogeneous, especially near the well and at large distances. High transmissivity value at W1 indicates that the aquifer material in the well vicinity is highly permeable. However, a relatively very small transmissivity value at W4 implies that the aquifer material becomes less permeable with distance.

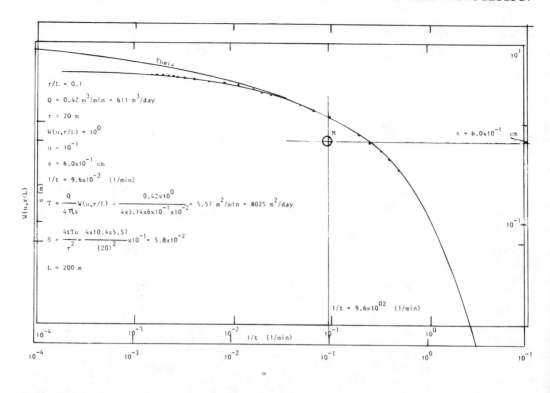

Figure 48 Leaky aquifer-type curve matching ($r = 20$ m).

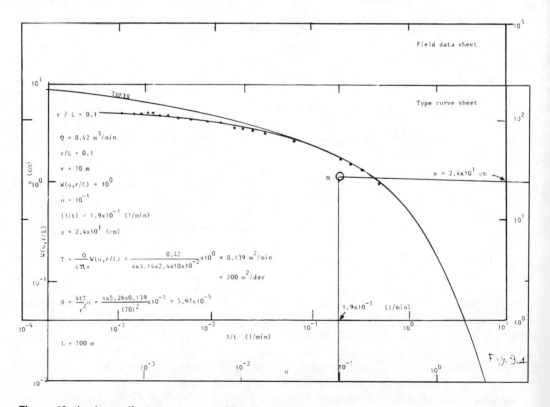

Figure 49 Leaky aquifer-type curve matching ($r = 70$ m).

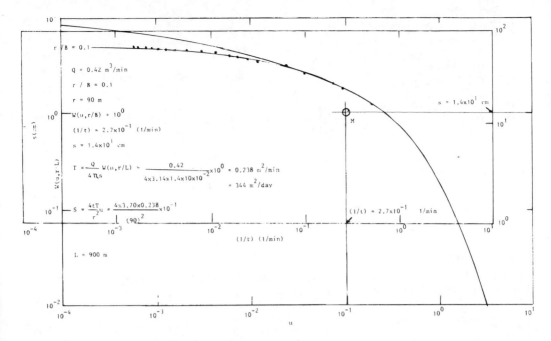

Figure 50 Leaky aquifer-type curve matching ($r = 90$ m).

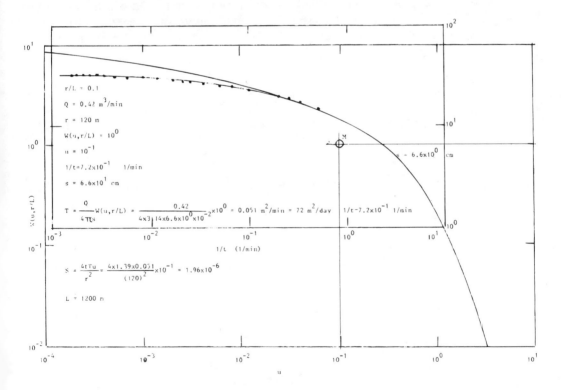

Figure 51 Leaky aquifer-type curve matching ($r = 120$ m).

Table 21 Leaky Aquifer Parameter Estimations

		Match point coordinates curve					Parameter estimates		Leakage factor
	Radial	Field data		Type curve					
Well no.	distance r (m)	s_M (cm)	$(1/t)_M$ (1/min)	u_M	$W_M(u,r/L)$	Label (r/L)	Equation 66 T (m²/d)	Equation 67 S	L (m)
W1	20	0.6	0.092	0.1	1	0.1	8025	5.8×10^{-2}	100
W2	70	24	0.190	0.1	1	0.1	200	6.0×10^{-5}	700
W3	90	14	0.270	0.1	1	0.1	344	4.4×10^{-5}	900
W4	120	6.6	0.720	0.1	1	0.1	72	2.0×10^{-5}	1200
						Averages	2165	1.5×10^{-2}	750

Note: Discharge $Q = 0.42$ m³/min.

B. APPROXIMATE MODELS

The Hantush-Jacob method does not give reliable results unless a sufficient number of time-drawdown data fall within the initial portion of type curve where the leakage effects are insignificant. When this portion is fit to the Theis-type curve in the best possible way, the data for large times will follow one of the family-type curves.

1. Inflection Point Model

Hantush (1956) observed that the initial time-drawdown data fall on the Theis-type curve for a period $t < t_i/4$ on the semilogarithmic paper where t_i is the time at which an inflection point occurs on this curve. Later, Hantush (1964) mentioned various properties of this point on a semilogarithmic time-drawdown plot and obtained the following relationships which are valid whatever the r/L value:

$$u_i = \frac{r^2 S}{4Tt_i} = \frac{r}{2L} \tag{69}$$

$$\Delta s_i = \frac{2.3Q}{4\pi T} e^{-r/L} \tag{70}$$

$$s_i = 0.5 s_m = \frac{Q}{4\pi T} K_0\left(\frac{r}{L}\right) \tag{71}$$

and

$$f\left(\frac{r}{L}\right) = e^{r/L} K_0\left(\frac{r}{L}\right) = 2.3 \frac{s_i}{\Delta s_i} \tag{72}$$

in which u_i, t_i, Δs_i and s_i are the dimensionless time factor, the time value, the slope of the semilogarithmic plot, and the drawdown, respectively, all at the inflection point. Equation 72 can be approximated by

$$\log\left(\frac{2L}{r}\right) = 0.251 + \frac{s_i}{\Delta s_i} \tag{73}$$

for $r/L < 0.01$ which corresponds to $s_i/\Delta s_i > 2$. Table 22 presents numerically the relationship between r/L and $K_0(r/L)$. This method is applicable if time drawdown data are available from a single observation well. The aquifer test durations should cover long times so as to attain

POROUS MEDIUM AQUIFER TESTS

Table 22 Steady-State Leaky Aquifer Parameter Relationship

x	$K_0(x)$	$K_1(x)$	$I_0(x)$	$I_1(x)$
0.010	4.7212	99.9739	1.0000	0.0050
0.020	4.0285	49.9547	1.0001	0.0100
0.030	3.6235	33.2715	1.0002	0.0150
0.040	3.3365	24.9233	1.0004	0.0200
0.050	3.1142	19.9097	1.0006	0.0250
0.060	2.9329	16.5637	1.0009	0.0300
0.070	2.7798	14.1710	1.0012	0.0350
0.080	2.6475	12.3742	1.0016	0.0400
0.090	2.5310	10.9749	1.0020	0.0451
0.1	2.4271	9.8538	1.0025	0.0501
0.2	1.7527	4.7760	1.0100	0.1005
0.3	1.3725	3.0560	1.0226	0.1517
0.4	1.1145	2.1843	1.0404	0.2040
0.5	0.9244	1.6564	1.0635	0.2579
0.6	0.7775	1.0283	1.0921	0.3137
0.7	0.6605	1.0503	1.1263	0.3719
0.8	0.5663	0.8618	1.1665	0.4327
0.9	0.4867	0.7165	1.2130	0.4971
1.0	0.4210	0.6019	1.2661	0.5652
1.5	0.2138	0.2774	1.6467	0.9817
2.0	0.1139	0.1399	2.2796	1.5906
2.5	0.0624	0.0739	3.2898	3.5167
3.0	0.0347	0.0402	4.8808	3.9534
3.5	0.0196	0.0222	7.3782	6.2058
4.0	0.0112	0.0125	11.3019	9.7595
4.5	0.0064	0.0071	17.4812	15.3892
5.0	0.0037	0.0040	27.2399	24.3356

the maximum drawdowns, s_M, and yield an apparent location for the inflection point. The semilogarithmic plot of time-drawdown data yields estimates of the maximum drawdown s_M and the slope, Δs_i, at the inflection point. The drawdown at this point is $s_i = s_M/2$ and corresponding t_i value is read from the logarithmic axis. Substitution of these values into Equation 72 yields $K_0(r/L)$, a value that leads to r/L from Table 22. Substitution of Δs_i and r/L values into Equation 70 gives an estimate of T. Subsequently, the substitution of T, t_i and r/L values into Equation 69 yields the S value. It is obvious that the calculations depend on the initial estimation of s_M and Δs_i.

Example 14—The field data in Table 20 are subjected to the analysis by the inflection point method. They are plotted on a semilogarithmic paper for each observation well as shown in Figure 52. After extrapolating the maximum drawdowns, s_M, from this figure the inflection point drawdown value, s_i, is found as $1/2 s_M$. Subsequently, the corresponding inflection point time and the slope, Δs_i, of the tangent at each inflection point are found. The parameter estimation calculations are summarized as in Table 23.

The accuracy of the calculated constants depends on the accuracy of extrapolation of the maximum drawdown value.

2. Distance-Slope Model

The purpose of this section is to provide a solution procedure for aquifer parameters if there is more than one observation well. Consideration of Equation 70 in the logarithmic domain leads to

$$r = 2.30 \, L \left[\log \frac{2.30Q}{4\pi T} - \log s_i \right] \tag{74}$$

It is obvious from this expression that a semilogarithmic plot of distance vs. logarithmic slope

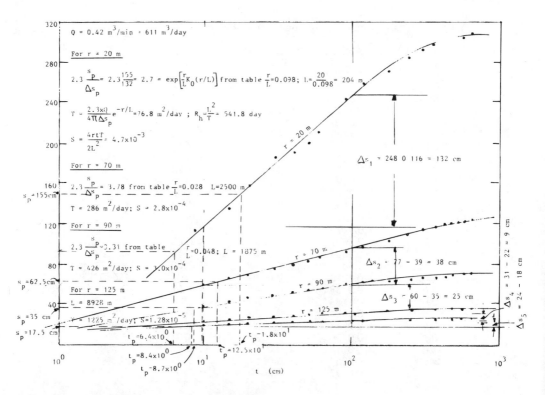

Figure 52 Inflection method on semilogarithmic paper.

Table 23 Inflection Point Method Parameter Determinations

Well no.	Radial maximum		Inflection point			Parameter estimations			
	Distance r (m)	Drawdown extrapolation s_m (m)	Drawdown $s_i = 0.5 s_m$ (m)	Time t_i (min)	Tangent slope s_i (m)	Equation 72	Table 9 r/L	Equation 70 T (m²/d)	Equation 69 S	L (m)
W1	20	3.15	1.57	18.0	1.32	2.70	0.098	76.8	4.7	204
W2	70	1.28	0.64	12.5	0.38	3.78	0.028	286	2.8	2500
W3	90	0.75	0.38	8.7	0.25	3.31	0.048	426	3.0	1875
W4	120	0.35	0.18	6.4	0.09	4.47	0.014	1225	1.3	8928
						Averages		503	1.3	3377

Note: Discharge $Q = 4.6$ m³/min

values appears as a straight line. The slope, Δr of this distance-slope line must be equal to the slope term in Equation 74. Hence,

$$\Delta r = 2.3 L \tag{75}$$

The intercept value, $(\Delta s_i)_0$, of the straight line has the analytical form

$$(\Delta s_i)_0 = \frac{2.3 Q}{4 \pi T} \tag{76}$$

The leakage index and transmissivity values can be estimated from Equations 75 and 76, respectively. Calculation of S value is obtained similar to the inflection point method as explained earlier.

Example 15—Aquifer test data presented in Table 20 are used in working out a numerical example. From Figure 52 the slopes of the straight-line portions of each curve are determined

POROUS MEDIUM AQUIFER TESTS

and in Figure 53 these slopes are plotted vs. the radial distances on a semilogarithmic paper and, finally, a straight line is fitted through these points.

The slope of this straight line is denoted by Δs_D and its intercept on the s axis is $(\Delta s_i)_0$. Their substitutions in Equations 75 and 76 yields L and T values. The calculations are shown explicitly on Figure 53.

3. Hantush-Theis Method

Under some restrictive conditions Hantush (1964) has proposed a type curve method by using the Theis curve for leaky aquifer parameter determination. He noticed that Equation 68 can be written with no loss of generality as

$$W\left(u, \frac{r}{L}\right) = 2K_0\left(\frac{r}{L}\right) - W\left(q, \frac{r}{L}\right) \tag{77}$$

where q is a dimensionless time factor defined for the leaky aquifers in terms of leakage factor.

$$q = \frac{tT}{SL^2} \tag{78}$$

If the period of test is long enough then a sufficient number of the observed data fall within the period $t > 4t_i$ and the scatter of these data on a semilogarithmic plot is such that the maximum drawdown can be reasonably extrapolated and a type curve method can be devised

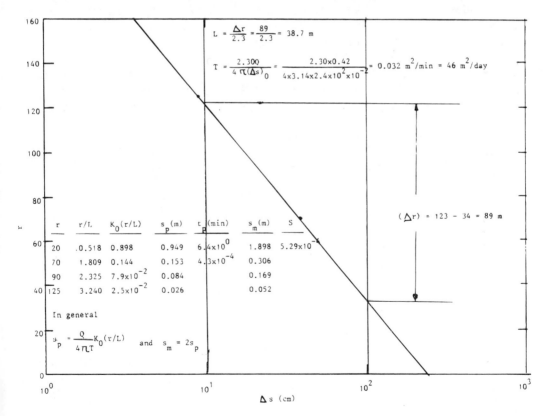

Figure 53 Distance slope method semilogarithmic paper.

to estimate the aquifer parameters. In addition, Hantush (1964) indicated that, for all practical purposes, if $q > 2(r/L)$, then Equation 77 can be approximated by

$$s_m - s = \frac{Q}{4\pi T} W(q) \qquad (79)$$

in which $W(q)$ is the well function for confined aquifer as proposed by Theis. Determination of three aquifer parameters, S, T, and L, can be obtained from Equations 73, 77, and 71, respectively, after the necessary matching procedure.

Example 16—To expose the numerical application of this method, the data given in Table 20 are used and the drawdowns are converted to $(s_m - s)$ values as shown in Table 24.

The calculated drawdown differences are plotted against times on double-logarithmic papers for the piezometers at each radial distance as shown in Figures 54 to 57.

In the same figures the field data are matched with the Theis-type curves. Consequently, reading the match point coordinates and their substitutions in the relevant equations yield the necessary aquifer parameter estimates. The calculations are expressed on each figure and the final results are summarized in Table 25.

C. NEUMAN-WITHERSPOON MODEL

There are two critical assumptions in the leaky aquifer models mentioned in the previous sections. First, the heads in the leakage-yielding aquifers do not change and, second, the rate of leakage is directly proportional to the hydraulic gradient across the semipervious layers that do not have any storage of water. However, Neuman and Witherspoon (1969a, 1969b) have provided complete solutions for leaky aquifers whereby the head changes in the unpumped aquifers as well as the storage capacity of the aquitards are taken into consideration.

The physical implication of no change in the hydraulic head within the unpumped aquifer holds approximately true for sufficiently small time values during which the pumped aquifer reacts as a confined aquifer and accordingly behaves similar to the Theis curve pattern on time-drawdown plot. Passage of time causes significant drawdowns even in the unpumped aquifer. On the contrary, ignorance of the aquifer storage leads to significant errors at small

Table 24 Drawdown Differences, $(s_m - s)$ (m)

Time, t (min)	$s_m = 315$ cm $r = 20$ m	$s_m = 128$ cm $r = 70$ m	$s_m = 75$ cm $r = 90$ m	$s_m = 37$ cm $r = 120$ m
2	280	103	57	21.5
3	261	94	52.5	19
4	249	89	49	18
5	237	85	46	17
10	195	69	38	13
15	177	63	31	11
20	158	58	28	10
30	130	52	26	9.5
40	125	48	24	9
50	115	44	21	8.5
60	103	40	20	8
90	70	35	15	7.6
120	55	29	14	5.5
180	45	27	13	5
240	31	17	10	4
300	22	12	9	4
360	17	10	8	3.5
420	15	7	7	3.0
480	11	5	5	3.0
570	8	4	4	2.5
650	5	3	3	2.0

POROUS MEDIUM AQUIFER TESTS

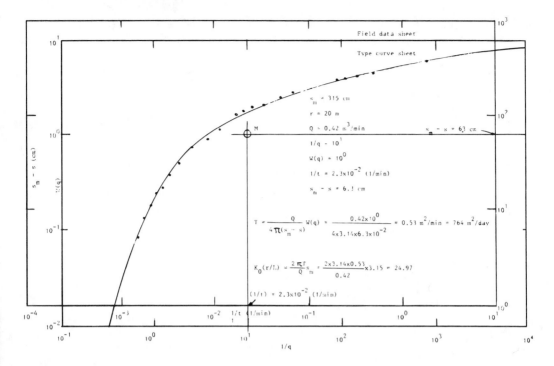

Figure 54 Hantush-Theis method ($r = 20$ m).

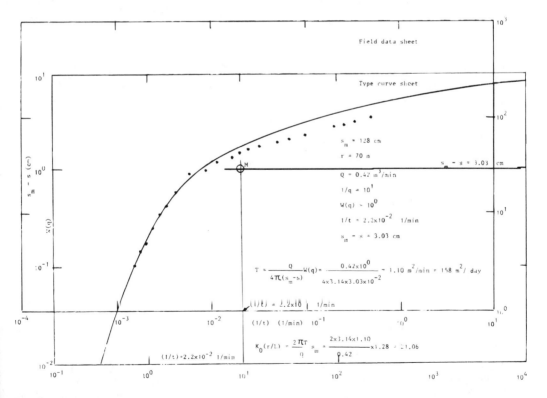

Figure 55 Hantush-Theis method ($r = 70$ m).

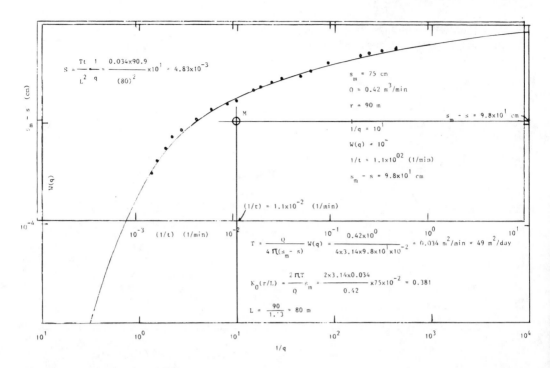

Figure 56 Hantush-Theis method ($r = 90$ m).

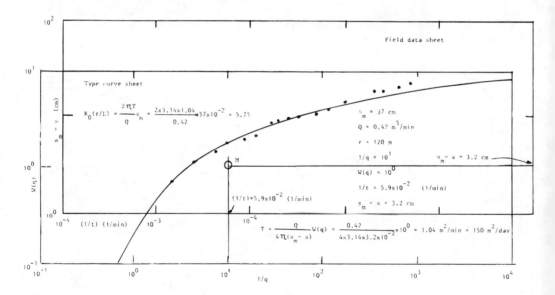

Figure 57 Hantush-Theis method ($r = 120$ m).

times holding true for large times. At every instant of time there are errors and sometimes with significant deviations from the Theis curve. A general solution of leaky aquifers with the release of these two assumptions was achieved first by Neuman and Witherspoon (1969a, 1969b). They applied the continuity and the Darcy equations with the proper initial and boundary conditions.

After involved mathematics they found out that the results can be expressed in terms of five dimensionless parameters that are lower and upper leakage functions.

$$\frac{r}{B_l} = r\left(\frac{K}{K_l m_l m_u}\right)^{1/2} \tag{80}$$

and

$$\frac{r}{B_u} = r\left(\frac{K}{K_u m_u m_l}\right)^{1/2} \tag{81}$$

respectively, and lower and upper storage-related leakage functions

$$L_l = \frac{r}{4m_l} = \left(\frac{KS}{K_l S_l}\right)^{1/2} \tag{82}$$

and

$$L_u = \frac{r}{4m_u} = \left(\frac{KS}{K_u S_u}\right)^{1/2} \tag{83}$$

and finally the dimensionless time factor u as

$$u = \frac{S_l r^2}{K_l t} \tag{84}$$

These five dimensionless parameters define the solution of leaky aquifer through a well function, written implicitly as

$$W\left(u, \frac{r}{B_u}, \frac{r}{B_l}, L_u, L_l\right) = \frac{4\pi T}{Q} s(r, t) \tag{85}$$

in which $s(r,t)$ is the drawdown in the pumped aquifer. The explicit form of Equation 85 is given by Neuman and Witherspoon (1969a) together with various special case solutions of the same equation. Tabulation of this well function would require many pages, therefore, the reader is referred the original reference.

Table 25 Hantush-Theis Method Parameter Estimates

Well no.	Field data Rad. dis. (m)	Field data $(s_m - s)_M$ (m)	Type curve $1/t_M$ 1/min	Type curve $1/q$	Type curve $W(q)$	Equation 70 T (m²/d)	Equation 62 K_0 (r/L)	Table 9 L (m)	L (m)	Equation 69 S
W1	20	0.63	0.023	10	1	764	24.97	—	—	—
W2	70	0.30	0.022	10	1	1588	21.06	—	—	—
W3	90	0.098	0.011	10	1	49	0.381	1.13	80	4.9 × 10⁻³
W4	120	0.032	0.0059	10	1	1504	5.750	—	—	—

Note: Disharge $Q = 4.6$ m³/min.

Table 26 Slope Matching Calculation for Leaky Aquifer

Time (day)	Draw-down (m)	Draw-down difference (m)	Slope	q	W(q)	T m²/day	K₀(r/L)	r/L	L (m)	S
2.43 × 10⁻²	0.069	0.078	−0.469	8.9 × 10⁻²	1.95	1514	1.837	0.183	492	1.710 × 10⁻³
3.06 × 10⁻²	0.077	0.070	−0.440	7.5 × 10⁻²	2.11	1824	2.214	0.126	715	1.455 × 10⁻³
3.75 × 10⁻²	0.083	0.064	−0.603	1.6 × 10⁻¹	1.41	1337	1.623	0.230	392	2.040 × 10⁻³
4.68 × 10⁻²	0.091	0.056	−0.480	9.2 × 10⁻²	1.90	2055	2.494	0.093	968	1.115 × 10⁻³
6.74 × 10⁻²	0.100	0.047	−0.746	2.5 × 10⁻¹	1.04	1345	1.632	0.228	395	2.324 × 10⁻³
8.96 × 10⁻²	0.1	0.038	1.026	4.9 × 10⁻¹	0.60	952	1.155	0.382	236	3.125 × 10⁻³
1.25 × 10⁻¹	0.120	0.027	1.399	7.8 × 10⁻¹	0.33	734	0.891	0.520	173	4.540 × 10⁻³
1.67 × 10⁻¹	0.129	0.018	2.243	1.7 × 10⁰	0.08	274	0.332	1.170	77	3.593 × 10⁻³
2.08 × 10⁻¹	0.136	0.011	3.295	3.7 × 10⁰	0.02	112	0.136	1.850	49	2.670 × 10⁻⁴
2.50 × 10⁻¹	0.141	0.006	1.174	5.9 × 10⁻¹	0.47	4765	5.783	—	—	—
2.92 × 10⁻¹	0.142	0.050	1.698	1.1 × 10⁰	0.20	2420	2.937	0.058	1551	2.670 × 10⁻⁴
3.33 × 10⁻¹	0.143	0.004								

Note: Extrapolated steady-state drawdown (maximum drawdown) is 0.147 m.

D. SLOPE METHOD

In fact, this method is the extension of confined aquifer slope matching method which has been explained in Section V.F. The slope, α_1, of the type curve for leaky aquifers can be found similarly to Equation 38 considering the Hantush-Theis method as

$$\alpha_1 = -\frac{e^{-q}}{W(q)} \tag{86}$$

By exchanging u with q, Table 12 can be used to find q vs. the slope. The procedure of aquifer parameter determination is similar to that of the confined aquifers. However, the calculation of the slopes from the field data should be as $\alpha_1 = Ln[s_m - s_{i-1})/(s_m - s_i)]/Ln(t_i/t_{i-1})$, for $i = 2,3,\ldots,n$. Due to the differences of drawdowns from the maximum drawdown, the slope value is always positive. Therefore, in the use of Table 12 for the leaky aquifer case, the sign is immaterial.

Example 17—The application of leaky aquifer slope method is presented for "Qude Korendijk" test data given by Kruseman and de Ridder (1979). The aquifer test data in an observation well at 90 m away from the main well are given in the first column of Table 26 together with the slope method calculations. These data were analyzed earlier by Kruseman and de Ridder (1979) using the Hantush method, by Rushton and Chan (1976) using a discrete numerical model, and by Chander et al. (1981a) using an iterated extended Kalman filter. From the aquifer test "Dalem" the data from piezometers at 30, 60, 90, and 120 m away from the main well were used for the numerical calculations of S, T, and L. The results are shown in Table 27 together with the results of other methods.

In this table the estimates of Bessel's function values are obtained from Equation 71. The values of the leakage, r/L, can be taken from tables of Bessel function values given in Table 22 leading to the calculation of leakage factor, L. The storage coefficient can be obtained from Equation 78.

Table 27 Aquifer Parameters by Different Methods

Model	Transmissivity (m²/d)	Storativity (×10⁻³)	Leakage factor (m)
Inflection	1665	1.70	600
Type curve	1729	1.90	900
Discrete	1680	1.50	850
Adaptive	1658	1.74	668
Slope	1576	2.40	505

Table 28 Confidence Limits of Parameter Estimates

Parameter	Lower confidence limit	Average value	Upper confidence limit	Standard deviation
S	1.512×10^{-3}	3.410×10^{-3}	3.300×10^{-3}	1.375×10^{-3}
T (m²/d)	787	1576	2365	1273
L (m)	200	505	810	465

The average parameter estimates are shown in Table 28 along with the corresponding upper and lower limits for 95% confidence interval. It is obvious that the upper and lower confidence limits include most of the parameter estimates obtained by the slope method.

E. VOLUMETRIC MODEL

Most often in the practical applications time-drawdown records are available from the main well, whereas models mentioned so far all require drawdown measurements in an observation well. As mentioned earlier in Section V.D, the Papadopulos and Cooper method provides determination of confined aquifer parameters from the main well time-drawdown record. In the case of leaky aquifers time-drawdown response will deviate from Papadopulos and Cooper curves especially at large times. Şen (1985) developed an approximate method to evaluate leaky aquifer parameters from the records in the main well only. The basis of his methodology is the successive volume variations of the depression cone that is intimately related to the storage and transmissivity properties of the aquifer and aquitard. The same set of assumptions as were for Hantush-Jacob method has been adopted in this approach. The pump discharge Q is composed of three components as

$$Q = Q_w(t) + Q_a(t) + Q_L(t) \tag{87}$$

in which $Q_w(t)$, $Q_a(t)$, and $Q_L(t)$ are time-dependent discharge contributions from the well storage, confined aquifer storage release at the well surface, and the total leakage rate over the depression cone, respectively. Considering the storage coefficient as the ratio of water volume taken from the pumped aquifer to the depression cone volume, Şen obtained type curve equation explicitly for the leaky aquifer as

$$u_w = \frac{1}{4\left(\dfrac{L}{r_w}\right)^2 \ln\left[1 + \dfrac{1 - e^{W(u_w)} - \dfrac{1}{S-1} W(u_w)}{4\left(\dfrac{L}{r_w}\right)^2}\right]} \tag{88}$$

in which u_w is the leaky aquifer dimensionless time factor defined as

$$u_w = \frac{r_w^2 S}{4Tt} \tag{89}$$

the well function $W(u_w)$ as

$$W(u_w) = \frac{4\pi T}{Q} s_w \tag{90}$$

Type curves for any S value can be obtained from numerical solution of Equation 88. Figure 58 shows the type curve set for $S = 10^{-4}$.

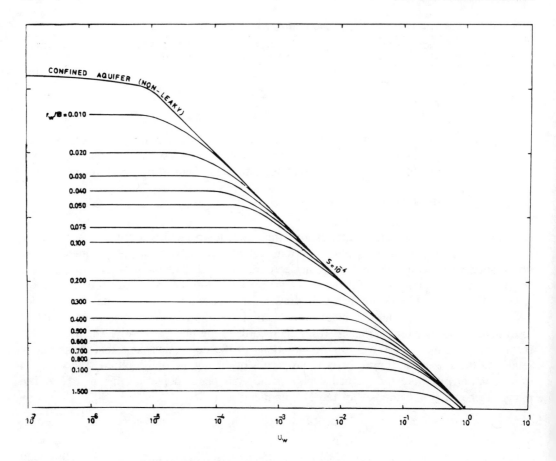

Figure 58 Volumetric method type curves ($S = 10^{-4}$).

Example 18—The numerical application of the methodology has been presented for the pumping test data obtained from a main well in the central Saudi Arabia. The well taps an aquifer in the Saq sandstone, which is overlain by a semipervious layer of shale. Above this shale layer, another layer of sandstone interbedded with shale extends to the surface. The well radius is 0.35 m, and the aquifer is pumped at a constant rate of 1.5 m³/min. On double-logarithmic paper, time-drawdown data are plotted as shown in Figure 59. After a series of comparisons with the type curve set one can see that the field points fall best along the curve for $r_w/L = 0.01$ and $S = 10^{-4}$. The match point chosen corresponds to $s_w(t) = 10^0$ and $1/t = 10^{-1}$ min^{-1}. On the type curve sheet, this point has coordinates as $W_M(u_w) = 1.3 \times 10^0$ and $u_w = 8.6 \times 10^{-6}$. Substitution of the appropriate numerical values into Equation 90 yields $T = 224$ m²/day, and Equation 89 leads to $S = 4.3 \times 10^{-4}$, which is fairly close to the initially chosen S value. The leakage factor can be found as $L = 35$ m.

It was observed that the field data can be easily fitted on two or three of the type curves and, therefore, it is not easy to find the best matching unless a complete steady state is reached.

VIII. BARRIER EFFECTS

Barriers may appear as a result of various hydrogeological situations. These are, in unconsolidated and semiconsolidated sedimentary rocks, pinching of aquifer layers due to change in facies; termination of an aquifer due to overlap; abutment of aquifer against impermeable

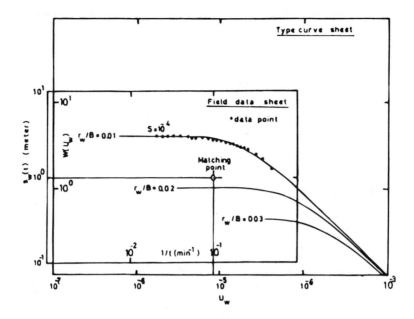

Figure 59 Large-diameter well leaky aquifer-type curve match.

layer; and barrier due to fault and deposition against a bed rock high. In consolidated sedimentary rocks barriers occur in abrupt thinning of weathered zones against massive rock; discontinuity in fracture or vesicular zones against massive rock; presence of dikes and veins of impermeable rock; and juxtaposition of fractured and weathered rocks against massive rocks due to faults or unconformity (Karanth, 1978).

In the preceding sections the discussion revolved around the wells in extensive aquifers. However, in practice extensive aquifers are rarely encountered, whereas most often different types of geological or hydrological discontinuities interfere with the groundwater flow such as dikes, faults, fractures, joints, seats, rock valleys, buried valleys, streams lakes, seas, etc. The existence of at least one of these discontinuities affects the groundwater movement and level fluctuations theoretically at all times. In general, their influence depends on abstraction duration and distance to the pumping well. Practically, short durations may not allow the depression cone to reach the discontinuity especially if it is far away. On the other hand, long duration of pumping leads to interference of depression cone with the discontinuity. The presence of at least one of these discontinuities in the vicinity of a well invalidates direct application of the above solutions. In literature, rather than seeking for new solutions, modification of existing aquifer solutions are preferred. To achieve such a goal the main assumption is that these continuities form effectively and fairly straight sections with full penetration. Only after such assumption does the method of images as suggested by Jacob (1950) become the most useful treatment of wells near the line of discontinuities.

From groundwater flow point of view, the discontinuities either prevent the continuity of flow or promote it. The former is referred to as the impervious barrier (Figure 60a, b, c, and d), whereas the latter acts as a source for recharge and is named as the hydrologic boundary (Figure 60e and f).

Physically, however, the impervious barriers appear as an accelerator of drawdown and on the contrary hydrologic barriers are decelerators of the drawdowns. Mathematically, to accelerate the drawdown, some additional water must be abstracted from the aquifer and deceleration is possible only by addition of water. In fact, the method of images has been devised by Ferris (1948) for the barrier effect treatment on the time-drawdown data in an observation well to determine the distance and location of barriers. According to this method

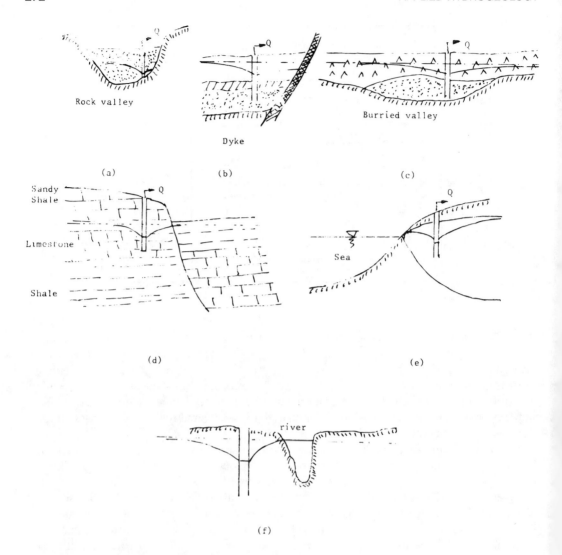

Figure 60 Pervious and impervious barriers.

the discontinuity is eliminated by a fictitious extensive aquifer that is composed of the real aquifer and its mirror extension from the discontinuity location. In such a mirror image the real well will have its image counter as in Figure 61. Depending on the acceleration or deceleration the image well will discharge or recharge the same quantity of water as in the main well.

Mathematically, the drawdown, s_T, in the main well after the depression cone touches the discontinuity will be summation (impervious barriers) or subtraction (hydrologic barriers) of individual drawdowns at the main well. In light of the aforementioned discussion the drawdown in a well near a discontinuity can be written in general as

$$s_T = s_r + \cdot \sum_{i=1}^{n} s_i \tag{91}$$

in which s_r and s_i are the drawdowns due to the real and imaginary wells. This expression is applicable in any type of aquifer with steady or unsteady flow situations.

POROUS MEDIUM AQUIFER TESTS

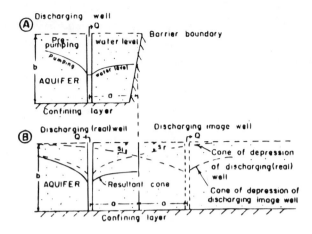

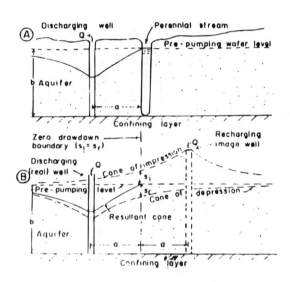

Figure 61 Mirror image of wells. (A) Actual well; (B) actual and mirror image well.

A. FERRIS METHOD

Ferris (1948) described a type curve-matching procedure by plotting r^2/t against drawdown, s, values in an observation well on a double-logarithmic paper and matching data with the Theis-type curve. Logically early time data are not affected by the barrier because of non-extensiveness of the depression cone. Therefore, Theis curve matching to these data yields estimations of S and T. Systematic departures for immediate and/or late time data from the Theis curve indicates the existence of a boundary effect. These departures are calculated for each r^2/t value and plotted on the same graph vs. drawdown to obtain departure curve. If the comparison of this departure data also show a boundary effect, the process is repeated until the last drawdown departure is dealt with. The observed data array can then be used to compare the distances between the observation and the image wells. For real and imaginary wells Theis equation gives

$$\frac{S}{4T} = \frac{u_r}{r_r^2/t_r} = \frac{u_i}{r_i^2/t_i} \qquad (92)$$

in which subscript r and i are for real and imaginary wells, respectively. If r_r^2/t_r and r_i^2/t_i values are selected on the early drawdown plot and first departure curves, respectively, such that $s_r = s_i$ then $u_r = u_i$ and Equation 92 gives

$$r_i = r_r \sqrt{\frac{t_i}{t_r}} \tag{93}$$

where r_i is the distance from the observation well to the image well and r_r is the distance from the real well to the observation well; t_r and t_i are the elapsed times at which the drawdown components are equal. Equation 93 becomes more practical after simple arrangement as

$$r_i = r_r \sqrt{\frac{r_r^2/t_r}{r_i^2/t_i}} \tag{94}$$

where r_r^2/t and r_i^2/t_i are r^2/t values from early drawdown and departure data, respectively.

B. WALTON METHOD

In the case of two or more image well effects Walton (1962) described a simplified method for boundary effect identification. Initially, as in the Ferris method early boundary intact time-drawdown data are matched with the Theis-type curve. Aquifer parameters are calculated with readings through a convenient matching point. The data points effected by the nearest boundary are then shifted to match the type curve extension. For more than one boundary the procedure is repeated. Two successive drawdown departures, s_{i1} and s_{i2} corresponding to times t_{i1} and t_{i2} are noted within the field of divergence of the type curve extension and the later data points. The times t_{p1} and t_{p2} at which the early drawdown data curve intersect drawdown values equal to these read at t_{i1} and t_{i2} are also read. The distances to the image wells are calculated by

$$r_{ij} = r_r \sqrt{\frac{t_{ij}}{t_{pj}}} \quad (j = 1, 2, \ldots, m) \tag{95}$$

where m shows the number of barriers. In using the Walton method the drawdown value at the second matching point must be twice the drawdown value at the first matching point, and the drawdown at third matching point must be three times the value of the first matching point and so on

C. STALLMANN METHOD

To solve a single boundary problem Stallman (1963) presented a set of modified Theis-type curves. Considering the total drawdown in Equation 91 he suggested that

$$s_T = \frac{Q}{4\pi T}[W(u_r) \pm W(u_i)] \tag{96}$$

where u_r and u_i are dimensionless time factors identical to the Theis method with different distances. The Theis ratio from Equation 10 becomes

$$u_i = \left(\frac{r_i}{r_r}\right)^2 u_r \tag{97}$$

Stallman proposed a set of type curves [$W(u_r) \pm W(u_i)$] vs. $1/u_r$ for several r_i/r_r dimensionless

POROUS MEDIUM AQUIFER TESTS

distance values as shown in Figure 62. The curves above the Theis curve are for discharging boundary effects whereas below the curve they are for recharging boundaries. Type curve-matching procedure between the field data and the most convenient curve from the set gives first the value of r_i/r_r and then from a match point coordinate readings are substituted into Equations 96 and 97 to solve for T and S provided that $r_i s$ are already calculated.

Example 19—For the application of the last three procedures time-drawdown data in an observation well given by Karanth (1978) are presented in Table 29. The bounded aquifer test is conducted at a constant discharge, $Q = 2.725$ m³/min. The distance of observation well to the main well is 30.48 m.

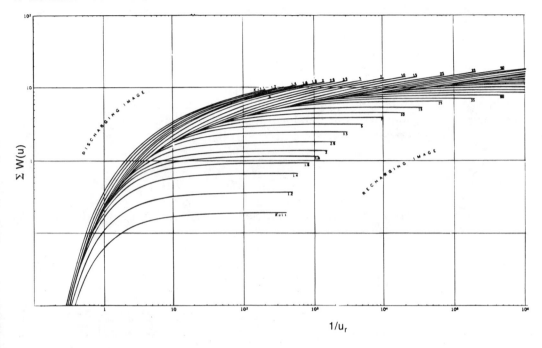

Figure 62 Stallman-type curves.

Table 29 Observation Well Time-Drawdown Data

Time since pump start (min)	Drawdown (m)	Drawdown on type curve trace (m)	Drawdown due to image well (s_i), (m)
1.0	0.34	0.34	
1.5	0.42	0.42	
2.0	0.51	0.51	
2.5	0.56	0.56	
3.0	0.61	0.61	
3.5	0.64	0.64	
4.0	0.66	0.66	
5.0	0.70	0.74	−0.04
6.0	0.72	0.78	−0.06
8.0	0.75	0.86	−0.11
10.0	0.77	0.92	−0.15
15.0	0.80	1.03	−0.23
20.0	0.81	1.10	−0.29
30.0	0.83	1.21	−0.38
40.0	0.84	1.30	−0.46
90.0	0.86	1.51	−0.63
150.0	0.86	1.70	−0.84
400.0	0.87	1.90	−1.03
700.0	0.88	2.10	−1.22
1440.0	0.89	2.20	−1.31

Stallman solution—The field data are plotted on a double-logarithmic paper as shown in Figure 63. Only the initial data points up to 4 min fit the Theis-type curve. The points after 4 min fall below the Theis curve, confirming the existence of recharge boundary. The values of drawdown on the Theis type curve, the value of observed drawdown, and the calculated values of drawdown departures, s_i, are shown in Table 29. The values of s_i are plotted vs. time on the same graph paper and matched with the type curve. No further boundary effect is apparent because all the points fall on the type curve trace. The values of $W(u)$, $1/u$, s, and t are recorded for the match points in the same graph. For an arbitrary drawdown value of 1 m time intercepts are obtained as 14 and 300 min. Substitution of these values into Equation 93, the distance of the image from the observation well, is computed as

$$r_i = 30.48\sqrt{\frac{320}{14}} = 146 \text{ m}$$

The time and drawdown data given in Table 29 are plotted on a semilogarithmic graph paper with time on the logarithmic axis as in Figure 64. The straight line passing through the initial points intercepts the time axis at 16 min for 1 m of drawdown. The second line intercept for 1 m drawdown difference due to image well is 760 min. Therefore,

$$r_i = 30.48\sqrt{\frac{760}{16}} = 210 \text{ m}$$

The aquifer parameters T and S can be calculated from the early time drawdown data by making use of the Theis or Jacob methods.

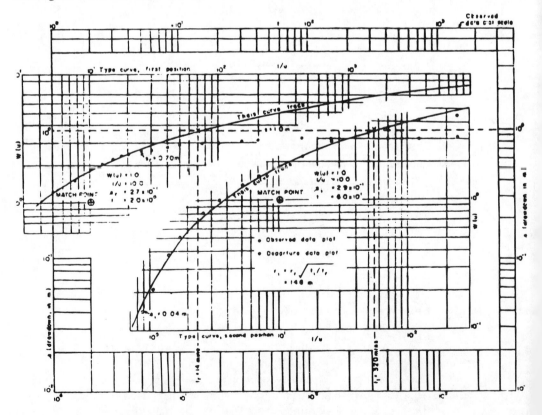

Figure 63 Bounded aquifer-matching procedure.

POROUS MEDIUM AQUIFER TESTS

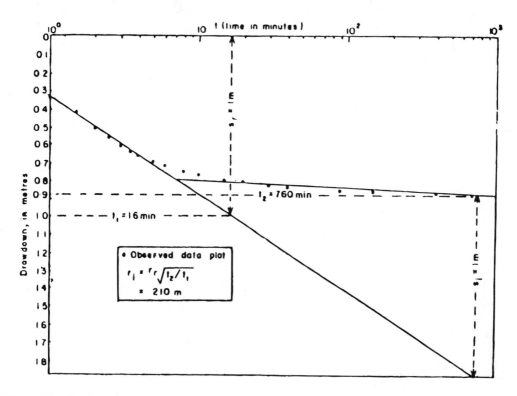

Figure 64 Semilog plot.

Walton solution—Figure 65 shows the early and late time-drawdown data matched with the convenient portion of the Theis-type curve. T and S are calculated by utilizing the match point coordinates of the first match point which are $W(u) = 5.0$, $1/u = 4 \times 10^0$, $s = 1.25$ m, and $t = 100$ min.

$$T = \frac{2.725 \times 60 \times 24 \times 5.0}{4 \times 3.14 \times 1.25} = 1250 \text{ m}^2/\text{day}$$

$$S = \frac{2.5 \times 10^{-3} \times 4 \times 1250 \times 100}{(30.48)^2 \times 1440} = 9 \times 10^{-4}$$

The distance to the recharge boundary is obtained from Equation 95 by substitution of the relevant values as

$$r_i = 30.48\sqrt{\frac{30}{1.3}} = 146 \text{ m}$$

The drawdown value (1.25 m) for the first match point (A in Figure 65) is nearly twice the drawdown value of 0.61 in the second match point.

Stallman solution—The time-drawdown data are plotted and matched with the Stallman's type curve set as shown in Figure 66. Aquifer parameters are calculated similar to the Theis procedure by substituting the match point coordinates $W(u) = 10^0$, $1/u = 10^0$, $s = 0.27$ m, and $t = 2$ min into relevant equations leading to

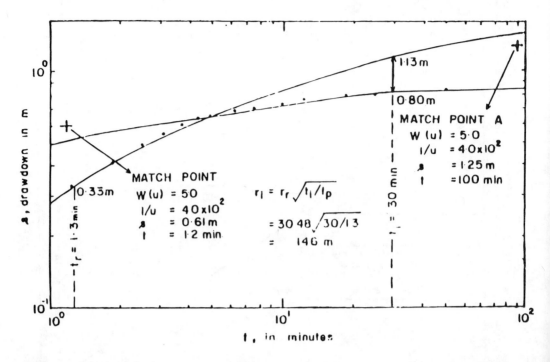

Figure 65 Walton method-type curve matching.

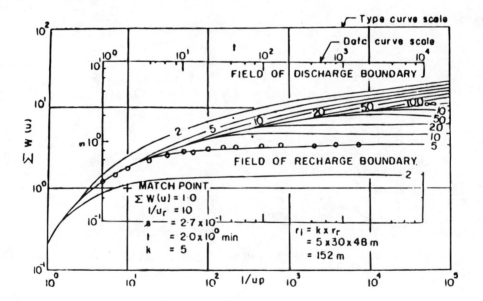

Figure 66 Stallman-type curve matching.

$$T = \frac{2.725 \times 60 \times 24 \times 1.0}{4 \times 3.14 \times 0.27} = 1157 \text{ m}^2/\text{day}$$

$$S = \frac{0.1 \times 4 \times 1157 \times 2}{(30.48)^2 \times 1440} = 6.9 \times 10^{-4}$$

For the selected type curve $r_i/r_r = 5$; therefore, $r_i = 5 \times 30.48 = 152$ m.

POROUS MEDIUM AQUIFER TESTS

D. RORABOUGH METHOD

Respective drawdowns s_{rp} and s_{r0} in pumping and observation wells near a recharge boundary have been expressed by Rorabough (1953) for steady-state flow as

$$s_{rp} = \frac{2.30Q}{2\pi T} \log \sqrt{4a^2 + r_r^2 - 4ar_r \cos\left(\frac{\phi}{r_r}\right)} \tag{98}$$

and

$$s_{r0} = \frac{2.30Q}{2\pi T} \log \frac{2a - r_w}{r_w} \tag{99}$$

where a is the distance from pumping well to the recharge boundary; ϕ is the angle between the time joining the pumping and image wells and another line joining pumping and observation wells (see Figure 67) and r_w is the radius of the main well.

IX. STORAGE COEFFICIENT DETERMINATION

Theoretically, Equation 9 in Chapter 8 represents a straight line on a semilogarithmic paper with drawdown vs. distance of observation wells from the main well. It is known from Equation 12 that the slope of this line gives the estimate of transmissivity coefficient. It is suggested herein that the revolutionary volume of the same straight line about the drawdown axis is related to the depression cone volume, V_D. If this volume is known then the storage coefficient can be estimated from Equation 11 in Chapter 4 as $S = Qt/V_D$. The necessary steps for the suggested procedure are as follows:

1. Plot the observed late drawdown at available distances on a semilog paper with times on horizontal logarithmic axis.
2. Draw the best-fitting straight line through the scatter points of data.
3. Extend this line on both directions until the intercept points with the vertical and horizontal axis.
4. Determine the slope, Δs, and then intercept values, S_L and r_U of this line on drawdown and distance axis, respectively. Of course, physically r_U corresponds to the radius of influence and the drawdown value beyond this distance is equal to zero everywhere within the aquifer. Besides, the drawdown intercept has the smallest distance, r_L, from the main well center (see Figure 68).
5. It is well known that the slope of the fitted line theoretically can be found from Equation 11 in Chapter 8.

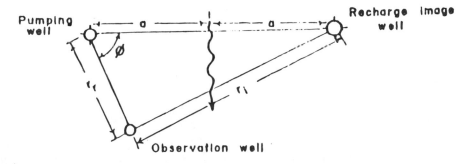

Figure 67 Rorabough method setup.

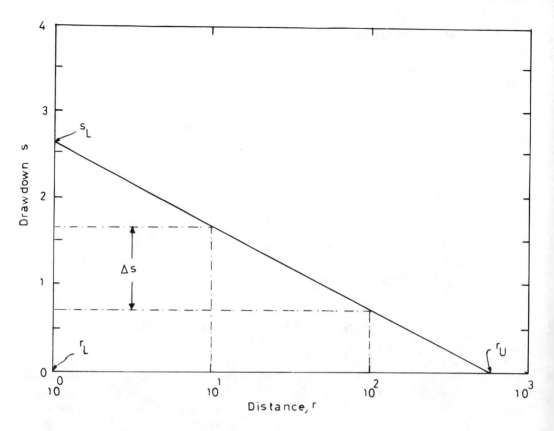

Figure 68 Schematic distance-drawdown plot.

6. Calculate the area, A, of the triangle constituted by the two axis and the fitted straight line in Figure 68:

$$A = \frac{1}{2}(r_U - r_L)s_L \tag{100}$$

which is logically related to the depression cone volume

7. Consider the revolution of this area about the drawdown axis but with back transformation of distances to linear scale which yields the depression cone volume estimate as

$$D = \pi \frac{(r_U - r_L)^2}{\ln\left(\dfrac{r_U}{r_L}\right)} s_L \tag{101}$$

8. The substitution of this volume expression into Equation 11 in Chapter 4 gives the desired result as

$$S = \frac{Qt}{\pi(r_U - r_L)^2 s_L} \ln\left(\frac{r_U}{r_L}\right) \tag{102}$$

Example 20—The validity of herein developed methodology for storage coefficient identification from quasi-steady-state flows in confined aquifers or for perfectly steady-state flows

POROUS MEDIUM AQUIFER TESTS

in leaky aquifers is presented through applications to field data presented by Kruseman and de Ridder (1979). First, Oude Korendijk aquifer test data are reproduced in Table 30. The pumping discharge is given as $Q = 788$ m^3/day and the quasi-steady-state flow is reached after $t = 830$ min.

The plots of distant-drawdown data on a semilogarithmic paper as shown in Figure 69 give two points that are connected by straight lines and the procedure in the previous section gives the relevant quantities as shown in the second part of the table. The storage coefficient estimates show very good agreement with Theis (1.6×10^{-4}) and Jacob (1.7×10^{-4}, 4.1×10^{-4}, and 1.7×10^{-4}) method estimates (Kruseman and de Ridder, 1979).

Another set of data for the application is also given by Kruseman and de Ridder (1979) in the case of transient flow. The plots of distance-drawdown values from two piezometer at distances 30 and 90 m after 140, 300, 600, and 830 min from the aquifer test start are shown in Figure 70 with four parallel straight lines. The implementation of steps in the previous section gives rise to relevant data as shown in Table 31 for various cases.

It is interesting to notice in this table that changes in the time duration lead almost to equal storage coefficient values that all have the same order of magnitude. Besides, they also compare well with the results obtained from other techniques.

Table 30 Aquifer Test Data and Results

r_1 (m)	r_2 (m)	s_1 (m)	s_2 (m)	Δs (m)	r_u (m)	r_L (m)	s_L (m)	V_D ($\times 10^3$ m^3)	T (m^2/d)	S ($\times 10^{-4}$)
30	90	1.088	0.716	0.64	810	0.1	3.00	687	370	6.69
0.8	30	2.236	1.088	0.73	1000	0.1	2.87	979	396	4.86
0.8	90	2.236	0.716	0.68	850	0.1	2.88	722	389	6.28
								Averages	385	5.94

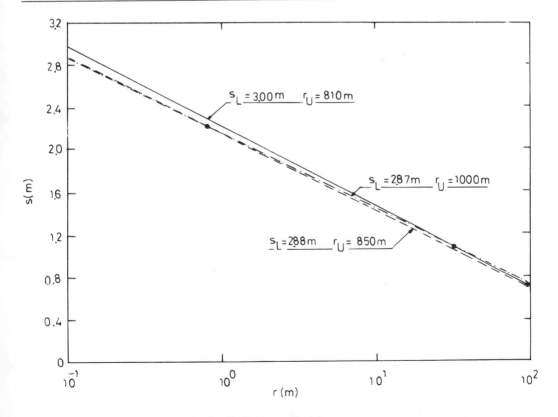

Figure 69 Distance-drawdown plot for Qude Korendijk data.

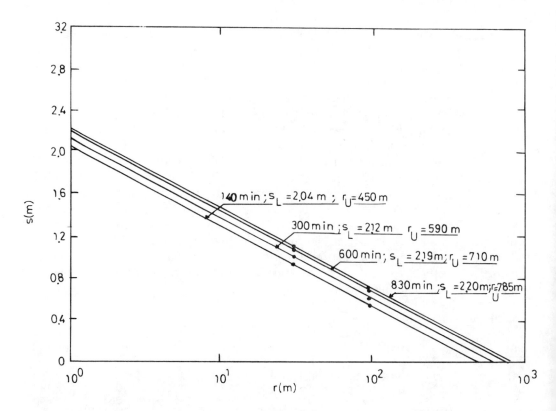

Figure 70 Distance-drawdown plot for the Jacob method.

A critical point in the application of the method appears in choosing the lower distance value. Logically, it might be taken as equal to the well diameter. However, any small distance choice does not make significant changes in the final storativity estimation. For this purpose, r_L is chosen as equal to 10 and the corresponding drawdown as well as storativity calculations are presented in the last two columns. It is obvious that, in spite of significant difference in r_L values, the storativities do not deviate significantly from each other.

Finally, the leaky aquifer steady-state flow data are also adopted from Kruseman and de Ridder (1979). The distance-drawdown plot is reproduced in Figure 71. The pump discharge is 761 m³/day and the steady state is reached after about 0.4 day from the pump start. The substitution of slope value from Figure 71 into Equation 15 yields that $T = 2020$ m²/day. Additionally, the necessary quantities for the storage coefficient calculation are $r_U = 1100$ m; $s_L = 0.32$ m; and $r_L = 6$ m. The substitution of these values into Equation 102 yields $S = 1.4 \times 10^{-3}$. This value compares very well with the storage coefficient estimates obtained from the application of other techniques such as the Hantush inflection point and Hantush-type curve methods that gave, respectively, 1.7×10^{-3} and 1.5×10^{-3} (Kruseman and de Ridder, 1979).

X. HYDROGEOPHYSICAL CONCEPTS

Aquifer test data analysis is an art leading to reliable hydraulic parameter identifications rather than a mechanical curve fitting. Most often aquifer test data processing is achieved by matching the data with a suitable type curve without detailed interpretations of deviations from this curve. In fact, relevant interpretations might yield valuable qualitative and quantitative

POROUS MEDIUM AQUIFER TESTS 283

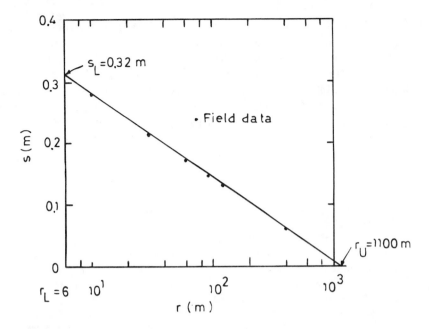

Figure 71 Leaky aquifer distance-drawdown plot.

features about the subsurface geological composition of the aquifer domain at least in the well vicinity. Available aquifer test data in terms of time- and/or distance-drawdown records help to identify the aquifer parameters such as the storage coefficient, transmissivity, hydraulic resistance, delayed yield factor, leakage factor, etc. However, any mechanical curve fitting without the qualitative interpretation of the data is an incomplete task. Type curve fittings devoid of physical reasoning lead to erroneous hydraulic parameter estimations and interpretation of the groundwater hydraulic mechanism. It is, therefore, advocated herein that, prior to application of any ready-type curves to available data, one should do one's best to extract physical interpretations from different types of paper (ordinary or semi- or double-logarithmic papers). The geology must not be forgotten in any hydrogeophysical reasoning.

With the advances in analytical and numerical modeling of groundwater flow toward wells, many scientific studies have discussed the response of an idealized aquifer geometry or flow regime under a set of simplifying assumptions with initial and boundary conditions. However, all the theories published so far and yet to be published should be viewed with their restrictive assumptions in any application. For instance, the Theis (1935) solution and its various modifications are derived on the basis of ideal initial as well as boundary conditions in addition to some assumptions not generally found in nature except in the average sense. In some cases, the deviations of the field data from the type curve are more significant than the use of the curve itself in calculating the aquifer parameter values. Hence, useful physical information should be obtained from the interpretation of these deviations by means of hydrogeophysical concepts. This term is coined herein distinctively from the classical terminologies of "geohy-

Table 31 Storage Coefficient Determination

Time (min)	Δs (m)	r_u (m)	r_L (m)	s_L (m)	S ($\times 10^{-4}$)	r_L (m)	S ($\times 10^{-4}$)
140	0.76	450	1.0	2.04	3.62	10.0	3.76
300	0.76	590	1.0	2.12	4.53	10.0	2.97
600	0.76	710	1.0	2.19	6.23	10.0	4.13
830	0.76	785	1.0	2.20	7.12	10.0	4.78
				Averages	5.35	10.0	3.91

drology", "hydrogeology", and "geophysics". It can be defined as a branch of earth sciences that deduces scientific information and interpretations about the physical behaviors of geological formations that are saturated with water. Hydrogeophysical concepts can be applied to aquifer test data only when there are deviations from the matched-type curves. Furthermore, deduction of information is achieved through the plot of drawdown vs. time or distance on various types of paper. It is, in fact, similar to a physicist trying to obtain information by examining X-ray sheets, likewise, the hydrogeologist should learn simple ways of qualitative field data sheet interpretations. Such simple interpretation techniques have been provided by Şen and Al-Baradi (1991). Hydrogeophysicists help to gather information about the following main points:

1. The geometrical type of streamlines that represents groundwater flow pattern within a certain flow domain. Among the most common streamline types are regular, such as radial, elliptical, linear, and spherical, or irregular patterns.
2. Hydraulic flow regimes that show the energy dissipation during the groundwater flow. In general, Darcian (linear) and non-Darcian (nonlinear) flows are two complementary alternatives. However, groundwater flow laws have been applied in aquifer analysis overwhelmingly for Darcian cases, (Theis, 1935; Hantush, 1956; Boulton, 1963, etc.) whereas non-Darcian flows are relatively new in groundwater flow toward wells, (Şen, 1986b, 1988, 1989).
3. Flow domain medium types are either porous or nonporous, such as fractured or karstic media. Again, most of the attention has been directed to porous medium, whereas others appear rather scarcely in the literature. This is due to difficulties in the analytical solutions of groundwater movement equation.
4. Hydrogeophysical concepts help to determine whether the aquifer material is homogeneous or not. In this connection it is also possible to know the spatial changes that might occur in the hydraulic parameters, e.g., the classical barrier effect, etc.
5. The checking of Jacob straight line validity is also among the most important hydrogeophysical concepts (Şen, 1989).
6. Identification of major aquifer types (confined, unconfined, or leaky) and related relevant interpretations are among the duties of hydrogeophysicists.
7. Identification of aquifer parameters from an unsteady groundwater flow record such as the aquifer test data. The best available average solution results from the suitable-type curve matching to these data usually on double-logarithmic paper. The local deviations from such a curve must be well documented. In some special cases similar graphical averaging may be accomplished by using semilogarithmic paper. It is among the scopes of the hydrogeophysical concepts to identify the local significant persistent deviations with plausible physical explanations.
8. The type curve methods assume that the aquifer is aerially unlimited, i.e., extensive. On the other hand, Dupuit assumption in unconfined aquifers implies that in most groundwater flows the slope of the phreatic surface is very small. The fact that these conditions do not occur in nature leads to deviations, i.e., errors hidden in the aquifer test data. Such differences do not make great harm to the quantitative accuracy in many cases but need further reasoning with hydrogeophysical concepts for additional and practically useful interpretations.
9. Due to elastic lag in confined aquifers and especially capillary fringe lag in unconfined aquifers, the storage coefficient calculation from short aquifer tests leads to comparatively smaller values than the asymptotic true value. Besides, due to aquifer material heterogeneity and/or thickness variations, the transmissivity might show fluctuations around an average value. Such systematic or erratic deviations in the parameter estimations may be dealt with the hydrogeophysical concepts. It is very difficult to determine the aquifer parameters in heterogeneous aquifers especially by means of conventional aquifer tests. In addition, there is also a need to address the uncertainties in the parameters being identified. Attempts have been made to study these by means of so-called stochastic continuum methods (Follin, 1992) as well as a discrete approach (Black, 1993). These studies show clearly that the presently used aquifer test methods based on various types of idealized flow patterns are incorporated (to various degree) because of problems with definition of the support scales, etc. It is in fact

already known beforehand that it is irrelevant to perform the analysis on the basis of such simplified solutions.

As a consequence, today many investigators try to apply statistical techniques (Marsily et al., 1983) combined with inverse modeling (LaVenue and RamaRao, 1992) to determine the hydrological parameters by matching observed and model curves of, for example, pumping tests, rather than to try to match standard type curves based on highly idealized flow patterns. This is particularly true in cases where it is obvious that observation data will deviate significantly from those of the type curves. Moreover, this approach provides measures on the uncertainties in the parameter data.

Nevertheless, the situation today is that many investigators (particularly in engineering problems) apply the "classical" aquifer test methods. Hence, it is of value to see what possible information could be extracted from these tests. The present paper contributes to this by demonstrating some ideas on how to explain various causes for the deviations from the "standard curves". The various explanations to the deviations will then guide the practicing engineer on how to proceed with the aquifer analysis.

It is not possible to present a complete list of what a hydrogeologist can interpret through the hydrogeophysical concepts but with practice he will be equipped with further insights into the meaningful physical aquifer test data interpretations.

Examples 21—The best way to gain appreciation in hydrogeophysical concepts is through the worked field examples. For this purpose the first example is taken from Kruseman and de Ridder (1979) as the aquifer test data from Oude Korendijk. The geological well site description, type curve application, and quantitative aquifer parameter estimations are presented by them in detail. For the sake of hydrogeophysical concepts discussion, their Figure 3.6 is reproduced herein as Figure 72. A close inspection of relative data points with respect to the Theis curve shows very clearly that there are significant deviations. Unfortunately, a mechanical

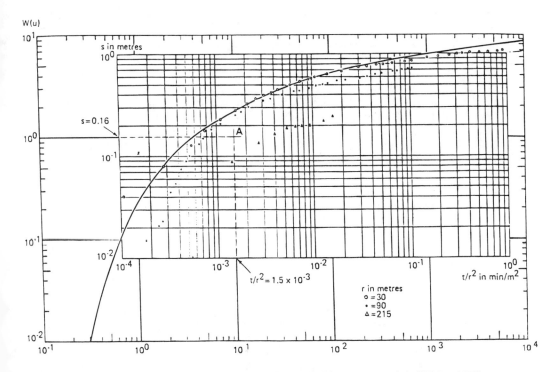

Figure 72 Type curve and field data from Qude Korendijk (Kruseman and de Ridder, 1990).

type curve matching to overall data gives only susceptive parameter estimates as average values equal to $S = 1.6 \times 10^{-4}$ and $T = 392$ m²/day. However, the following hydrogeophysical interpretations can be made concerning the data.

1. The type curve matching has been mechanical without any further concern since all of the field data fall below the Theis curve. However, to have average aquifer parameters the field data should lie in a rather balanced manner above and below the type curve, if possible. Nevertheless, the first part of the type curves ($r = 30$ m) seems to fit the Theis curve but afterward deviations occur perhaps due to leakage.
2. It is not possible to match the field data in Figure 1 with the Theis-type curve as shown but other suitable type curves must be tried in to have representative aquifer parameter estimates.
3. For small r/t^2 values, i.e., either for small times or large distances the field data lie consistently below the Theis-type curve. It means physically that the field data pattern has steeper slope than the type curve for early portions of time-drawdown record and smaller slope for late portion. However, for moderate data portions this pattern has almost the same slope as is evident from Figure 79. An increase in slope implies excessive energy dissipation than the laminar (Darcian) flow. Such an interpretation is indicative of non-Darcian flow which, in turn, reminds us that the flow domain might be nonporous.
4. Among the classical nonporous media are the media of coarse or very fine porous medium, fractured medium, or karstic medium. It has been already shown by Şen (1989) that the flow regime is non-Darcian for these data since the Reynolds number is 27 which is far greater than the upper limit (10) of Darcian flow.
5. It is interesting that the initial portion of field plot (about ten data points) confirms very clearly the existence of a straight line. It implies that

$$\log W(s) \alpha \log\left(\frac{t}{r^2}\right) \qquad (103)$$

or

$$\log W(u) \alpha \log\left(\frac{1}{u}\right) \qquad (104)$$

in which α is the proportionality sign. Such linear relationships on double-logarithmic paper have been observed in groundwater flow toward wells in the case of large-diameter well aquifer tests (Papadopulos and Cooper, 1967). However, the slope in their case is equal to 1 whereas herein it is equal to almost 2. Herein, the large diameter is out of question but still there is a straight line appearance with greater slope that one. Increase in the slope might show that the aquifer material around the well is composed of coarse grains; therefore, there is rather an easy (almost without resistance) entrance of water to the well.
6. The aquifer test data from piezometer at 215 m away fall further below the other field data from other piezometers. This discrepancy indicates systematic heterogeneity in the aquifer material composition, i.e., changes in hydraulic properties with distance from the well.
7. Only moderate data from piezometer $r = 90$ m and early as well as moderate data from other piezometers follow type curve rather closely. However, late data patterns from both piezometers show systematic downward deviations. This indicates recharge to the aquifer from adjacent layers. In fact, the geological cross-section presented by Kruseman and de Ridder for the aquifer test site shows that the main aquifer is composed of coarse sand and gravel and overlain by a rather thick fine-sand layer that gives rise to leakage.
8. The field data points are not haphazardly different from each other and, therefore, it can be concluded that the aquifer has regional homogeneity.
9. The field data plot from the piezometer at far distance has a sort of S-shape that might indicate delayed recharge effect.

The second example is also from Kruseman and de Ridder (1979) concerning the Jacob straight line fit to field data from piezometer $r = 30$ m during an aquifer test at Oude Korendijk. Once again, the mechanical fitting (this time, a straight line) is very obvious. In theory, the straight line is valid only for the late time-drawdown data. On the contrary, in Figure 73 the straight line is fitted to early time data that is against the basic theoretical principle. It is interesting to note that the late time-drawdown data come along with another straight line which is correct at least theoretically. On the basis of this line the aquifer parameters are estimated as $T = 572$ m²/day and $S = 3.0 \times 10^{-5}$ which are significantly different from the ones given on the basis of early straight line as $T = 385$ m²/day and $S = 1.7 \times 10^{-4}$. Hence, there is 32% underestimation in the transmissivity and 82% overestimation in the storage value.

Furthermore, two straight lines with different slopes for the same aquifer data imply physical existence of boundary effect in the aquifer configuration. To check whether the Jacob method application is valid even with the late time-drawdown data, dimensionless time-drawdown data are calculated according to the procedure presented in Section V.B.4.4. This dimensionless data plot is presented in Figure 74 for late time. The important point is that the dimensionless straight line slope is equal to 1.5 which is significantly smaller than the standard Jacob slope of 2.3. It is known that the small slope implies recharge and this conclusion is consistent with type curve deductions in the previous paragraph.

The third data set is from the Saq sandstone aquifer which lies in the northwestern part of Saudi Arabia. The aquifer is of the leaky type and the conventional type curve has been matched to field data as shown in Figure 75. There is no question that the type curve represents fairly well the field data leading to average aquifer parameters. The deviations from the type curves give domain for the application of hydrogeophysical concepts. The main point of this stage is that, since there appears rather erratic variations in the drawdown especially for large times, one may suspect that the aquifer material is not homogeneous. Consequently, these variations give rise to local variations in the aquifer parameters. The effect of these erratic variations on the parameter estimations can be obtained from the slope-matching method

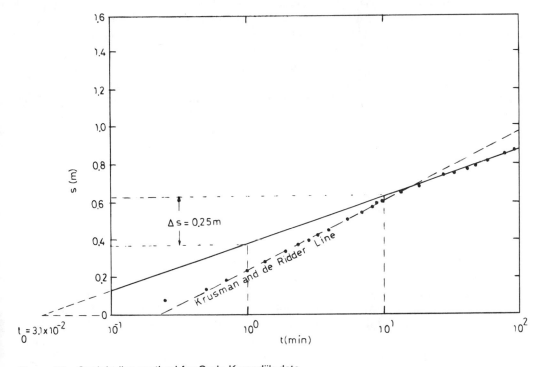

Figure 73 Straight line method for Qude Korendijk data.

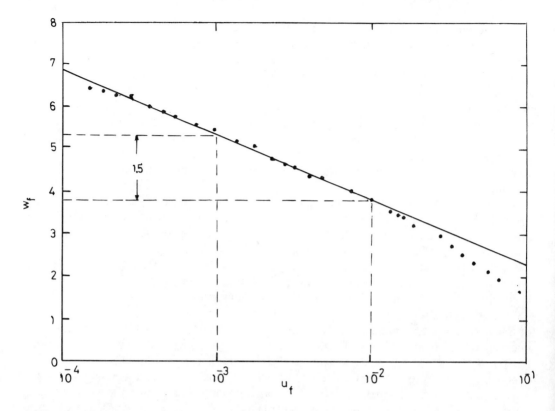

Figure 74 Dimensionless time-drawdown data.

presented in Sections V.F and VII.D. The direct application of this method yields a series of transmissivity and storage coefficient estimates. Their plots are shown in Figures 76 and 77, respectively, with averages represented by horizontal straight lines. It is very obvious that the aquifer parameters change from one constant value to a new constant value as the depression cone expands with time. The reader may make many useful interpretations from these figures concerning the aquifer domain in the well vicinity.

Last but not least, a final hypothetical example is considered where all of the field data lie on the type curve without any deviation, (see Figure 78). Nonexistence of erratic or systematic deviations hinder the employment of hydrogeophysical concepts. Besides, such a situation never appears in nature and its existence implies that all of the underlying assumptions in type curve derivation are valid exactly.

XI. AQUIFER HETEROGENEITY AND SAMPLE FUNCTIONS

An adaptive slope-matching method is used to evaluate the aquifer heterogeneity from constant rate aquifer test data. The basic aquifer parameters such as the storativity and transmissivity are considered to vary spatially and the effects of such changes appear as temporal variations during an aquifer test period. It is stressed that this is due to the depression cone expansion with time around the pumping well. The plots of individual parameters vs. time are referred to as the sample functions. These functions have erratic variations, indicating that the aquifers considered in this study have local heterogeneities. Furthermore, a moving average technique is applied to each parameter sample function for the purpose of finding

POROUS MEDIUM AQUIFER TESTS

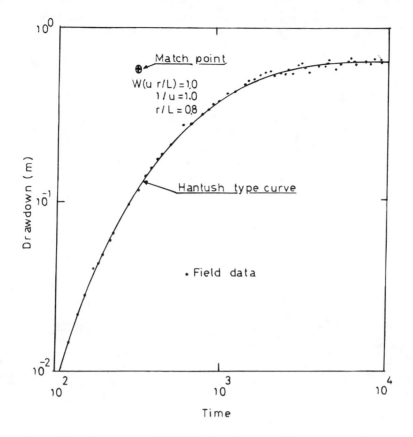

Figure 75 Leaky aquifer-type curve and field data from Saq sandstone.

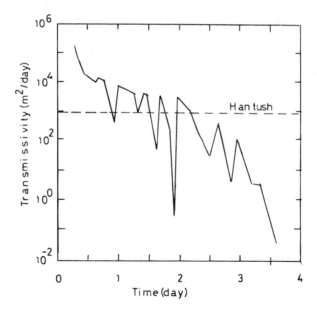

Figure 76 Variation in transmissivity.

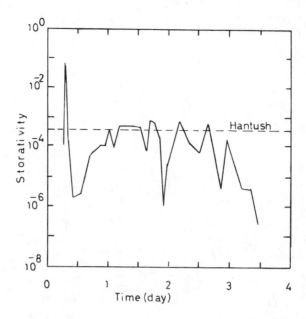

Figure 77 Variation in storage coefficient.

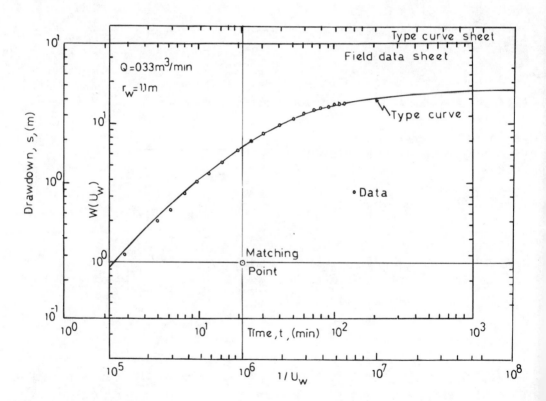

Figure 78 Perfect match of type curve to field data.

regional trends in the parameter variations. It is interesting to note that the arithmetic average values of sample functions yield aquifer parameter values as obtained by the classical type curve-matching techniques that assume that the aquifer concerned is homogeneous. Therefore, the use of the sample function concept is recommended for the detailed interpretation of the aquifer heterogeneity (Şen and Al Baradi, 1991).

Usually there is an apparent lack of consistency between the parameters such as the hydraulic conductivity, transmissivity, and storage coefficient determined by means of the aquifer tests and those obtained in the laboratory. This is due to the fact that the scale of observations is different and a natural aquifer may be considered as heterogeneous if sampled at different points by small sample volumes. However, in any aquifer test the same aquifer can represent homogeneity at least in the statistical sense. In other words, regional properties differ from local heterogeneous subunit properties. Hence, a natural aquifer can be idealized either as an equivalent homogeneous aquifer at a macroscale or as a random field at a microscale for any parameter considered.

Likewise, expansion of the depression cone with time during an aquifer test leads to the idea that initially this cone will have a small volume and it will attain its maximum expansion after a long time when the quasi-steady-state flow is reached. Therefore, during such an expansion the depression cone moves across different heterogeneous subunits. As a result, the heterogeneity affects the drawdown and consequently, the aquifer heterogeneities appear as hidden components within the recorded time-drawdown data in the field. It is essential to develop a technique to clearly detect and define aquifer heterogeneities and this is one of the most attractive aspects of modern aquifer testing interpretations. Testing can also indicate hidden anomalies that might not have been detected by other aquifer-modeling techniques. Numerous kinds of heterogeneities influence the time-drawdown response during aquifer testing. Lack of interpretation models hindered our ability to detect most heterogeneities.

Almost all of the aquifer test data evaluation techniques are based on the homogeneity assumption that is simplification in the analytical solution of groundwater flow problems. None of these techniques is capable of detecting the hidden components but rather they yield a single average value for the parameter estimate. There are few techniques that deal with the heterogeneous aquifer parameters but all of them consider systematic heterogeneities in the forms of either horizontal layering or vertical geological facies change or hydrologic barriers (see Section VIII). Field experiences, since the original work of Theis (1935), have indicated that always there are local discrepancies between the type curves and the time-drawdown data plots. At times, if the time drawdown plot deviates systematically from the type curve, that is expected theoretically, then the interpretation may be that either an impermeable or constant-head boundary may exist; if the deviations are in the form of a random pattern, the results of the analysis may be labeled as unreliable with the observation that the deviations may be due to either heterogeneous aquifer properties or variations of these parameters with time and/or space or both. The graphs that show the change of parameters with time are referred to as the parameters sample functions in this book.

Hence, it is among the objectives to estimate the spatial variations of aquifer parameters around the pumping well from observed time-drawdown measurements by means of the sample functions. The local heterogeneities are represented as individual deviations from the regional trend hidden within the sample function. Then, by using different orders of moving averages, the regional trends of these functions are depicted. It is observed that the overall arithmetic mean value of any sample function corresponds to the parameter estimates obtained from classical type curve-matching procedure.

Hydraulic parameters such as the transmissivity, storativity, hydraulic conductivity, leakage factor, etc., are related to the material and geometric properties of the aquifer domain. Almost in any analytical treatment of groundwater movement the material is assumed to be isotropic and homogeneous. Consequently, a single value for each parameter emerges after the application of the classical type curve-matching methods that are available. It is a known fact that, even though the aquifer might be isotropic and homogeneous, the values of the hydraulic conductivity and storage coefficient affect the depression cone around the pumping well; the smaller the hydraulic conductivity the narrower is the depression cone as shown in Figures 1 to 3. It is then physically plausible to ask whether some erratic changes during an aquifer test in the drawdown measurements are due to the expansion of depression cone with time? The answer to this question is affirmative because, for instance, any change in the geological facies such as the existence of an impervious boundary gives rise to an increase of drawdown more than usual after a certain time. It is by now accepted that such an increase of drawdown or effect of the barrier starts to occur at a time when the extent of the depression cone reaches this barrier, that is, the aquifer parameters obtained from aquifer test data, in fact, reflect the properties of material changes (heterogeneities) within the effective domain of depression cone. The question is then if erratic changes appear in S or K (i.e., heterogeneous aquifer material), will there be reflections of such a situation in the drawdown measurements? No doubt, there will be significant effects but none of the existing analytical solution methods of groundwater flow toward wells accounts for such local variations. Logically, the increment, ds, in the drawdown during a time interval, dt, means there is an increment in the depression cone volume, dV, and also some additional aquifer domain is sampled, dr, within the depression cone dominance (see Figure 79). If there exists any heterogeneity during such an increment the drawdown will be affected accordingly. For instance, local permeability variations due to a difference in compaction, aquifer thickness changes over short distances, random impermeable lenses, etc., all give rise to local erratic variations in the time-drawdown measurements. The similar argument is valid for distance-drawdown data. Hence, the drawdown increment with time (or distance) might be a good indication of aquifer heterogeneity. Derivation of Equation 7 with respect to distance yields after some algebra

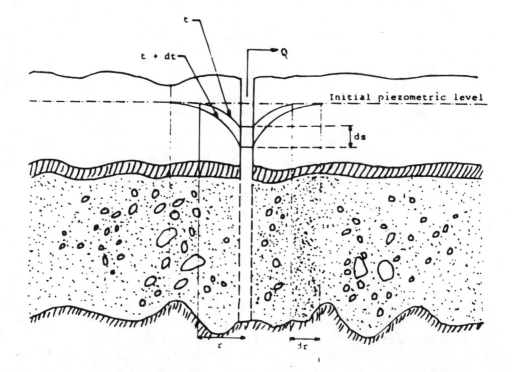

Figure 79 Depression cone expansion.

$$\frac{ds(r,t)}{dt} = -\frac{Q}{2\pi T} \frac{e^{-r^2 S/4tT}}{r} \tag{105}$$

This expression represents the radial expansion of the depression cone around the pumping well. It also indicates theoretically that the hydraulic gradient, $dh(r,t)/dr$, is inversely and nonlinearly related to the transmissivity as well as the storativity. Equation 105 is the basis of the classical Theis equation, provided that the aquifer is completely homogeneous and isotropic which gives way to its integration from r to ∞ leading to the Theis expression. However, if the aquifer modeled is not homogeneous, then the integration procedure is not applicable due to the expected spatial changes in T and/or S. In such situations, Equation 105 can be used in obtaining the local parameter estimations. To obtain the temporal changes in the drawdown during an aquifer test one can differentiate Equation 7 with respect to time according to Leibnitz's rule leading to

$$\frac{ds(r,t)}{dt} = \frac{Q}{4\pi T} \frac{e^{-r^2 S/4tT}}{t} \tag{106}$$

This expression is similar to Equation 105 but it represents the expansion of the depression cone with time. The left hand side of Equation 106 represents the slope of time-drawdown plot and theoretically this slope is also inversely and nonlinearly related to the aquifer parameters. Although in Equation 106 T and S are the only unknowns and the other terms are all known from an aquifer test performed in the field, unfortunately the solution of S and T is not uniquely possible. However, this difficulty can be overcome by the slope-matching method which estimates the local changes in the transmissivity and storativity values.

Local deviations of drawdown measurements from the relevant type curve might mean local changes either in the transmissivity and/or in the storativity or in any other hydraulic property. Because the transmissivity is a compound value of hydraulic conductivity, K, multiplied by the aquifer thickness, m, their individual or simultaneous changes also affect the drawdown variations during a pumping test. This is tantamount to saying that, even though the aquifer is homogeneous from a hydraulic conductivity point of view, the changes in the aquifer saturation thickness with the radial distance from the well imply corresponding changes in the transmissivity and storativity estimations. As mentioned earlier the time durations of such increases are also important.

Example 22—The heterogeneity of aquifer is sought for the Saq sandstone formation which lies in the northwestern part of Saudi Arabia. The surficial deposits covering the area consist mainly of silt and below with variable proportions of pebbles from quartz gravel and pebble of limestone and rocks of basement complex. The overlying Tabuk formation is a thick sequence of shale, clay, and sandstone units of marine and continental origin. Its thickness is about 500 m in the study area and age ranges from Ordovician to Devonian. At places the Tabuk formation provides confining layers for the Saq sandstone; otherwise, it is semipervious, constituting a leaky aquifer setup with the Saq formation. Three aquifer tests (two leaky, A1 and A2, and one confining, A3) are considered as the field time-drawdown data which are presented in Figures 80 through 82 with the most suitable type curve model matchings (Theis, 1935 and Hantush, 1956). It is obvious from these figures that there are erratic deviations from the type curves that indicate the heterogeneity of the Saq sandstone aquifer. These heterogeneities are accounted by the slope-matching procedure and consequently the series of storage coefficients, S, and transmissivities, T, are obtained with the results presented in Figures 83 and 84. A general inspection of the main values in these figures reveals that the storativity and the transmissivity values have rather random patterns of ups and downs. At times significant deviations appear from the constant parameter values obtained from the type curve matching. Compared to other well locations A3 has less heterogeneity because the

294 APPLIED HYDROGEOLOGY

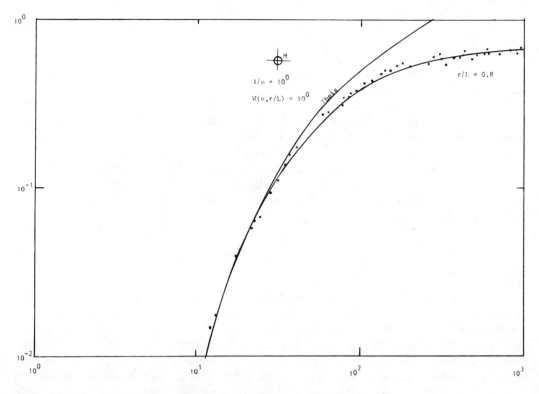

Figure 80 Hantush-type curve matching for well A1.

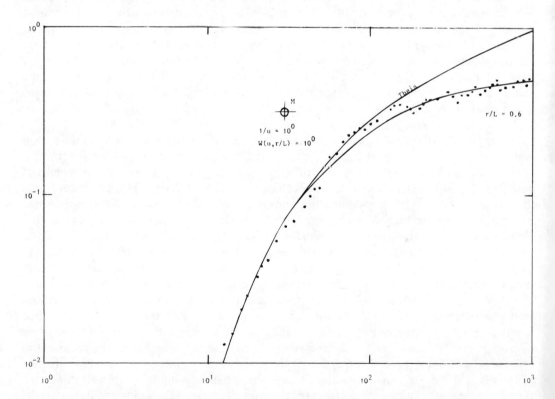

Figure 81 Hantush-type curve matching for well A2.

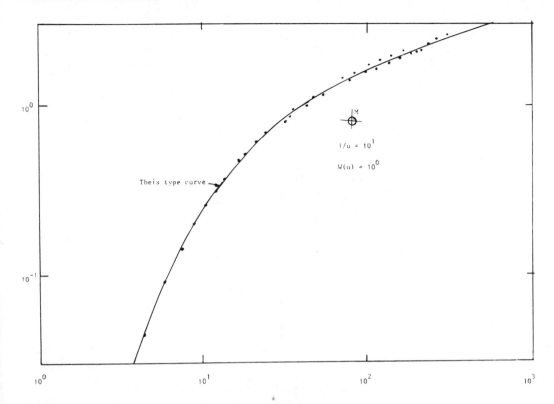

Figure 82 Hantush-type curve matching for well A3.

general fluctuations of the parameter values have relatively smaller amplitudes. The comparison of field data points and the type curve in Figure 85 confirms this statement. Furthermore, these fluctuations occur around more or less constant trends. The storativity sample functions in all other well locations occur around rather constant levels, whereas the transmissivity sample functions expose obvious trends in the forms of transmissivity decrease with time, i.e., with the expansion of the depression cone. This is tantamount to saying that, as the cone of depression expands, the aquifer material covered is less permeable. Within the depression cone the specific discharge becomes greater toward the pumping well and therefore fine particles are washed out more and more in the well vicinity, giving rise to the permeability increment toward the same well. To be able to detect a trend more explicitly, the moving average scheme is applied to each sample function and the results are shown on the same figures for moving average orders of three and ten. As the order increases the amplitude of fluctuations decrease and then it is possible to deduce whether the underlying trends are linear or nonlinear. The physical significance of moving average method application is to be able to depict the hidden trends, jumps, or functional spatial changes in the aquifer parameters around the pumping well. In Figure 84a there is a trend of exponential decrease in the transmissivity values which is very obvious with the moving average of order ten. However, the storativity sample function in Figure 84b for the same aquifer indicates a composite pattern of trends where initially there appears a sharp decrease and then an increase of storativity

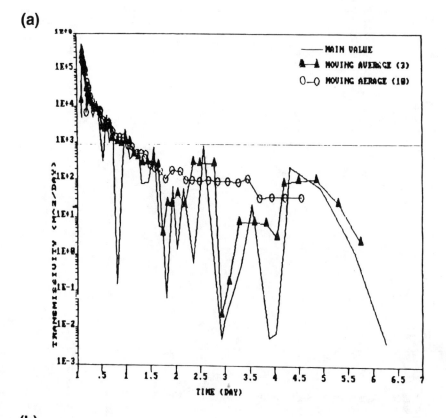
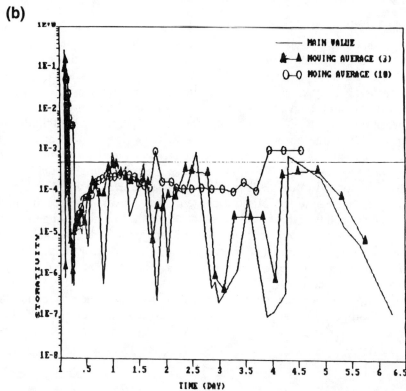

Figure 83 Well A1 sample functions. (a) Transmissivity; (b) storage coefficient.

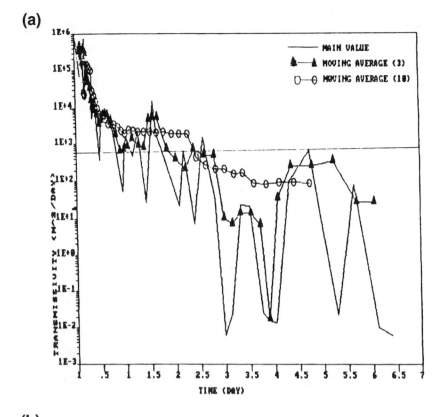
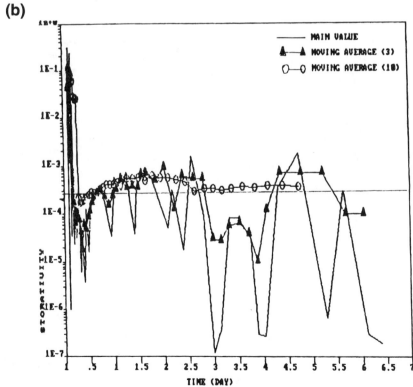

Figure 84 Well A2 sample functions. (a) Transmissivity; (b) storage coefficient.

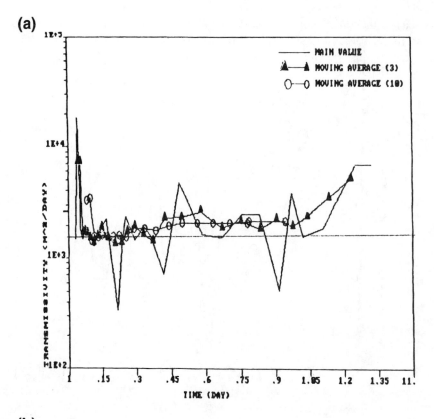
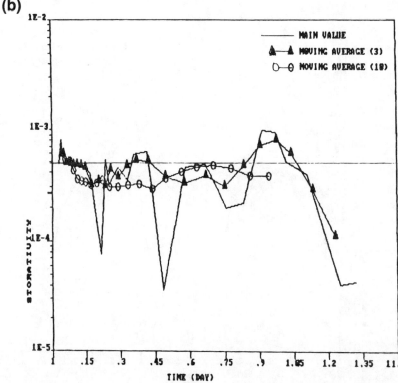

Figure 85 Well A3 sample functions. (a) Transmissivity; (b) storage coefficient.

takes place and after some stable level it decreases again. The sample function in Figure 85a indicates an initial decrease that is followed by a rather stable level in the moving average values of order ten but the storativity sample function in Figure 85b also has a stable trend.

XII. PITFALLS IN AQUIFER TEST ANALYSIS

A pitfall may be defined as taking of a false-logical path that may lead to absurd conclusions, a hidden mistake capable of destroying the validity of an entire argument. It is in fact a conceptual error into which analysts frequently and easily fall if they do not have sound plausible physical reasoning about the phenomenon concerned. Unfortunately, in many parts of the world aquifer test analysts overlook any geological, physical, or configurational prospects of the test environment, depending mechanically on derived-type curves such as mathematical treatment of time-drawdown data that might lead easily to unrealistic and unrepresentative aquifer parameter estimations and consequently to over- or underestimations of groundwater potentiality of an area. Many analytical models may not even have examples of actual data to fit. However, one of the benefits from the multiplicity of such models is that they give rise to similarities that may exist among the performance of completely different systems.

To reach reliable identification of aquifer parameter estimations it is useful to be aware of pitfall types that might occur either before the beginning of actual field data collection or during the data collection or after the type curve matching. The following points are among the pitfalls in an aquifer test.

1. The analysts must be familiar not only with different type curves but know the underlying assumptions' validity for the data at hand and field environment.
2. Starting data collection too early before the confirmation of steady-state flow condition.
3. Not looking for other plausible explanations for the outcomes.
4. Overdependence on statistical significance, particularly at the expense of practical significance.
5. Focusing only on the overall (average) results and not considering the local changes.
6. Not indicating the assumptions, uncertainties, and other limitations of the evaluation when presenting the findings.
7. Poor presentation of findings so that potential users cannot really understand them.
8. Late time- and distance-drawdown plots may not necessarily mean that the Jacob method is applicable.
9. In the literature it is stated that actually the data lie along straight lines only for sufficiently large time, or sufficiently small distances, (Bear, 1981, page 469). The second part of this statement is not plausible physically because the smaller the distance the greater is the hydraulic gradient and consequently there is a possibility of nonlinear flow for which the definition of transmissivity is not valid. On the other hand, most of the line sink or source solutions assume that the well diameter is equal to zero! Consequently, theoretical drawdown in such a well is extremely big. However, in physical reality the drawdowns are finite; therefore, theoretical drawdowns cannot be compared with actual ones in the well vicinity.
10. How does one distinguish the physical reality if the type curves are similar to each other? For instance, leaky aquifer curves are similar to exponentially decreasing discharge

REFERENCES

Bear, J., 1981. *Hydraulics of Groundwater*, McGraw-Hill, New York.
Black, J. H., 1993. Hydrogeology of Fractured Rocks—A Question of Uncertainty about Geometry, Int. Assoc. Hydrogeology XXIVth Congress, Norway, 783.
Boulton, N. S., 1963. Analysis of data from nonequilibrium pumping test allowing for delayed yield from storage, *Proc. Instn. Civ. Eng. (London)*, 26, 469.

Carslaw, H. S. and Jeager, J. C., 1939. *Conduction of Heat in Solids,* 2nd ed., Oxford University Press, London.

Chander, S., Kapoor, P. N., and Goyal, S. K., 1981a. Aquifer parameter estimation using Kalman filters, *J. Irr. Drain. Div.,* ASCE, Proc. Paper 19101, 107(IR1), 25.

Chander, S., Kapoor, P. N., and Goyal, S. K., 1981b. Analysis of pumping test data using Marquardt algorithm, *Ground Water,* 19(3), 275.

Chow, V. T., 1952. On the determination of transmissibility and storage coefficients from pumping test data, *Trans. Am. Geophys. Union,* 33, 397.

Cooper, N. N. and Jacob, C. E., 1946. A generalized graphical method for evaluating formation constants and summarizing well field history, *Trans. Am. Geophys. Union,* 27, 526.

Ferris, J. G., 1948. Ground Water Hydraulics as a Geophysical Aid, Geol. Surv. Div., Tech. Report 1, Michigan Department of Conservation, 16.

Follin, S., 1992. On the interpretation of double-packer tests in heterogeneous porous media: numerical simulations using the stochastic continuum analogue, SKB TR 92-36, Swedish Nuclear Fuel and Waste Management Co., Stockholm.

Hantush, M. S. and Jacob, C. E., 1955. Non-steady radial flow in an infinite leaky aquifer, *Trans. Am. Geophys. Union,* 36(1), 95.

Hantush, M. S., 1956. Analysis of data from pumping test in leaky aquifers, *Trans. Am. Geophys. Union,* 37(6), 702.

Hantush, M. S., 1964. Hydraulics of wells, in *Advances in Hydroscience,* Vol. 1, Chow, V. T., Ed., Academic Press, New York, 281.

Jacob, C. E., 1940. On the flow of water in an elastic artesian aquifer, *Am. Geophys. Union Trans.,* 72 (Part II), 574.

Jacob, C. E., 1944. Notes on Determining Permeability by Pumping Test under Water Table Condition, U.S. Geol. Survey, Open File Report.

Jacob, C. E., 1950. Flow of groundwater, in *Engineering Hydraulics,* Rouse, H., Ed., John Wiley & Sons, New York, 321.

Kalman, R. E., 1962. A new approach to linear filtering and prediction, *J. Basic Eng.,* 82, 34.

Karanth, K. R., 1978. *Ground Water Assessment. Development and Management,* Tata McGraw-Hill, New Delhi.

Kruseman, G. P. and de Ridder, N. A., 1979. Analysis and Evaluation of Pumping Test Data, Int. Inst. for Land Reclamation and Improvement, Wageningen, The Netherlands.

LaVenue, A. M. and RamaRao, B. S., 1992. A Modeling Approach to Address Spatial Variability within the Culebra Dolomite Transmissivity Field, AND92-7309, Sandia Laboratories.

Marino, M. A. and Luthin, J. N., 1982. *Seepage and Groundwater,* Elsevier, Amsterdam.

Marsily, G. D., 1986. Quantitative hydrogeology, in *Groundwater Hydrology for Engineers,* Academic Press, New York.

Marsily, G. Labedan, M. Boucher and G. Fasanino, 1993. Interpreting the interference test in a well field using geostatistical techniques to fit the permeability distribution in a reservoir model, in *Geostatistics for Natural Resources Characterization,* Second NATO Advanced Study Institute, California, 831.

Muskat, M., 1937. *The Flow of Homogeneous Fluids through Porous Media,* McGraw-Hill, New York.

Neuman, S. P., 1972. Theory of flow in unconfined aquifers considering delayed response of the water table, *Water Resour. Res.,* 8(4), 1031.

Neuman, S. P., 1973. Saturated-unsaturated seepage by finite elements, *J. Hydraul. Div.,* ASCE, HY12, 2233.

Neuman, S. P., 1975. Galerkin method for analyzing flow in saturated-unsaturated porous media, in *Finite Elements in Fluids,* Gallagher, G., et al., Eds., John Wiley & Sons, London, chap. 10.

Neuman, S. P., 1987. On methods of determining specific yield, *Ground Water,* 25(6), 679.

Neuman, S. P. and Witherspoon, P. A., 1969a. Theory of flow in a confined two-aquifer system, *Water Resour. Res.,* 5(4), 803.

Neuman, S. P. and Witherspoon, P. A., 1969b. Applicability of current theories of flow in leaky aquifers, *Water Resour. Res.,* 6(4), 817.

Neuman, S. P. and Witherspoon, P. A., 1972. Field determination of the hydraulic properties of leaky multiple aquifer systems, *Water Resour. Res.,* 8, 1284.

Nwankwor, G. I., Cherry, J. A., and Gilham, R. W., 1985. A comparative study of specific yield determination for a shallow sand aquifer, *Ground Water,* 22(6), 764.

Papadopulos, I. S. and Cooper, H. H., 1967. Drawdown in a well of large diameter, *Water Resour. Res.,* 3(1), 241.

Papadopulos, I. S., 1967. Drawdown Distribution around a Large Diameter Well, Proc. Mat. Symp. Groundwater Hydrology, San Francisco.

Rorabaugh, W. I., 1953. Graphical and theoretical analysis of step drawdown test of artesian well, ASCE Proc. No. 362, Vol. 79.

Rushton, K. R. and Chan, Y. K., 1976. A numerical model for pumping test analysis, *Proc. Inst. of Civ. Eng.,* 61, 241.

Shultz, E. F., 1973. *Problems in Applied Hydrology,* Colorado State University, Ft. Collins.

Stallman, R. W., 1963. Type curves for solutions of single boundary problems, in Short Cuts and Special Problems in Aquifer Tests, U.S. Geol. Survey, Water Supply Paper, 1545-C, 45.

Streltsova, T. D., 1972. Unsteady radial flow in an unconfined aquifer, *Water Resour. Res.,* 8, 1059.

Streltsova, T. D., 1973. Flow near a pumped well in an unconfined aquifer under nonsteady conditions, *Water Resour. Res.,* 9(1), 227.

Şen, Z., 1983. Large diameter well evaluations and applications for the Arabian Shield, in *Symp. Water Resour. in Saudi Arabia,* Vol. 1, King Saud University Press, Riyadh, A-157.

Şen, Z., 1984. Adaptive pumping test analysis, *J. Hydrol.,* 74, 259.

Şen, Z., 1985. Volumetric approach to type curves in leaky aquifers, *J. Hydraul. Div.,* 111(3), 467.

Şen, Z., 1986a. Determination of aquifer parameters by the slope matching method, *Ground Water,* 24(2), 217.

Şen, Z., 1986b. Discharge calculation from early drawdown data large diameter wells, *J. Hydrol.,* 83, 45.

Şen, Z., 1988. Dimensionless time drawdown plots of late aquifer test data, *Ground Water,* 26(5), 615.

Şen, Z., 1989. Non-linear flow toward wells, *J. Hydraul. Div.,* 115, 193.

Şen, Z. and Al-Baradi, A., 1991. Sample functions as indicators of aquifer heterogeneities, *Int. J. Nordic Hydrol.,* 22, 37.

Theis, C. V., 1935. The relation between lowering of the piezometric surface and the rate and duration of discharge of a well using ground water storage, *Trans. Am. Geophys. Union,* Part 2, 519.

Walton, W. C., 1962. Selected Analytical Methods for Well and Aquifer Evaluation, Illinois State Water Survey Bull. 49.

Walton, W. C., 1970. *Groundwater Resources Evaluation,* McGraw-Hill, New York.

CHAPTER 10

Fractured Medium Aquifer Tests

I. GENERAL

An unfractured rock mass is effectively impermeable to water from engineering and hydrogeological points of view. Louis (1969) has shown that the water permeabilities of most intact rocks vary between 10^{-10} and 10^{-15} m/s. All rocks, particularly those close to the earth surface, include fractures of various sizes that provide major conducting pathways to water movement. Development of fractures due to different geological conditions such as tectonics and thermal and chemical activities in the porous or nonporous rocks changes the original strength and hydraulic properties. Many productive freshwater aquifers as well as petroleum and geothermal reservoirs are found in fractured igneous or sedimentary rocks such as sandstones and limetones. Crystalline rocks are heterogeneous and anisotropic water-bearing formations and their permeability varies markedly in short distances. Consequently, hydraulic behaviors of fractured rocks depend upon the number, extension, direction, size, aperture, and degree of connectedness of the fractures. Similarly, the amount of extractable water depends on the same features penetrated by the bore hole of the well. The existence of a connected network of fractures facilitates extraction of fluids (water, oil, and gas) from these formations by man-made structures such as vertical (well) and horizontal (gallery) shafts. Fractures play a multiple role in aquifer property alteration by changing the porosity or the permeability or both. In addition, fractured formations provide a potential underground space for nuclear waste repository siting.

It is often observed in the field that fractured rock masses contain sets of discontinuities with a geometry that resembles that shown in Figure 1. The existence of fractures provides an extra ability for the rock mass to transfer water easier than the original intact forms and also the water-storage capacity may increase significantly. Furthermore, the fractures cause any tight formation to be broken up into irregular shape of blocks, referred to as a matrix.

In general, fractured aquifers are complex, anisotropic, and heterogeneous. Prior to any mathematical treatment for the purpose of evaluating their hydraulic properties, it is necessary to render this complex system into an idealized medium by keeping the basic physical phenomena as closely as possible to reality in nature. The theory of groundwater flow in fractured reservoirs is, however, not yet fully developed, mainly owing to the difficulties in defining the topology (geometry) of the fracture system, basic physical flow phenomenon within the fractures or blocks, and block-to-fracture or fracture-to-well or fracture and block-to-well connections.

Methods presented in the previous two chapters primarily for homogeneous and isotropic porous medium become inadequate in interpretation and evaluation of fractured medium reservoirs. Most often either new approaches are found for the fractured reservoir solutions or for some simple cases existing conventional procedures are extended and/or modified to provide a reasonable explanation for the fractured medium peculiarities.

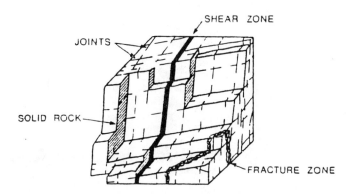

Figure 1 Naturally fractured rock.

Models presented in this chapter are based on the double-porosity theory, provided that the medium does not have any dominant horizontal or vertical major fracture. Methods for single vertical or horizontal fracture are also presented in this chapter with practical examples. Due to the complexity of the theoretical mathematical expressions for the groundwater flow through fractured medium, only basic expressions and procedures for practical applications are presented with their physical meanings.

II. FRACTURED MEDIUM IDEALIZATION MODELS

Naturally geometry of fractures is rather complex and detailed features are difficult to grasp. Therefore, some idealizations are necessary prior to any treatment of the flow in naturally fractured rock media. In general, there are three types of conceptual ideas to simplify the fractured medium geometry to apply continuity and flow law equations. These are referred to as the "double-porosity" concepts.

1. The first systematic approach in determining fractured medium aquifer parameters is due to Barrenblatt et al. (1960), who conceptualized the fractured reservoir as two coexisting and interacting media, namely, a medium of primary porosity of blocks with low permeability but high storativity and a medium of secondary porosity of fractures with high permeability but low storativity. It is assumed that any small representative volume element includes numerous random fractures and porous blocks. Such a conceptualization is referred to herein as the "random double-porosity" model (see Figure 2). Generally, each medium acts as a classical porous medium, however, their interaction is different from any homogeneous and isotropic porous medium concept due to the hydraulic interaction between them especially at moderate time periods during water withdrawal. Herein, there is no regular Euclidean geometry of fractures or blocks.
2. The second systematic approach considers regular Euclidean block geometry either in the forms of equal parallelepipeds surrounded by orthogonal fracture pattern or as alternate

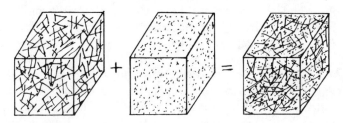

Figure 2 Coexistince of two porosities as random double porosity.

horizontal layer successions with significantly different hydraulic conductivities (Figure 3). Individual fractures are described by various parameters such as aperture size, roughness, orientation, dipping, frequency, and hydraulic radius. A fractured aquifer is regarded as a result of a certain pattern of fractures or the configuration of a few finitely or infinitely extended fractures. This is referred to "regular double-porosity" model and used by Warren and Root (1963) and Kazemi (1969).

3. The third approach is based on defining a fictitious continuum which in some limited sense is equivalent to the actual fractured medium. This approach has resulted from establishing whether or not the fractured media behaves like porous media. It is called the "equivalent double-porosity" model. First studies toward this purpose have been presented by Snow (1965, 1969) where fracture orientations and apertures intersected by a borehole were determined in the field. The fractures were assumed to be infinite in length, and an equivalent porous medium permeability was then computed as an accumulation of individual fracture permeability.

Sagar and Runchal (1982) have extended the Snow (1965, 1969) theory for the permeability of fractured systems in an attempt to account for the finite size of fractures. The equivalence is established in terms of flow rates. Once an equivalent continuum is defined, existing porous medium flow laws become usable. The equivalent continuum model provides correct answers for the fractured medium of equivalence only. Otherwise, aquifer parameter estimations will be erroneous. Simulation of fractured aquifers by using equivalent groundwater flow models of porous media has been justified as reasonable for many cases involving large areas. This

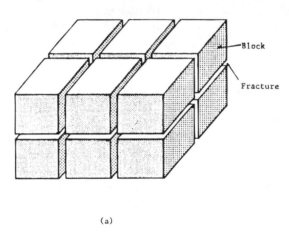

(a)

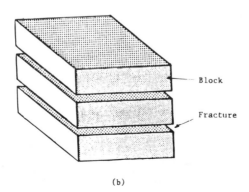

(b)

Figure 3 Regular double porosity madia. (a) Orthogonal and (b) horizontal fracture system.

is because the area of interest is sufficiently large to smooth out heterogeneities resulting from the presence of numerous fractures.

Even after these conceptual idealization and simplifications the theoretical mathematical models developed for depicting the behavior of fractured reservoirs are complex and lead to complicated-type curve expressions with several parameters and hence their exhaustive tabulations are not practical. Perhaps this is one reason these models are not preferable in practical applications. Practicing engineers or hydrogeologists prefer simple expressions that lead to speedy results. However, if the type curves are available, then procedures and matching techniques similar to single-porosity medium aquifers are applicable for aquifer parameter determinations.

III. QUALITATIVE PHYSICAL INTERPRETATION

It is always preferable to have some insight into the logical and physical evaluation of the phenomenon concerned before any hasty mathematical elaborations. Such insight will give the ability of interpreting intermediate stages and final results obtained from models in a meaningful manner. In a naturally fractured rock saturated with water prior to pumping, static piezometric levels both for the fracture and block media coincide with each other and, physically, there is no water exchange between fractures and blocks. Any water withdrawal from a double-porosity aquifer results in fracture response faster than blocks because of the permeability differences. Consequently, the fracture hydraulic head, h_f, deviates promptly from the static piezometric level, implying horizontal groundwater flow toward the well within the fractures only. At small time periods the double-porosity medium response is similar to single-aquifer behavior as the only active conveyors of water. Logically, the time-drawdown curve of this period is expected to comply with the Theis-type curve (see Figure 4a). The whole depression cone area behaves as a single-porosity medium with fracture flow only. With the passage of time the intermediate stage becomes dominant where the block hydraulic head also starts to deviate from the initial piezometric level as shown in Figure 4b. In the meantime the head difference between fractures and blocks causes groundwater recharge from the blocks to surrounding fractures. Note that the block depression cone is over the fracture depression cone. The common area of the two cones on a horizontal plane shows the influence of the double-porosity behavior. Outside of this area single porosity of fractures prevails. Physical

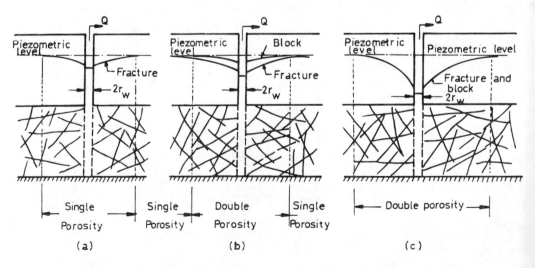

Figure 4 Fractured aquifer response to pumping. (a) Early, (b) transition, and (c) final stages.

reasoning indicates that recharge from blocks reduces the rate of fracture drawdown increment. If this recharge balances the amount of water delivered by the fractures to the well, then a quasi-steady state flow appears after some time during this stage. It is logical to expect that the time-drawdown pattern of this stage will fall below the classical Theis curve. The aquifer behavior during this stage is strictly a function of the block geometry and interporosity flow. Toward the end of this stage the head difference between the blocks and fractures become negligible (see Figure 4c) and again the aquifer behaves as a single formation. Of course, the response in the form of a time-drawdown pattern again follows the Theis curve but with a position different from the first-stage case. In the late stage both block and fracture elements are active simultaneously with fractures dominating in terms of permeability and the blocks in storativity. At the same time, the flow is horizontal in both fractures and blocks. The response of fractured aquifer to water withdrawal appears as a characteristic S-shaped curve with an inflection point. This response is very similar to time-drawdown behavior of unconfined aquifer as explained in Chapter 9, Section VI. Delayed yield in unconfined aquifers resembles the block-to-fracture interporosity effect in fractured media. The following useful information results from the discussions so far.

1. In general, double-porosity medium-type curves are expected to start and end in the form of the Theis-type curve but transition from early to late time includes either a horizontal zone (equilibrium) or an inflection point during moderate times.
2. At large radial distances from the well deviations of fracture and block hydraulic heads might not be very distinctive. It means that the double-porosity effect appears only within a certain region around the well.
3. Considerations from the previous chapter indicate that, when the time drawdown data with double porosity effect are plotted on a semilog paper then the result is expected to appear as two parallel straight lines each corresponding to the late-time Theis-curve portion with a curved middle part connecting them. This situation occurs if complete double-porosity time-drawdown data are available.
4. Observation wells far away from the main well might not catch the double-porosity behavior of the aquifer. The measurements in these observation wells are responses to pumping and yield information concerning the third stage only, i.e., the combined hydraulic properties of blocks and fractures. Hence the drawdown behavior is that of an equivalent homogeneous, isotropic confined aquifer representing both the fracture and block hydraulic properties.
5. If the fractures have rather high permeability, then their response to the pumping will not continue for sufficiently long time to show the early time Theis behavior. However, late-time Theis behavior will always appear, provided that the pumping time is long enough.
6. If the geological setup of the aquifer is not known prior to the aquifer test, then one might erroneously conclude from the time-drawdown response that the aquifer is a homogeneous, isotropic unconfined type with delayed yield. Gringarten (1978) in petroleum reservoir evaluation has already shown that double porosity type curves are identical to type curves for a single porosity unconfined aquifer with delayed yield as presented by Boulton (1963) in the groundwater domain.
7. If the fractured aquifer test does not continue for sufficiently long time then it is not possible to observe the third stage where equilibrium exists between block and fracture flows.

Early fractured aquifer models have appeared in the petroleum reservoir engineering domain and all of the mathematical results were presented on semilogarithmic paper drawdown vs. logarithm of time. In light of the aforementioned physical reasoning in all cases, it is agreed that the transition period develops that is strictly a function of the matrix geometry and matrix-to-fracture flow properties. Figure 5 shows a typical semilog plot depicting the transition period and the parallel lines for early and late times, respectively. In such a plot the early time straight line reflects the hydraulic parameters of the fracture set only, whereas the late time straight line is a combined characteristic of fracture and block features in the

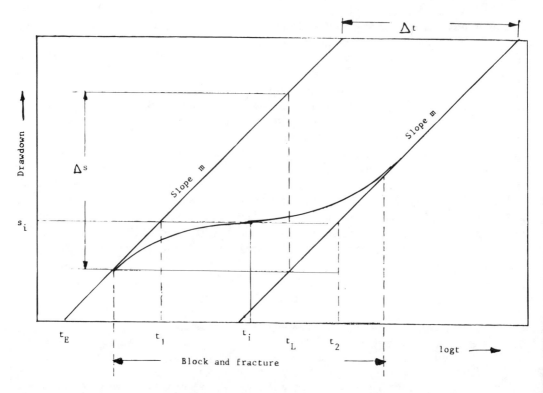

Figure 5 Semilogarithmic plot of drawdown vs. time.

equilibrium state. However, the transition period depends on the matrix geometry as well as the matrix-to-fracture flow properties. It is already known from single porous medium studies in the previous chapters that any straight line provides two graphical values: the slope and the intercept corresponding to zero drawdown. These two variables play the key role in the determination of fractured aquifer properties. In fact, in a single-porosity medium two aquifer parameters, namely, storativity and transmissivity, are enough for complete hydraulic behavior assessments. Similarly, in the double-porosity model each of the medium hydraulic properties can be determined from straight lines on semilog plots. A close inspection of Figure 5 gives further graphical measures for the double-porosity behavior. These additional quantities might be

1. The drawdown difference, Δs, between the straight lines along any vertical line.
2. The time difference, Δt, between the two parallel lines along any horizontal line.
3. The coordinates of inflection point if any, (t_i, s_i).
4. The average slope of transition period, Δs_{tr}.
5. This average slope is always smaller than the slope of early (late) straight lines.
6. The intercept, (t_E and t_L) of straight lines with the time axis (zero drawdown).
7. The slopes of early and late straight lines.
8. The time coordinates (t_1 and t_2) of the horizontal straight line intercepts with the parallel straight lines.

These graphical points provide information for combined effects of matrix geometry and interporosity flow properties and they give an opportunity to determine some other combined variables that are defined during the mathematical modeling.

The main purpose of model development is to find relevant equations that relate these graphical values to the model parameters. Although the physical basis and field observations give S-shaped time-drawdown pattern, its representation by a suitable mathematical model requires a set of simplifying assumptions.

IV. SIMPLIFYING ASSUMPTIONS

The geometric idealizations of blocks and fractures presented in the previous section are not enough for the mathematical treatment of groundwater flow in fractured rocks. It is necessary prior to any modeling scheme to set up simplifying assumptions concerning the spatial geometry of the aquifer itself, well configuration, and flow exchange law between the fractures and blocks.

Different researchers established fractured medium aquifer test solutions by considering a set of these assumptions. There are deviations from Jacob straight line for pumping response data measured in fractured rock aquifers and these deviations from the expected linear relationship indicate the need for a new mathematical model to characterize flow in the fractured rocks. The difference in the analytical treatment of groundwater movement equation in different double-porosity approaches does not lie in the mathematical solution technique but rather in further assumptions that are stated as the models are explained in the subsequent sections of this chapter.

The following set of common assumptions is necessary in any case for reaching practically valid simple solutions. First, the assumptions related to the geometry of the aquifer and well configuration are

1. That the aquifer is confined, horizontal, and areally extensive.
2. That the thickness of aquifer is uniform over the depression cone effective area.
3. That the well is vertical and fully penetrates the aquifer and recharge to the well is through the fracture medium only.
4. That there is neither well skin nor storage effects and the well has infinitesimally small diameter.

The assumptions concerning the groundwater flow within the fractures and blocks are

1. That the flow in fracture medium has a horizontal component only while in the blocks the flow direction is toward surrounding fractures.
2. That the flow from blocks to fracture may be proportional to either the hydraulic head difference between matrix and fracture or to the averaged hydraulic head gradient throughout the matrix block. The former assumption was introduced by Barenblatt et al. (1960) and employed by Warren and Root (1963). It has an advantage of simplifying the mathematical analysis of the flow problem and a disadvantage of not representing correctly either the mechanism of hydraulic head adjustment between block and fracture by time-variant crossflow or the aquifer hydraulic head response during the transitional time. The pseudo-steady-state matrix-to-fracture flow equation is given in general as

$$q_b = \alpha(h_b - h_f) \qquad (1)$$

where q_b is the rate of flow from block into fractures; α is a constant that is dependent on the block geometry; and h_b and h_f are the block and fracture hydraulic heads, respectively. This equation implies that the matrix flux is independent of spatial position which can be true only at a state of equilibrium or at a pseudo-steady-state time, i.e., practically for large times after pump start. This flow assumption, therefore, is often referred to as a "pseudo-steady-state" or "lump-parameter" matrix-to-fracture flow assumption.

On the other hand, the averaged gradient assumption on matrix-to-fracture crossflow, although leading to a somewhat complicating mathematical analysis, has an advantage of more correctly describing the pressure equilibrium process that occurs during the transitional period. The expression of gradient flow is given by Streltsova (1983) as a one-dimensional diffusion equation:

$$\frac{\partial^2 h_b}{\partial z^2} = \beta \frac{\partial h_b}{\partial t} \qquad (2)$$

in which z is the distance from matrix edge; t is the time; and β is a coefficient that represents the spread of pressure within the matrix blocks. The average gradient assumption is referred to as "distributed-parameter" flux assumption as Streltsova (1976) stated; the gradient flow model for matrix-to-fracture flow predicts transitional periods which when plotted vs. logarithm of time gives a linear segment that connects the early and late time drawdown curves without an inflection point. This type of flux equation has been adopted in models by Kazemi (1969), de Swan (1976), and Najurieta (1980). It is pointed out by some researchers that the gradient flow model is more realistic in describing the matrix-to-fracture flow (Streltsova, 1982; Serra et al., 1983).

3. That no flow enters the blocks.
4. That the change in the volume of water due to the compressibility of the fractures is negligible when compared with the change in water volume caused by the flow from blocks. This is tantamount to saying that the storativity of fractures is negligibly small.
5. That the change of volume due to the recharging of the blocks is negligible when compared with the change of volume due to the water expansion in the blocks. It implies that the hydraulic conductivity in the blocks is negligible.
6. That the flow toward the well is in unsteady state.
7. That before the pump start the piezometric level is horizontal both in fractures and blocks.
8. That the aquifer response to water withdrawal is instantaneous.

Finally, assumptions concerning the material composition of the medium are

1. That two porous media with different hydraulic characteristics exist within the fractured aquifer domain.
2. That individually each porous medium is homogeneous and isotropic but collectively they constitute a heterogeneous medium.

V. DOUBLE-POROSITY MODELS

A. BARRENBLATT ET AL. METHOD

The first study toward well hydraulic problems in fractured rocks is due to Barenblatt et al. (1960), who considered the rock as naturally broken up into various sizes and irregular shapes of blocks. They assumed that any infinitesimally small volume includes large numbers of block fragments and fractures of random aperture, orientation, and length in addition to their connectedness. This assumption enables the use of Darcy law for the flow movement within the fractured medium.

Under the assumptions mentioned in the previous sections this method has an advantage in simplifying the mathematical analysis of the flow problem but a disadvantage of not correctly representing either the mechanism of piezometric level readjustment between block and fracture flow or the aquifer piezometric response during the transitional time. The random double-porosity aquifer model has numerous random (in length and direction) fractures coexisting with porous medium as shown in Figure 2.

The solution of the problem by using Laplace transformation leads to fractured medium well function as

$$W\left(u_f, \frac{r}{B}\right) = \frac{4\pi T_f}{Q} s(r,t) \qquad (3)$$

in which u_f is the dimensionless time factor defined for the fracture as

$$u_f = \frac{r^2 S_b}{4t T_f} \qquad (4)$$

and B_f is the dimensionless fracture factor given by

$$B^2 = \frac{K_b}{\alpha} \qquad (5)$$

where α is a characteristic of fractured rock with dimension $[T]^{-1}$ as appears in Equation 1. Herein, T_f, K_b, and S_b are the fissure transmissivity, block hydraulic conductivity, and storage coefficient, respectively. The necessary values of $W(u_f,r/B)$ and u_f are given in Table 1 for a set of r/B values.

Streltsova (1976) provided fractured media type curves using the assumptions of Barenblatt et al., but considering storage coefficient, S_f, of the fissured medium also. Her analytical solutions lead to a well function similar to Equation 3 but with the storage coefficient ratio, η, defined as

$$\eta = \frac{S_b}{S_f} + 1 \qquad (6)$$

If $S_f \cong 0$ then $\eta \to \infty$ and type curves given in Figure 6 are valid. The relationship between the well function dimensionless time factor and η are presented numerically in Table 2. Corresponding type curves are plotted on double-logarithmic graph paper as shown in Figure 7 for $\eta = 10^1$.

B. WARREN-ROOT METHOD

The first identification method was proposed by Warren and Root (1963), who gave a mathematical solution for the time-drawdown data behavior obtained from the pumped well. They solved the fractured medium problem with the similar sets of assumptions as was for the Barrenblatt model with the alteration of two basic assumptions:

1. Instead of random distribution of fractures their model assumes a set of uniformly distributed blocks that are identical rectangular parallelepipeds as shown in Figure 3a in the form of orthogonal fracture systems. Hence, the fracture network consists of three diagonal, continuous, uniform sets of fractures and each fracture is parallel to one of the principal axis of hydraulic conductivity.

Table 1 Values of Fractured Media Well Function, $W(u_f, r/B)$

	r/B					
$1/u_f$	0.05	0.1	0.2	0.5	1.0	10
1×10^0	0.0039	0.0121	0.0349	0.0115	0.0150	0.0165
1×10^1	0.0388	0.1202	0.3379	0.1138	0.3032	0.2346
1×10^2	0.3794	1.0101	2.5136	0.9805	1.4952	1.7641
5×10^2	1.7133	3.8947	5.3619	3.8455	3.8874	4.0318
1×10^3	3.085	5.5050	6.2162	5.6062	3.6316	5.6368
2×10^3	4.8987	6.7377	6.9690	6.3151	6.3277	6.3294

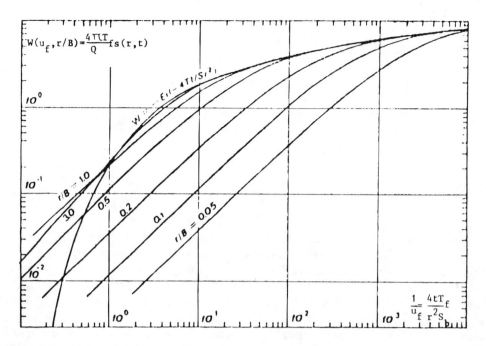

Figure 6 Barenblatt et al. type curves, ($\eta = \infty$).

2. The fracture compressibility is not negligible but plays a significant role in the total storage coefficient of the fractured medium.

Warren and Root evaluated approximate forms of their pseudo-steady-state interporosity flow and found that the result appears as two parallel lines on a semi-log paper similar to Figure 5. They concluded that two parameters were enough to characterize the behavior of the fractured rock aquifers. One dimensionless parameter, w, relates the water storage capacity of the fracture, S_f, to that of the whole fractured medium and it is defined as

$$w = \frac{S_f}{S_f + \beta S_b} \qquad (7)$$

in which S_b is the block storage coefficient and β is a parameter that assumes the value of zero for early straight line but the values of 1.3 or 1 for the late straight line according to whether the blocks have cube (Figure 3a) or strata shape (Figure 3b), respectively.

The second dimensionless parameter, λ, is the interporosity flow coefficient and relates to the degree of heterogeneity of the fractured medium. In terms of well radius, r_w, the block, K_b, and the fracture, K_f, hydraulic conductivities of the anisotropic medium are expressed as

Table 2 Values of Fractured Media Well Function, $W(u_f, r/B)$, for $\eta = 10$

	r/B											
	0.05		0.1		0.2		0.5		1.0		10	
	η		η		η		η		η		η	
$1/u_f$	10^1	10^2	10^1	10^2	10^1	10^2	10^1	10^2	10^1	10^2	10^1	10^2
1×10^0	0.0006	0.0019	0.0024	0.0074	0.0093	0.0268	0.0033	0.0070	0.0010	0.0140	0.0042	0.0159
1×10^1	0.0178	0.0311	0.0697	0.1099	0.2444	0.3279	0.0498	0.1049	0.1262	0.1950	0.1925	1.2290
1×10^2	0.2963	0.3650	0.9815	1.0864	2.3494	2.5014	0.8478	0.9689	1.3264	1.4777	1.6146	1.7676
5×10^2	1.5801	1.6845	3.7324	3.8696	5.1574	3.3476	3.6406	3.8286	3.7882	3.9762	3.8411	4.0135
1×10^3	2.8827	2.9959	5.3244	5.4755	6.0105	6.2019	5.3981	5.5890	5.4203	5.6102	5.4442	5.6183
2×10^3	4.6057	4.8189	6.5515	6.7066	6.7631	6.9546	6.1067	6.2979	6.1163	6.3063	6.1365	6.3109

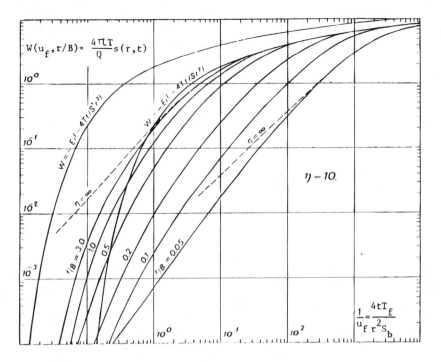

Figure 7 Barenblatt et al. type curves, ($\eta = 10$).

$$\lambda = \alpha r_w^2 \frac{K_b}{K_f} \tag{8}$$

where α is the block geometry shape parameter for the heterogeneous region and has the dimension $[L^{-2}]$.

Essentially identical solutions to the Warren and Root approach were presented later by Odeh (1965) and Kazemi et al. (1969). Well storage and skin effects were analyzed with the pseudo-steady-state interporosity flow solution by Movar and Cinco (1977). Their conclusion was that time-drawdown plot on a semilogarithmic paper of a well produced from a fractured reservoir may be characterized by two parallel and linear segments. This solution was extended by Bourdet and Gringarten (1980) to account for interporosity flow and to generate type curves useful for the analysis of double-porosity system field data. Necessary procedural steps for the application of the Warren and Root model are deduced from a concise study by Gringarten (1984).

Naturally fractured rock time-drawdown data are processed conventionally on a semilog paper. The Warren and Root (1963) identification procedure can be explained first through a hypothetical time drawdown graph as in Figure 5. The first straight line represents homogeneous radial flow in the fracture portion of double-porosity system, whereas the last line corresponds to radial flow in the total aquifer. These two parallel straight lines are connected through a transition period during which pressure tries to stabilize. Warren and Root indicated that w can be expressed in terms of the vertical drawdown difference, Δs, between the two straight lines and their slope, m, as

$$w = 100^{-\Delta s/m} \tag{9}$$

On the other hand, λ from the time intersection of the horizontal line is drawn through the middle (inflection point) of the transition curve with either the first, t_1, or the second, t_2, semilog straight lines. This was shown by Bourdet and Gringarten (1980) as

$$\lambda = \frac{S_f r_w^2}{1.78 \ K_f t_1} = \frac{(S_f + \beta S_b)}{1.78 \ K_f t_2} \tag{10}$$

The values of t_1 and t_2 are only approximately readable from the figure and consequently λ obtained by this equation is not very accurate but usually remains within the order of magnitude of the correct value.

Furthermore, the drawdown at which the transition occurs from the dominant fracture flow to combined flow from fracture and matrix block is given by the same authors as

$$s_i = \frac{Q}{2\pi T_f} K_0(\sqrt{\lambda}) \tag{11}$$

where $K_0(\sqrt{\lambda})$ is the modified Bessel function of the second kind and of zero order. This last expression for $\lambda < 0.01$ becomes

$$s_i = \frac{2.30Q}{4\pi T_f} \log \frac{1.26}{\lambda} \tag{12}$$

Type curve solution procedure—Bourdet and Gringarten (1980) have presented type curve solution of fractured aquifer with the regular double-porosity model based on the Warren and Root concept. The well function, $W(u^*, w, \lambda)$ and the dimensionless time factor, u^* expressions are given for an observation well as

$$W(u^*, \lambda, w) = \frac{4\pi T_f}{Q} s \tag{13}$$

and

$$u^* = \frac{T_f t}{(S_f + \beta S_b) r^2} \tag{14}$$

with w and λ parameters already defined in Equations 7 and 8 but calculated from time-drawdown plot on semilogarithmic graph paper by using Equations 9 and 10. As given by Serra et al. (1983) typical values of w and λ fall within the ranges of 10^{-1} to 10^{-4} and 10^{-3} $(r_w/r)^2$ to 10^{-9} $(r_w/r)^2$, respectively. Provided that different values of w and λ are given, type curves of $W(u^*, w, \lambda)$ vs. u^* are presented by Bourdet and Gringarten from available tables not presented in this book. They indicated that the double-porosity behavior occurs as long as $\lambda < 1.78$. The elegancy of Equation 13 is that it reduces to classical single porous medium Theis-type curve with dimensionless time factor defined as

$$u = \frac{(S_f + \beta S_b) r^2}{4 T_f t} \tag{15}$$

This provides an opportunity to determine the early and late aquifer response parameters without any resort to complex and numerous type curve-matching procedure. With this information at hand, part by part matching procedure steps for fractured aquifer parameter determination are as follows:

FRACTURED MEDIUM AQUIFER TESTS

1. Plot observed time-drawdown data from an observation well on a double logarithmic graph paper with the same scale as a ready Theis-type curve sheet.
2. Try to find the best matching position with the early time-drawdown values.
3. Choose a match point E (E signifies early data) and read the four values from the type curve and field data sheets. These are $W_E(u)$, $1/u_E$, s_E, and t_E.
4. Calculate the fracture transmissivity as $T_f = QW_E(u)/(4\pi s_E)$; here Q is the constant pump discharge.
5. Substitute this transmissivity value into Equation 15 by considering $\beta = 0$ and calculate $S_f = 4T_f t_E u_E/r^2$ where r is the radial distance of observation well to the main well.
6. Notice for either a horizontal transition portion or an inflection point in the field data and read the corresponding drawdown value, s_i. The substitution of this value into Equation 12 yields the estimate of λ as $\lambda = 1.26 e^{-\frac{4\pi T_f}{Q} s_i}$.
7. Find the best matching position of the late time-drawdown data with the Theis curve.
8. Choose a matching point L and note its coordinates from the graphs as $W_L(u)$, $1/u_L$, s_L, and t_L.
9. Calculate $(S_f + \beta S_b)$ value from Equation 15. For the Warren and Root model by considering $\beta = 1.3$ and knowing the value of S_f from step (5) one can find the value of S_b.

Because the magnitude of w affects the transition period duration for relatively small values of w, this period is very long and consequently matching the Theis curve to late time drawdown may not be possible. On the other hand, for high values of λ which implies large radial distances, the drawdown in an observation well will no longer reflect the double-porosity behavior and the type curve analysis fails for the early time matching.

Semilogarithmic paper calculations—In naturally fractured aquifers, procedure based on semi-log paper similar to the Jacob straight line method yields the aquifer parameters, provided that the conditions for the description of straight lines are satisfied as suggested by Movar and Cinco (1977). For early and late time straight line occurrences, the necessary conditions are

$$u^* < \frac{w(1-w)}{3.6\lambda} \tag{16}$$

and

$$u^* > \frac{1-w}{1.3\lambda} \geq 100 \tag{17}$$

respectively. With the satisfaction of these conditions the following procedural steps lead to aquifer parameter estimates.

1. Find the slope, m, and early as well as late time intercepts t_E and t_L on the time axis for zero drawdown.
2. Calculate the fracture transmissivity $T_f = 2.3 Q/4\pi m$ where Q is the constant pump discharge.
3. Calculate the fracture medium storage coefficient as $S_f = 2.25 T_f t_E/r^2$ with r being the radial distance from the main well to observation well.
4. Calculate the combined storativity value as $(S_f + \beta S_b) = 2.25 T_f t_L/r^2$ by considering the convenient value of β which is equal to $1/3$ for the Warren and Root model.
5. Calculate the separate values of S_f and S_b.

C. KAZEMI ET AL. METHOD

Later, Kazemi (1969) was stimulated by the Warren-Root work and solved the same problem numerically by considering a model consisting of a finite circular aquifer layer with a well located at the center and the two distinct porous regions for the blocks and fractures as shown in Figure 8. Hence, the main change in the assumptions of Warren-Root model is the geometry of the aquifer, blocks, and fractures. For layer type of blocks geometry $\beta = 1$ and the procedures in the previous section must take this change into consideration. Furthermore, he assumed block-to-fracture flow as gradient flow Equation 2, i.e., unsteady state.

Kazemi's results are similar to those obtained by Warren and Root except for a smooth transition zone in the sense that there is no inflection point as a result of unsteady-state flow in the blocks (see Figure 9). Kazemi's observation was that for such a system the transition period develops as a straight line with no inflection point. His statement was supported by de Swan (1976) studies. It is observed that the only significant difference between Kazemi's and Warren-Root curves occurs in the transition zone, otherwise, early and late time straight lines coincide in the two models. Kazemi concluded that the fractured reservoir characterizations of Warren and Root were applicable to cases where fracture distribution was uniform and the contrasts between fracture and block flow were large. When this contrast was small only one straight line appears as the model implying a single-porosity medium.

Streltsova (1982) and Serra et al. (1983) emphasized the transient nature of flow from blocks to fracture and pointed out the development of a unique slope ratio between the transition straight line and the early (late) time lines' slopes. The straight line shape of the transition period develops a slope that is numerically one half the slope of the parallel straight lines corresponding to the early and late time data. This property provides a chance for the characterization of naturally fractured reservoirs from aquifer test data as presented by Ershanghi and Aflani (1985). They mentioned that the half-slope occurs around the dimensionless time factor of

$$u_H = \frac{5w}{e^\gamma \lambda} = 2.58 \frac{w}{\lambda} \tag{18}$$

with storage capacity ratio w and interporosity flow coefficient λ defined as in Equations 7 and 8.

Several previous studies have indicated that the block shape geometry (Figure 3a) considered in the Warren-Root model under pseudo-steady or transient (gradient) state interporosity flow conditions results in the formation of an inflection point during the transition period. On the other hand, the strata geometry of the blocks tends to linearize the transition period in half-slope. Identification of half-slope is very different for $w > 0.1$. The slope at the point

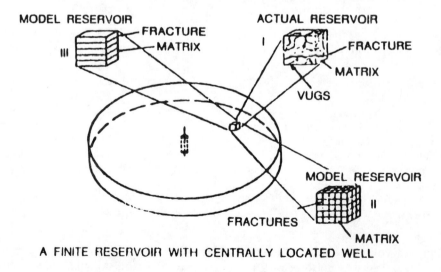

Figure 8 Kazemi model of fractured rocks.

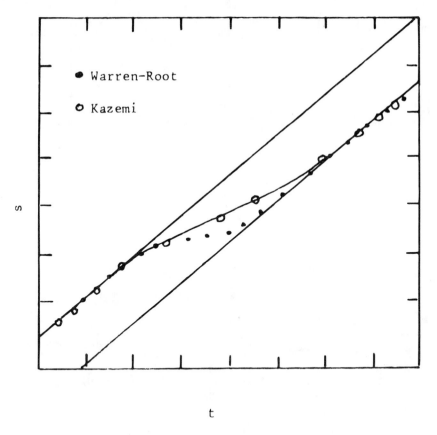

Figure 9 Comparison of Warren-Root and Kazemi models.

of inflection of the theoretical Warren-Root model is also a function of w. At this point of inflection

$$u_L^* = -\frac{wLn(w)}{\lambda} \qquad (19)$$

and the equation for the slope ratio at this point is

$$m = [1 + w^{1/(1-w)} - w^{w/(1-w)}] \qquad (20)$$

The numerical solution of this equation is presented by Ershanghi and Aflani (1985) as shown in Figure 10. It is obvious from this figure that a slope ratio of 1/2 occurs at $w = 0.227$.

Later, Kazemi et al. (1969) showed that the drawdown expression valid for the main well in the Warren-Root model is equivalently applicable to observation wells. The aquifer parameter estimation of the Kazemi model can be achieved with the similar type curve or straight line procedures as explained in the previous section.

D. BOULTON-STRELTSOVA METHOD

All the aforementioned methods and models are developed by petroleum engineers for characterizing the fluid flow in the fractured petroleum reservoirs. The first appearance of quantitative studies directed toward the groundwater flow was due to Boulton and Streltsova (1977a,b).

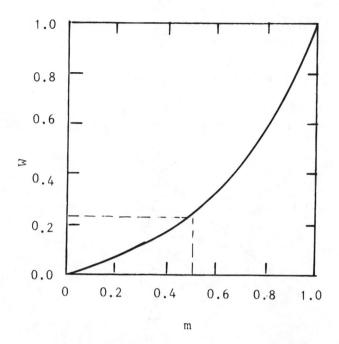

Figure 10 Slope ratio change with w.

It is by now rather clear from the discussions in the previous sections that, to account for recharge flow from blocks to fractures, specification of the block geometry is necessary. Boulton and Streltsova adopted the assumptions of the Barenblatt et al. (1960) model with modifications in the block geometry and interporosity flow as follows.

1. The fractured rock mass is idealized as alternating layers (slabs) of blocks and fractures where the thickness of the blocks and the aperture of fissures represent average fracture spacings and apertures as shown in Figure 11. The block layers representing the rock mass have thickness, $2H$, which is assumed to be equal to the average thickness of the actual blocks whereas the fractures have a thickness, $2H$, equal to the average thickness of the actual fractures.
2. Strictly one-dimensional flow is assumed in the blocks. Simultaneous consideration of flows in the blocks led to account for the variation that is expected to occur in the hydraulic head in the blocks.

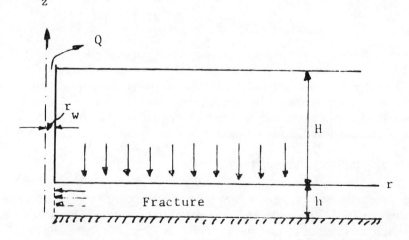

Figure 11 Block-fracture unit.

The analytical solutions of rather involved mathematics lead succinctly to the well function for fractures

$$W_f\left(u_f, \frac{r}{B_f}\right) = \frac{4\pi T_f}{Q} s_f \quad (21)$$

in which the dimensionless time factor, u_f, in terms of fracture hydraulic parameters is

$$u_f = \frac{r^2 S_f}{4t T_f} \quad (22)$$

and B_f is a dimensionless hydraulic conductivity ratio defined as

$$B_f^2 = \frac{K_f h}{K_b H} \quad (23)$$

They also provided the well function for blocks,

$$W_b\left(u_f, \frac{r}{B_f}\right) = \frac{4\pi T_f}{Q} s_b \quad (24)$$

in which s_b is the drawdown in the blocks. Fracture and block type curves as presented by Streltsova (1976) are given in Figures 12 and 13, respectively.

It is obvious that fracture type curves have inflection points similar to the Warren-Root model, whereas the block-type curves are without any inflection and resemble the homogeneous and isotropic medium curves. Streltsova stated that this approach is applicable only during the late time periods of pumping and only when the size of the block-to-fracture units forming the aquifer have finite block and fissure thicknesses.

E. ŞEN MODEL

The review of literature on naturally fractured aquifer test analysis indicates discomfort in their applications and interpretations of the results obtained. The sources of these difficulties are mainly due to the following points:

1. Practicing engineer and hydrogeologist alike are confused with the multitude of approaches. In fact, most of the analytical solutions are essentially identical. Apparent differences stem from the definition of various parameters used in the derivations.
2. High-level mathematics involved in the analytical derivations at times hinder understanding of basic physical concepts which causes difficulties in final result interpretations. Sometimes it is not even possible to grasp the mathematical expositions and the solutions are possible after very restrictive assumptions.
3. The results of analytical solutions are given in lengthy table forms for a specified set of parameters that are most often different from what is required in practical applications.

It is, therefore, preferable by any practicing personnel to use simple empirical relationships, if available, or even to resort to approximate but simple mathematical formulations. Şen (1988) formulated the problem with least restrictive assumptions. In his approach the geometric and material composition assumptions by Barrenblatt et al. (1960) have been adopted. On the other hand, the following releases in the groundwater flow assumptions are made.

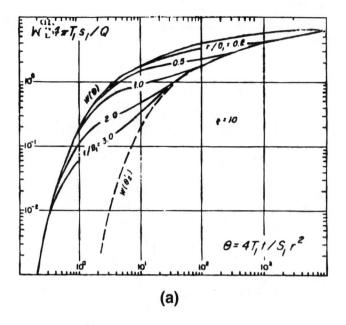

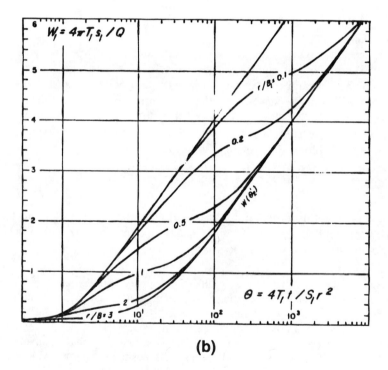

Figure 12 Boulton-Streltsova fracture-type curves.

1. The flow in fractures and blocks can take place in any possible direction.
2. Neither the storage capacity of fractures nor the transmissivity of the blocks are negligible.

Idealized representation of this model with the significant parameters are shown in Figure 14.

FRACTURED MEDIUM AQUIFER TESTS

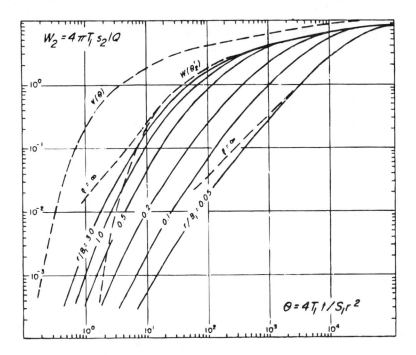

Figure 13 Boulton-Streltsova block-type curves.

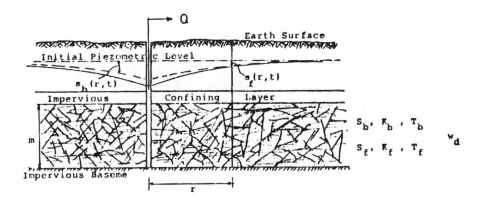

Figure 14 Fracture medium representation.

Although the solution is an approximation to fractured aquifer flow, it leads to comparatively simple type curve expression which does not require any table and besides it is possible to assess the effect of different block and fracture parameters on the type curve.

In the analytical treatment of the problem the continuity and flow law equations are written separately for the fractures and the blocks, in addition to interporosity flow according to pseudo-steady-state law.

After the necessary analytical treatment fracture well function is given as

$$W\left(u_f, \frac{r}{B_f}\right) = \int_{u_f}^{\alpha} \frac{e^{-xC(x)}}{x} dx \qquad (25)$$

in which the fracture dimensionless time factor is

$$u_f = \frac{r^2 S_f}{4t T_f} \qquad (26)$$

$C(x)$ function is defined explicitly as

$$C(x) = \frac{1 + \dfrac{S_b}{S_f}[1 - e^{-\beta(x)}]}{1 + \dfrac{T_b}{T_f}[1 - e^{-\beta(x)}]} \qquad (27)$$

where the dimensionless shape factor, $\beta(x)$, is defined as

$$\beta(x) = \frac{1}{4}\frac{S_b}{S_f}\left(\frac{r}{B}\right)^2 \frac{1}{x} \qquad (28)$$

and the fracturation ratio, B, is defined as

$$B^2 = \frac{T_f}{\alpha} \qquad (29)$$

in which α is a characteristic factor as appeared in Equation 1 and related to block-to-fracture characteristics.

In the foregoing discussions, the well is assumed to take water from the fractures only, i.e., the well is screened along against the fractures. However, this is a difficult task to accomplish in practice and consequently it is preferable to design the well such that the seepage into the well appears along the whole saturation thickness of the aquifer. To this end the new boundary condition is considered as

$$\lim_{r \to 0}\{2\pi m r[q_f(r,t) + q_b(r,t)]\} = Q \qquad r \to 0 \qquad (30)$$

where $q_f(r,t)$ and $q_b(r,t)$ are the fracture and block specific discharges, respectively. The solution for this case becomes

$$W'\left(u_f, \frac{r}{B}\right) = \int_{u_f}^{\alpha} \frac{e^{-C(x)x}}{x\left\{1 + \dfrac{T_b}{T_f}[1 - e^{-\beta(x)}]\right\}} dx \qquad (31)$$

or implicitly as

$$W'\left(u_f, \frac{r}{B}\right) = \frac{4\pi T_f}{Q} s'(r,t) \qquad (32)$$

in which $s'(r,t)$ is the drawdown in a piezometer for this new situation. Solutions of Equation 25 or 31 with the given definition of all the terms can be achieved numerically for any desired

FRACTURED MEDIUM AQUIFER TESTS

set of parameters. Some representative type curves are presented in Figures 15 and 16. The following common useful interpretations are noticeable from these figures.

1. Given a pair of storativity and transmissivity ratios all type curves for small and large time periods merge to an early and late Theis-type curve. As $t \to 0$ ($u \to \infty$), $\beta(u) \to 0$, and therefore Equation 27 yields $C(u) \to 1$. Under these circumstances Equation 25 reduces to early Theis-type curve with dimensionless time factor defined with the fracture transmissivity and storativity. Likewise, as $t \to \infty$ ($u \to 0$), $\beta(u) \to \infty$, and accordingly $C(u) \to (T_f/S_f)[(S_b + S_f)/(T_b + T_f)]$ and finally these asymptotic limitations reduce Equation 25 again into the Theis curve but this time with different dimensionless time factor defined as

$$u_L = \frac{r^2(S_f + S_b)}{4t(T_f + T_b)} \qquad (33)$$

On the other hand, for very finely fractured natural rocks $\alpha \to 0$ which implies that $B \to \infty$ and from Equation 28 $\beta(u) \to 0$; hence, again a Theis curve appears for this limiting case. As the degree of fracturation increases the whole aquifer behaves similar to a homogeneous and isotropic single porosity medium.

2. For transition (moderate) times apart from the limiting cases mentioned in (1), the fractured media type curves increasingly deviate from the Theis curves. Transition from an early Theis curve to a late one occurs along the transition period with an inflection point as shown in Figures 15 and 16.

3. Comparison of type curves in Figures 15 and 16 among themselves indicates another important property that the transition period duration changes with changes in storativity and transmissivity ratios. The smaller these ratios are the shorter becomes the duration. However, the storativity ratios are more effective than the transmissivity ratios.

4. The duration of early Theis curve is rather very short and, therefore, during an actual aquifer test extra care must be given to detect this portion which may continue for a few minutes in

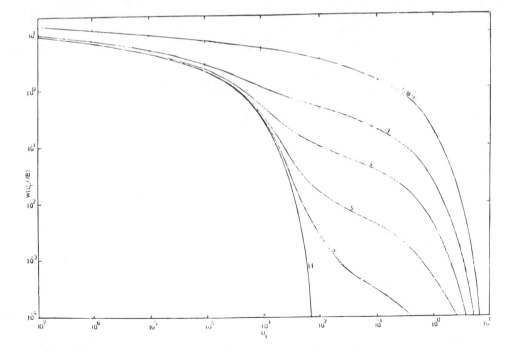

Figure 15 Type curve with fracture to well discharge ($S_f/S_b = 10^{-4}$, $T_f/T_b = 2$).

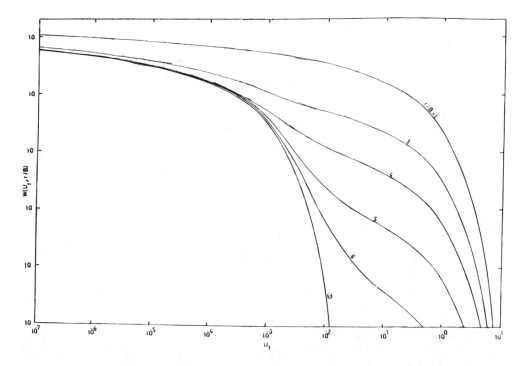

Figure 16 Type curve with block and fracture-to-well discharge ($S_f/S_b = 10^{-4}$, $T_f/T_b = 2$).

many cases. As explained in Section III.B this early part is very significant in estimating the fracture hydraulic parameters.

In naturally fractured rocks with large block sizes the direct contribution of flow in the blocks to the pumping well gains importance. When the fracture and block transmissivities are close to each other it becomes significant even for moderate times. However, a non-negligible contribution does always exist for large times. Equation 32 is valid for such a situation in general but as a limiting case for early times ($t \to 0$) it reduces to a modified Theis curve as

$$W'(u_f) = \frac{1}{1 + \dfrac{T_b}{T_f}} \int_{u_f}^{\infty} \frac{e^{-x}}{x} dx \tag{34}$$

or the relationship between the two cases becomes

$$W'(u_f) = \frac{1}{1 + \dfrac{T_b}{T_f}} W(u) \tag{35}$$

Significance of this last expression increases especially in fractured coarse-grained sandstones and limestones.

VI. INDIVIDUAL FRACTURE MODELS

These models study a single vertical or horizontal excessive fracture existing within the reservoir and intersecting with the well. Fractures have relatively higher hydraulic conductivities than the surrounding host rock material. The water moves easily through fractures. Intersection of a well by a major fracture increases the productivity of a well producing from low permeability formations. It is, therefore, a desirable task to search for potential major fractures in an area and locate the wells such that these fractures are intercepted. This is an inexpensive way to increase the water production from a well. The success of many marginal wells can be attributed directly to the intersection of fractures. Figure 17 shows various possible intersections of vertical, horizontal, or inclined individual fractures with the wells. In any individual fracture model development, generally the following simplifying assumptions are considered.

1. The aquifer is infinite in horizontal directions, consisting of a single layer with uniform thickness surrounded from above and below by horizontal impervious layers.
2. The well intersects the fracture in its midpoint and the vertical fracture penetrates completely the whole thickness of aquifer.
3. The fracture is planar and its aperture is infinitesimally small which implies zero storage coefficient for the fracture whereas its length is finite.
4. The groundwater flows from the matrix into the fracture at a constant rate.
5. The water level in the vertical fracture at any time during pumping is uniform along the entire length of fracture and the groundwater enters the well through the fracture only.
6. The groundwater flow is laminar and the Darcy law is valid.
7. The pump discharge is constant.
8. The fracture has an infinite hydraulic conductivity.
9. The formation is isotropic, that is, the vertical and horizontal hydraulic conductivities are not different from each other.
10. Initially the hydraulic pressure is constant everywhere in the aquifer.
11. Well losses and storage effects are all negligible.

Groundwater flow in some fractured rocks with a dominant fracture will not be radial but rather linear especially in the vicinity of the fracture itself. The well and the production surface will be referred to jointly as an extended well.

A. GRINGARTEN-WITHERSPOON MODEL

The first study to evaluate aquifer parameters in single fracture-well interferences was performed by Gringarten and Witherspoon (1972), who derived type curves for determining the hydraulic parameters from aquifer test response. In fact, their model is related to the work in the petroleum reservoir engineering by Gringarten and Ramey (1974), who developed techniques for evaluating fracture effects on oil and gas wells.

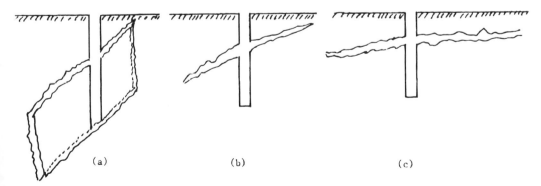

Figure 17 Well-fracture intersection. (a) Vertical, (b) horizontal, and (c) inclined fractures.

The time-drawdown relationship for the problem under consideration was obtained by using Green's function product solution technique presented by Gringarten and Ramey (1974). The solution is presented for time-drawdown variations in an observation well near a finite length vertical fracture during pumping from an extended well. The dimensionless well function for vertical fracture, $W_v(u,d)$ at a point with coordinates (x,y) and distance, d, from the pumping well is given explicitly as

$$W_V(u,d) = \frac{\sqrt{\pi}}{2} \int_0^u \left[\Phi\left(\frac{1-d}{2\sqrt{\pi}}\right) + \Phi\left(\frac{1+d}{2\sqrt{\tau}}\right) \right] \frac{d\tau}{\sqrt{\tau}} \qquad (36)$$

in which $\Phi(\cdot)$ is the error function of the argument and the definition of $W_v(u,d)$ is

$$W(u,d) = \frac{4\pi T}{Q} s \qquad (37)$$

where

$$u = \frac{Tt}{Sx_f^2} \qquad (38)$$

and

$$d = \frac{\sqrt{x^2 + y^2}}{x_f}$$

herein x_f is the half length of the fracture; x and y are the relative coordinates with respect to the pumping well (see Figure 18) The general solution in Equation 36 can be reduced into various practically useful forms as follows:

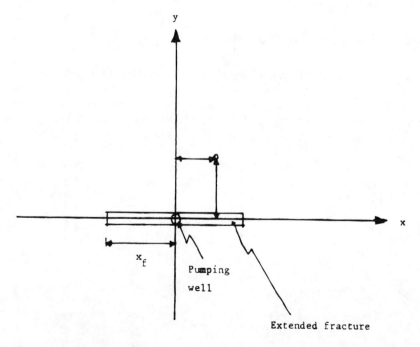

Figure 18 Finite vertical fracture pattern.

1. If observation well is located along the x axis, i.e., $y = 0$ $(d = x/x_f)$ then the well function becomes

$$W_V(u,d) = \frac{\sqrt{\pi}}{2} \int_0^u \left[\Phi\left(\frac{1-d}{2\sqrt{\tau}}\right) + \Phi\left(\frac{1+d}{2\sqrt{\tau}}\right) \right] \frac{d\tau}{\sqrt{\tau}} \tag{39}$$

The graphical representation of resulting type curves is given in Figure 19.

2. For an observation well located on y axis $(d = y/x_f)$ the well function is

$$W_V(u,d) = \sqrt{\pi} \int_0^u \Phi\left(\frac{1}{2\sqrt{\tau}}\right) e^{-\frac{d^2}{4\tau}} \frac{d\tau}{\sqrt{\tau}} \tag{40}$$

Necessary type curves are presented in Figure 20.

3. When the observation well lies along 45° line $(x = y)$ and $d = \sqrt{2}x/x_f = \sqrt{2}y/x_f$ that crosses through the pumping well (see Figure 21), then the well function takes its form as

$$W_V(u,d) = \frac{\sqrt{\pi}}{2} \int_0^u \exp\left(-\frac{d^2/2}{4\tau}\right) \left[\Phi\left(\frac{1-\frac{d}{\sqrt{2}}}{2\sqrt{\tau}}\right) + \Phi\left(\frac{1-\frac{d}{\sqrt{2}}}{2\sqrt{\tau}}\right) \right] \frac{d\tau}{\sqrt{\tau}} \tag{41}$$

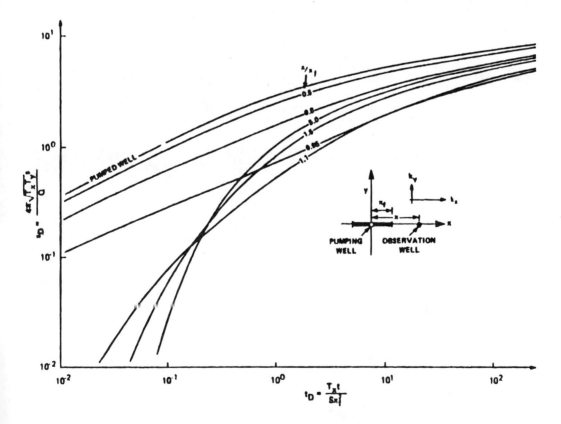

Figure 19 Vertical fracture-type curves for observation well along fracture axis.

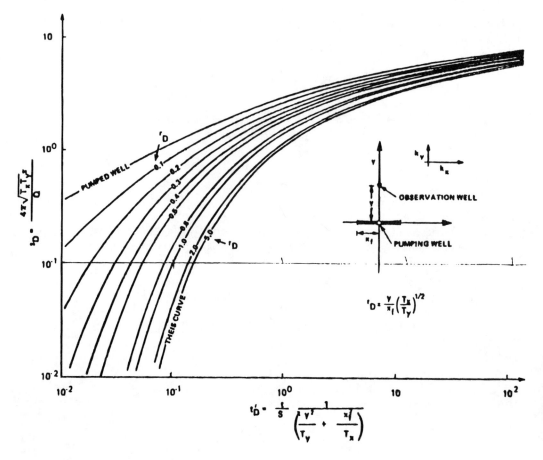

Figure 20 Vertical fracture-type curves for observation well perpendicular to fracture axis.

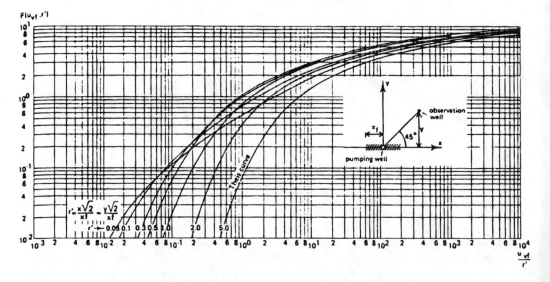

Figure 21 Vertical fracture-type curves for observation well along 45° straight line with the fracture direction.

For the sake of comparison the Theis-type curve is also shown in Figures 19 to 21. It is clear that the drawdown responses in an observation well differ from each other significantly at small times and/or distances. As the distance becomes smaller the initial portions of the curve approach a straight line with slope 1/2 on a double-logarithmic paper. At large distances, however, the drawdown response becomes asymptotically close to the Theis type curve. Finally, whatever the observation position the well function value is always greater than the Theis well function at any given time. This is equivalent to saying that the drawdown response in an observation well due to an extended well response is smaller than a point sink response.

For the application of method it is necessary to know the geometry and the configurations of the main and observation wells in addition to the fracture direction. Otherwise, a trial-and-error matching procedure is necessary for application by considering all three sets of type curves. This tedious procedure should be continued until finding the most suitable matching position that yields approximations of the fracture location, dimensions, and estimates of the aquifer parameter consistent with all the available geological and field measurement information in the study area. The multiplicity of vertical fracture-type curves resulting from this method increases the potential for obtaining a spurious fit between field data and type curve, so it is important to have a good conceptual picture of the aquifer system or to work with data from several observation wells to confirm that a valid solution is reached (Smith and Vaughan, 1985).

Type curve expression for the pumped well located in a vertical fracture is presented by Gringarten and Ramey (1974) as

$$W_{VF}(u) = 2\sqrt{\pi u}\,\Phi\left(\frac{1}{2\sqrt{u}}\right) + \int_0^{1/4u} \frac{e^{-x}}{x}\,dx \qquad (42)$$

At early times when linear flow dominates in the surrounding porous media Equation 42 reduces to $W_{VF}(u) = 2\sqrt{\pi u}$ which indicates a straight line with slope 1/2 on a log-log paper. The duration of this straight line depends on the aquifer and/or vertical fracture features. In low-permeability aquifers and elongated fractures the linear flow period may continue for a considerably long time. Gringarten et al. (1974) showed that transition to pseudo-radial flow starts at $u = 2$ and, consequently, the Theis-type curve can be used. In such a situation an approximate drawdown equation is given by Gringarten and Ramey (1974) as

$$s = \frac{2.30Q}{4\pi T}\log\frac{16.59T_t}{Sx_f^2} \qquad (43)$$

This expression provides a basis for determining the aquifer parameters through a procedure similar to the Jacob straight line method on a semilog paper with transmissivity and storativity calculations from $T = 2.30Q/4\pi\Delta s$ and $S = 16.59Tt_0/x_{f2}$, respectively.

B. JENKINS-PRENTICE METHOD

The solution of vertical fracture problem has been described by Jenkins and Prentice (1982) as an extreme condition where a homogeneous aquifer is assumed to be bisected by a single fracture with an extensive length. All of the remaining assumptions are adopted as stated above. Under these considerations flow in the aquifer is linear at all times toward the fracture. Their conceptual model is shown in Figure 22. In this model drawdowns in the observation wells in the rock matrix are directly related to distance, from the vertical fracture only. The drawdown, $s(d,t)$, at an observation well with its distance, d, to the fracture at time, t, is given as

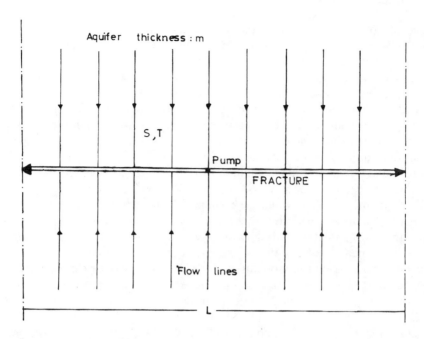

Figure 22 Extensive single vertical fracture.

$$s(d,t) = \frac{Q}{2LT}\left\{\frac{4tT}{S}\exp\left(-\frac{d^2}{4tT}\right) + \left[\Phi\left(\frac{d^2S}{4tT}\right) - 1\right]\right\} \quad (44)$$

where $\Phi(\cdot)$ is the error function, S and T are the storage coefficient and the transmissivity of the rock matrix, and L is the fracture length which is assumed to be very long. Finally, Q is the constant discharge from the well. Jenkins and Prentice did not provide any type curve but rather investigated late time or small distance behaviors of the time-drawdown variation during an aquifer test. For instance, the drawdowns in the pumping well ($x = 0$) or large time cases become from Equation 44 independent of distance as

$$s(t) = \frac{\sqrt{\pi}Q}{2L\sqrt{TS}}\sqrt{t} \quad (45)$$

which leads to conclusion that the drawdown vs. time plot on a log-log paper yields to a straight line with slope 1/2 indicating that the Jacob straight line is not valid even at large times for extensive fractures similar to the method in the previous section.

The same authors developed a model for the recovery drawdown following the pumping period. They have used the image theory and assumed that after the pump stop a recharge well with the same flow had been introduced at the same point where the drawdown was measured. Hence, the recovery equation for linear flow is

$$s(d,t,t') = \frac{Q}{2LT}\left\{\frac{4tT}{S}\exp\left(-\frac{d^2S}{4tT}\right) - \frac{4t'T}{S}\exp\left(-\frac{d^2S}{4t'T}\right)\right\} \quad (46)$$

where t is time since pumping began and t' is the recovery period time since pumping stopped. This expression which becomes independent of distance has been simplified to

$$s(d,t,t') = \frac{Q}{2LT}\frac{4T}{S}\sqrt{(t-t')} \quad (47)$$

Similarly, recovery data plotted as $s(t,t')$ vs. $\sqrt{t-t'}$ will give a straight line in a linear flow system. The point differs from a similar plot of data in a radial flow system where the drawdown decreases rapidly at small values of t or t' and tends to stabilize at large values of t.

Jenkins and Prentice stated that at small values of time and distance from an extended well, Equations 42 and 43 for linear flow describe the drawdown in the trough of depression that surrounds the extended well. However, at large values of time and distance from a finite extended well in an infinite aquifer, flow becomes radial. The time and distance at which radial-flow equations become valid are functions of the fracture length, aquifer storage coefficient, and transmissivity. These points will be elaborated in the next section.

In the case of extremely high conductivity vertical fracture if the orientation is known and the host rock transmissivities on both sides are the same, then two piezometers on the same sides of the fracture are required to determine the perpendicular distances between the piezometers and the trace of vertical fracture as in Figure 23a. The piezometer closest to the pumped well is not the piezometer closest to the fracture. Regardless of distances the drawdown will be greater in the piezometer closest to the fracture. On the other hand, if the precise orientation of the fracture is not known then two piezometers will be needed. As can be seen from Figure 23b, if x_1 is small relative to x_2, two orientations are possible because x_1 may be on either side of the fracture. More piezometers must then be placed to find the orientation.

C. ŞEN MODEL

A mathematical description of groundwater flow toward vertical fractures has been presented by Şen (1986) with the purpose of obtaining type curves that were not available in the

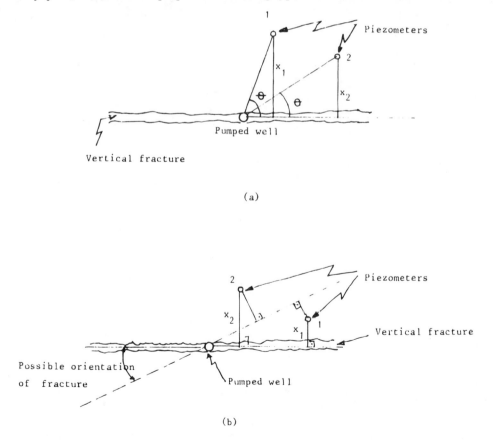

Figure 23 Piezometer arrangement near a fracture of (a) known and (b) unknown orientation.

Jenkins-Prentice study. On the basis of the same set of assumptions as in the Jenkins and Prentice model, he obtained the type curve expression succinctly as

$$W_V(u) = \frac{1 - \Phi(\sqrt{u})}{\sqrt{u}} \qquad (48)$$

where explicit definition of the well function for an extensively long vertical fracture is given as

$$W_V(u) = \frac{2TL}{Qd} s(d,t) \qquad (49)$$

here Q, L, and d have the same meanings as in the Jenkins-Prentice model. The plot of $W_v(u)$ vs. u on a double-logarithmic paper gives a type curve for the extended well composed of infinitely long vertical fracture (in Figure 24). Radial flow type curve (Theis) is also shown in the same figure. For small times the two curves are rather close to each other but at large times the linear flow curve diverges from the radial flow curve. On the other hand, for small u values, linear flow has greater drawdown than the radial flow. Asymptotically, it becomes a straight line as $u \to 0$, $\Phi(0) = 1/2$, and hence Equation 48 becomes

$$W_V(u) = \frac{1}{2\sqrt{u}} \qquad (50)$$

The slope of log-log paper straight line is equal to 1/2. The recovery type curve expression

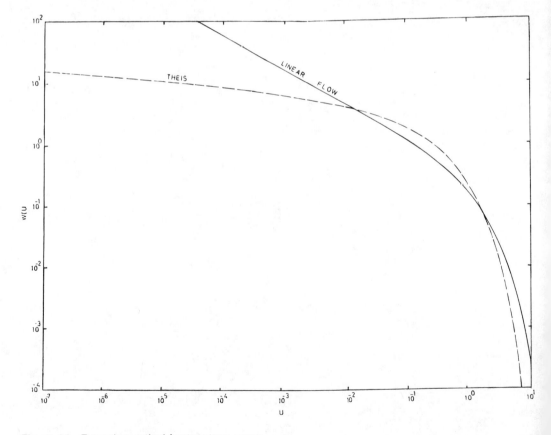

Figure 24 Extensive vertical fracture-type curve.

has also been obtained also by Şen (1986) as

$$W_{VR}(u) = \frac{\Phi[\sqrt{u/u_\tau}] - \Phi(\sqrt{u})}{\sqrt{u}} \tag{51}$$

Herein, $W_{VR}(u)$ is the extended well function for recovery period. It is interesting to note from this expression that the complete drawdown is possible theoretically for $u \to 0$; however, in practice, the time duration needed for complete recovery is larger than the pumping period. Type curves resulting from Equation 51 are shown in Figure 25 for different u_τ values. It is obvious graphically that the recovery period is always longer than the pumping period. In addition, for small u_τ values, the initial recovery rate is very rapid compared to the late recovery and initial recoveries for larger u_τ values. Furthermore, most of the recovery period shows a straight line relationship between u and $W(u)$ after the initial recovery along a curvature portion of the type curves.

One of the most important features of particular interest is that the drawdown continues to increase in the observation well after pumping is stopped provided that u_τ is large ($u_\tau > 1$). The reason for this phenomenon is that, although water is no longer pumped, the aquifer continues to supply water to refill the well (Rushton and Holt, 1981). Their field observations showed that such a behavior is not apparent in all tests. Their explanation is low transmissivity values of the aquifer. Low transmissivities cause large u_τ values; therefore, increase in drawdown after pump-stop is evident in type curves in Figure 25 for $u_\tau > 1$.

So far none of the above-mentioned individual fracture models considered the fracture storage. The assumption that a fracture can have negligible resistance to flow while having

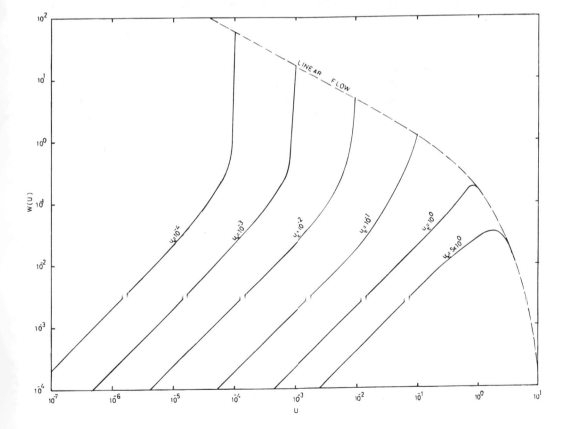

Figure 25 Recovery-type curve for vertical fractures.

no storage capability is an obvious conflict. Such an unrealistic assumption is released by Şen (1986), who developed type curve expressions for the vertical fracture of aperture, $2a$, as

$$W_{VS}(u) = -\frac{S}{2\sqrt{\pi}} exp\left[-\int_u^\infty A(x)dx\right]dx - \int_u^\alpha \frac{exp\left[\int_{u\alpha}^\infty A(x)dx\right]}{x^2} dx \qquad (52)$$

where

$$A(x) = -\left\{\frac{\sqrt{\pi}}{A} x^{3/2} e^x [1 - \Phi(\sqrt{x})]\right\}^{-1} \qquad (53)$$

and

$$u = \frac{Sa^2}{4tT} \qquad (54)$$

and $W_{vs}(u)$ is the extended well function with fracture storage which is defined similar to Equation 49. A set of type curves for $S = 10^{-1}$, 10^{-2}, and 10^{-3} is given in Figure 26. Each one of these curves has initial and final portions as straight lines. As the curvature portion starts, the aquifer contributes to the pump discharge and the fracture storage ceases to contribute to the pump discharge along the late straight line.

Another point worth noticing is that, in all the vertical fracture models, the fracture length was considered to be very long and as a result the laminar flow existed at every time and point. Kohut et al. (1983) indicated that a pure linear flow pattern implies an infinite fracture length that cannot be applicable to field tests where most of the time the fracture length is of finite extent and accordingly the equipotential lines in the fracture vicinity are neither completely linear nor radial but elliptical. This fact leaves the method of Gringarten and Witherspoon as the only tool in determining the aquifer parameters in the case of pumping from a finite length fracture. However, there appears practical complications in its applications

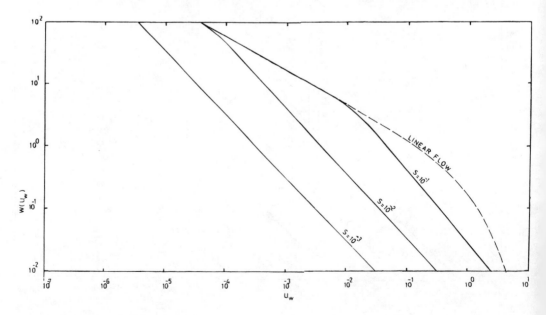

Figure 26 Large vertical fracture-type curves.

FRACTURED MEDIUM AQUIFER TESTS

as were pointed out by Kohut et al. (1983). Şen (1992) proposed an approximate but simplified solution for finite length vertical fractures by simplifying the flow pattern as composed of linear flow toward the fracture but semi-radial flow patterns toward the end points of the fracture as shown in Figure 27. In deriving the necessary formula all the assumptions in Section IV are adopted except that the fracture has a finite length, $2x_f$. The fracture well function and the dimensionless time factor for observation well are given as

$$W_{VF}(u,v) = \frac{2T}{Q} s(d,t) \tag{55}$$

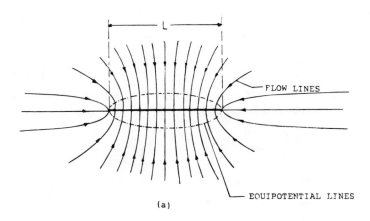

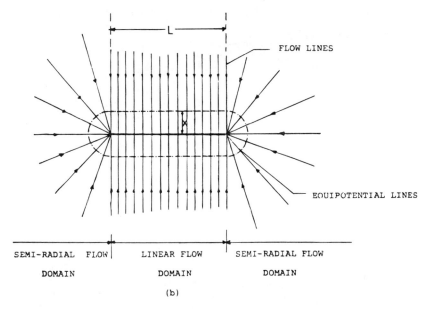

Figure 27 Extended well. (a) Theoretical and (b) idealized flow geometry.

and

$$u = \frac{d^2 S}{4tT} \tag{56}$$

respectively. Herein, v is a dimensionless distance factor defined as

$$v = \frac{2x_f}{d} \tag{57}$$

in which d is the perpendicular distance of the observation well to the vertical fracture. The explicit form of the well function is

$$W_{VF}(u,v) = \frac{\left(\int_u^\infty \frac{e^{-x}}{x} dx\right)[1 - \Phi(\sqrt{x})]}{2\pi\alpha[1 - \Phi(\sqrt{x})] + \frac{v\sqrt{u}}{\sqrt{\pi}} \int_u^\infty \frac{e^{-x}}{x} dx} \tag{58}$$

The numerical solution of this expression leads to type curves presented in Figure 28. Herein, the expression $\Phi(\sqrt{x})$ is the error function as

$$\Phi(\sqrt{x}) = \frac{1}{2\sqrt{\pi}} \int_{-\infty}^{\sqrt{x}} e^{-y^2/2} dy \tag{59}$$

which may be taken from normal probability distribution function tables given in any standard textbook on statistics. Prior to the numerical solution of Equation 58 it is illuminating to consider its special cases as follows

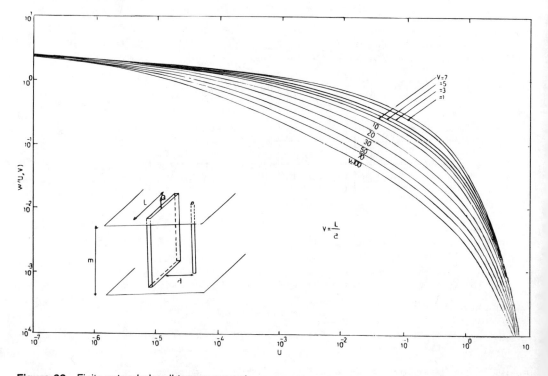

Figure 28 Finite extended well-type curve set.

1. If the fracture length is zero, i.e., the radial flow case, $v = 0$ where $\alpha = 1$, then its substitution into Equation 58 leads to the classical Theis solution as

$$2\pi W(u,v) = \int_{-\alpha}^{u} \frac{e^{-x}}{x} dx \qquad (60)$$

Notice that consideration of Equation 55 gives $2\pi W(u,v) = (4\pi T/Q)s(d,t)$ which is equivalent to the Theis well function, $W(u)$. In general, when $v \to 0$, the Theis solution becomes applicable for the extended well situation. For such a condition the observation well location must be far away from the fracture itself, i.e., $2x_f << d$. As a rule of thumb, in practical studies $v < 0.1$ can be acceptable for the validity of the Theis solution.

2. For the linear flow only, it is sufficient to substitute $\alpha = 0$ into Equation 58 which leads to

$$\frac{2x_f}{\sqrt{\pi d}} W(u,v) = \frac{1 - \Phi(\sqrt{u})}{\sqrt{u}} \qquad (61)$$

in which the left-hand side by substitution of $W(u,v)$ from Equation 55 becomes equivalent to Equation 48.

3. As $u \to 0$, $(t \to \infty)$, $\Phi(\sqrt{u}) \to 1/2$, and hence Equation 58 reduces to the Theis expression as given in Equation 60. This is tantamount to saying that, in a finite length fracture (extended well), the drawdown distribution at large times will comply to the Theis-type curve, and accordingly the Jacob straight line method can be used. However, as explained in the Jenkins-Prentice method, the linear flow case stipulated that the vertical fracture is of infinite length. The late time drawdowns are defined by a straight line on an arithmetic scale, not semilogarithmic, with the plot of $s(r,t)$ vs. $\sqrt{t}$. This is one of the major differences of finite and infinite length fracture behavior in the case of water withdrawal through them.

The calculation of the right-hand side in Equation 58 is achieved by the use of Simpson's numerical calculation rule on high-speed digital computers, and the results are shown in Figure 28 as a set of curves for different v values. It is obvious from this figure that the initial and final portions of the extended well (with finite fracture length) type curves merge with the classical Theis-type curve. Hence, deviations from the Theis-type curve occur at moderate time instances only. However, the deviations from this basic-type curve increase with the increase of the dimensionless fracture length, v.

1. Application

The field time-drawdown data for the application of the methodology developed herein have been adopted from Smith and Vaughan (1985). They conducted a low-rate aquifer test in a folded and fractured limestone formation in the Conasauga group of eastern Tennessee. The water was pumped through a well in hydraulic connection with a major fracture for 24 h at an average rate of 3.29 l/min. Water-level response in many observation wells has shown a through-like depression, indicating the effect of the vertical fracture on the equipotential and flow lines. The layout of the wells and the equipotential lines are presented in Figure 29. The best way to determine the aquifer parameter values has been suggested by the Gringarten and Witherspoon (1972) procedure. However, the major difficulty in their procedure was determining the well location coordinates with respect to the fracture midpoint. To overcome such a difficulty Smith and Vaughan assumed the observation well locations as having either abscissa zero ($x = 0$) or ordinate zero ($y = 0$). In fact, the methodology developed herein does not suffer from such a requirement and the distance in Equation 56 is taken either as the perpendicular distance to the fracture orientation for observation wells within the linear flow domain or the radial distance to the nearest fracture end in the semiradial flow domain. For

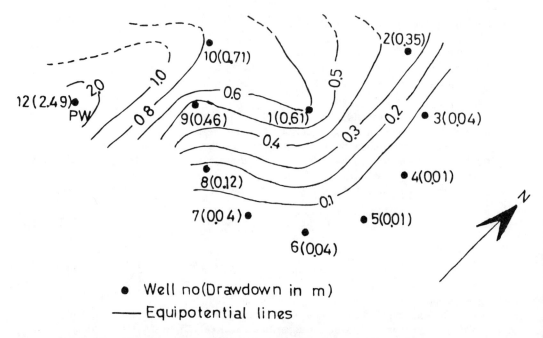

Figure 29 Well locations and equipotential lines (after Smith and Vaughan, 1985).

instance, the application of the methodology requires the perpendicular distances irrespective of the y values along the fracture which is important in the Gringarten and Witherspoon method because there the distances to the main well which is located in the center of the fracture are important. The following steps should be followed for an effective application of the methodology in estimating the aquifer parameters:

1. Plot the field data on the same scale of double-logarithmic graph paper as for the type curves.
2. Find the best matching curve to these data among the set of type curves presented in Figure 28. Observation well data are matched individually after a trial-and-error method with the best type curve as shown in Figure 30a to e. Hence, v values can be read easily from the matched type curves.
3. In each figure the matching point has the extended well function value as $W(u,v) = 10^1$ and $1/u = 10^0$ except in Figures 30a and e where $1/u = 10^{-1}$ and 10^{-2}, respectively.
4. The perpendicular distance, d, of each well to the fracture is adopted from the work of Smith and Vaughan. The well numbering, match-point coordinates, and fracture distances are presented in the first part of Table 3.
5. The substitution of the relevant values from the first part of Table 3 into Equation 55 gives the transmissivity values, whereas the fracture length estimation is obtained from Equation 57. The results are shown in the second part of Table 3. Due to the anisotropy of the aquifer, the transmissivities should be regarded as the equivalent transmissivities to the anisotropic media. As stated by Freeze and Cherry (1979), the two equivalent representations of the same problem can be written as $T = \sqrt{T_x T_y}$ in which T_x and T_y are the transmissivity values in the x and y directions. Hence, comparison of $\sqrt{T_x T_y}$ values of Smith and Vaughan with T values calculated herein indicates a good agreement between the two sets of values (Table 4).

During the calculations, it has been noted that the storage coefficient determination requires a precise estimation of the fracture (extended well) length. The effect of fracture length on storage coefficient can make a difference of about an order of magnitude. Because the transmissivity values are not dependent on the fracture length, they are more reliable than the storage coefficient values calculated from this method.

FRACTURED MEDIUM AQUIFER TESTS

VII. HORIZONTAL FRACTURES

Unfortunately, studies on the horizontal fractures to evaluate the aquifer parameters lag very much behind the vertical fractures. Hydraulic properties of the horizontal fractures are different from the vertical fractures along the following main points:

1. Horizontal fractures are subject to confined conditions more than any other type of fracture. They do not extend to the surface and they are often immersed into the water.
2. Two-dimensional flow occurs in the horizontal fractures depending on the well fracture combination and fracture surface roughness. Figures 15 and 16 in Chapter 5 indicate different

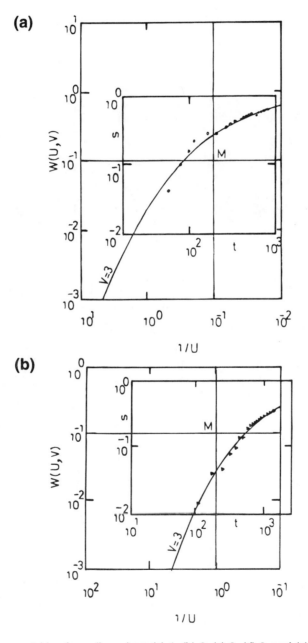

Figure 30 Type curve matching for well numbers (a) 1, (b) 2, (c) 8, (d) 9, and (e) 10.

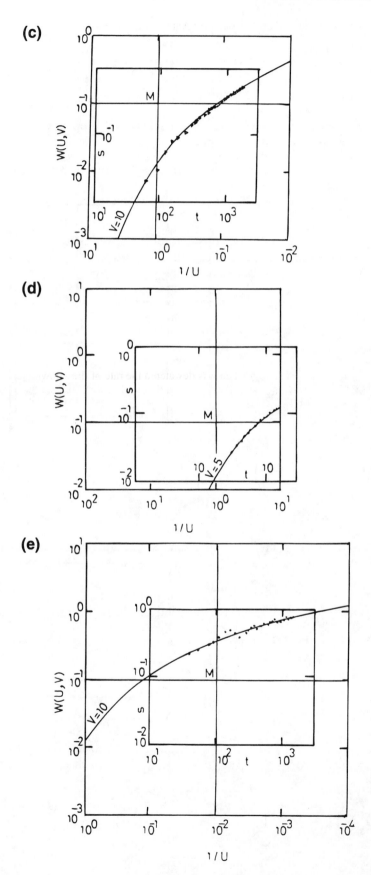

Figure 30 Continued.

Table 3 Match-Point Coordinates and Aquifer Parameters

Well no.	Dimensionless distance (V)	Match point t (min)	Match point s (m)	Fracture distance	T (m²/d)	S	L (m)
1	3.0	2.2×10^2	1.2×10^{-1}	3.5	1.97	9.8×10^{-3}	10.5
2	3.0	1.4×10^2	1.8×10^{-1}	4.0	1.32	3.2×10^{-2}	12.0
8	5.0	1.9×10^2	8.0×10^{-2}	9.0	2.90	7.9×10^{-2}	45.0
9	10.0	9.5×10^1	3.1×10^{-1}	2.5	0.76	8.4×10^{-2}	25.0
10	10.0	1.1×10^2	9.0×10^{-1}	0.2	2.60	2.0×10^{-3}	2.0

types of flow patterns. In the one-dimensional flow case the flow lines are horizontal and parallel, provided that the fracture walls are horizontal and parallel; otherwise, one-dimensional but nonparallel flow takes place within the fracture. Generally, the flow in a horizontal fracture has two dimensions as in Figure 16 in Chapter 5 but if water is withdrawn from or injected into fracture through a well then the flow becomes radially symmetrical similar to the porous medium. Such a situation does not occur in a vertical fracture.

3. Horizontal fracture hydraulic conductivity and roughness play a dominant role in the groundwater flow toward wells. However, as it was explained in the previous section, the vertical fractures are conveyors of groundwater to the well without any significant effect from its hydraulic conductivity or roughness on the well discharge.
4. Horizontal fractures have more areal extent than the vertical fractures and as a result resistance of horizontal fracture wells to the groundwater flow is relatively more than vertical fractures.
5. If the piezometric level falls below the horizontal fracture, then the discharge from this fracture to the well is independent of the drawdown in the pumping well. Any horizontal fracture acts as a reservoir and when it is dewatered the rate of drawdown increases rapidly. However, the effect of vertical fractures, intersecting with a well, is continuous without sudden drops in the drawdown measurement.

A. MULTI-HORIZONTAL ROUGH FRACTURES

In the crystalline rock environments of the world, increasing demand on groundwater supply has necessarily focused attention on granitic rocks which have considerable extent and sheet structures giving rise to the horizontal fractures subparallel to the rock surface. Under favorable conditions artesian pressure may be built in lowland sheeted openings where intersheet zones are rather continuous and are crossed by few fractures. They act as confining beds for the groundwater moving under hydrostatic pressure down the sheeted planes as explained by LeGrand (1949).

In hard igneous and metamorphic rocks the fractures are the only significant means of transmitting and starting water movement. Let the number of horizontal fractures be n each with different storativities, S_i, and transmissivities, T_i, fracture apertures, $2a_i$, and roughness, r_i, where $i = 1,2,\ldots,n$). The flow law within a fracture in terms of the fracture aperture, $2a$, and the relative roughness, r_r, was explained in Chapter 5. Şen (1989) applied these flow equations to problems of well behavior in the horizontally fractured rocks leading to a set of

Table 4 Hydraulic Parameter Estimations

Well no.	Fracture length (m)	Smith and Vaughan T (m²/d)	Smith and Vaughan S	Equations 17 and 19 T (m²/d)	Equations 17 and 19 S
1	15	1.43	2.9×10^{-3}	1.97	9.8×10^{-3}
2	20	1.53	9.1×10^{-2}	1.32	3.2×10^{-2}
8	20	4.40	1.4×10^{-2}	2.00	2.1×10^{-2}
9	15	0.52	4.7×10^{-3}	0.76	4.8×10^{-3}
10	18	1.05	2.9×10^{-3}	2.60	5.2×10^{-3}

type curves. A set of horizontal fracture with different hydraulic and geometric properties has been conceptualized as shown in Figure 31.

In the case of fractured aquifers where there is effectively no primary porosity, very little work, if any, has been done on the effect of fracture system on the behavior of a well completed in such a system. The problem of laminar flow in a single horizontal fracture has been investigated in the laboratory (Louis, 1969; Snow, 1965; Witherspoon et al., 1980) as an analogy of parallel planar plates to represent the fractures. It was proved that a cubic law is valid for the discharge, q_i, through a fracture with aperture, $2a_i$,

$$q_i = C_i(2a_i)^3 \tag{62}$$

where C_i is a constant for the ith fracture which in the case of radial flow toward a well is given by

$$C_i = \frac{\pi \rho g [h_{wi} - h_i(r)]}{6\mu \ln \dfrac{r}{r_w}} \tag{63}$$

in which ρ is the fluid density; g is the acceleration gravity; and μ is the fluid viscosity; $h_i(r)$ is the hydraulic head of ith horizontal fracture. The following set of assumptions is considered for the horizontal fracture system:

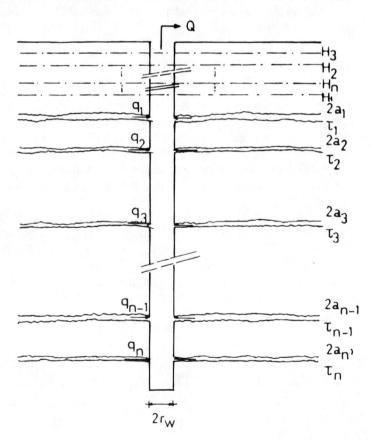

Figure 31 Multihorizontal fracture well.

1. The fractures are horizontal and areally extensive enough for practical purposes, i.e., there is no hydrologic or geologic boundary effects within the depression cone.
2. The fractures are of uniform thickness and the well is perpendicular to them.
3. The well diameter is finite which implies that the well storage plays an active role especially at the initial portions of pumping.
4. The fracture response to the water withdrawal from the well is not instantaneous.
5. Initially each horizontal fracture has different static piezometric level as H_i ($i = 1,2,\ldots,n$).
6. In the fractures the so-called "cubic law" given in Equation 62 is valid.
7. Fractures have rough surfaces.
8. The fractures are individually and collectively compressible, i.e., they are elastic.
9. Each fracture has different thickness, aperture, storativity, and transmissivity which are spatially and temporally invariant.

It is also assumed that the groundwater is included only in the horizontal fractures and the intact zones do not include any water. An approximate solution for the multihorizontal fracture well has been obtained by using the volumetric approach the fundamentals of which were presented in Section VII.E in Chapter 9. The relationship between the well function and the dimensionless time factor are

$$u_w = \left\{ e^{W_H(u_w)} + W_H(u_w) \left(\frac{1}{\sum_{i=1}^{n} S_i} - 1 \right) - 1 \right\}^{-1} \tag{64}$$

in which explicitly by definition

$$u_w = \frac{r_w^2 \sum_{i=1}^{n} S_i}{4t \left[\dfrac{g}{12v} \sum_{i=1}^{n} \dfrac{(2a_i)^3}{(1 + 8.8 r_i^{1.5})} \right]} \tag{65}$$

and

$$W_H(u_w) = \frac{4\pi \left[\dfrac{g}{12v} \sum_{i=1}^{n} \dfrac{(2a_i)^3}{(1 + 8.8 r_i^{1.5})} \right]}{Q} s_w \tag{66}$$

These two last expressions reduce to simpler forms for the following four cases of theoretical and practical interest.

1. If the fracture apertures and roughnesses are almost identical, i.e., $a_i = a$ and $r_i = r$ ($i = 1,2,\ldots,n$) then Equations 65 and 66 become

$$u_w = \frac{3 v r_w^2 (1 + 8.8 r^{1.5})}{t g n (2a)^3} \sum_{i=1}^{n} S_i \tag{67}$$

and

$$W_H(u_w) = \frac{\pi g n (2a)^2}{2vQ(1 + 8.8 r^{1.5})} s_w \tag{68}$$

However, this is an ideal model that cannot be practically encountered in nature. It may have some interest from the academic point of view.

2. If there is only one single horizontal fracture along the well bore, then Equations 67 and 68 are valid provided that $n = 1$. Such a case can be of theoretical interest to reveal the relationship between the aperture and roughness of a single natural fracture and the hydraulic quantities such as the storativity, drawdown, discharge, and time.

3. If the fracture wall roughness is assumed to be zero ($r_i = 0$), i.e., parallel plate analogy, then the dimensionless time factor and well function will have the following forms:

$$u_w = \frac{3\nu r_w^2 \sum_{i=1}^{n} S_i}{tgn \sum_{i=1}^{n} (2a_i)^3} \tag{69}$$

and

$$W_H(u_w) = \frac{\pi g \sum_{i=1}^{n} (2a_i)^3}{3\nu Q} S_w \tag{70}$$

4. In practice, there may appear multihorizontal fractures along the well borehole but one of them may have relatively thicker (at least ten times larger) fracture aperture than others. Because of the cubic power of the aperture in the aforementioned formulations the difference becomes even greater, e.g., 1000 times, and therefore for all practical purposes small fracture contributions to the summations in Equations 65 and 66 can be ignored. This is the most likely and practically observed situation in nature among the simplified alternatives. The dimensionless time factor and the well function for such a case become

$$u_w = \frac{3\nu r_w^2 (1 + 8.8 r_i^{1.5})}{tg(2a_i)^3} \sum_{i=1}^{n} S_i \tag{71}$$

and

$$W_H(u_w) = \frac{\pi g (2a)^3}{3\nu Q (1 + 8.8 r_i^{1.5})} S_w \tag{72}$$

in which $2a_i$ and r_r correspond to the major fracture aperture and relative roughness, respectively. Substitution of each u_w, $W_H(u_w)$ pair into Equation 64 gives type curve expressions for the cases considered above.

However, to separate fracture aperture and relative roughness as independent factors, the following dimensionless definitions are made especially for a single dominant horizontal fracture as in case (4). The dimensionless fracture aperture factor is defined as

$$\beta = \frac{g(2a)^3}{6\nu^2} \tag{73}$$

The horizontal fracture dimensionless time factor, u_f, and the well function, $W(u_f, \beta)$ are

FRACTURED MEDIUM AQUIFER TESTS

$$u_f = \frac{r_w^2(1 + 8.8r_i^{1.5})\sum_{i=1}^{n} S_i}{tv} \tag{74}$$

and

$$W(u_f, \beta) = \frac{\pi v}{(1 + 8.8r_i^{1.5})Q} s_w \tag{75}$$

respectively. Furthermore, u_w and $W(u_w)$ can be written in terms of these last definitions as

$$u_w = \frac{u_f}{2\beta} \tag{76}$$

and

$$W_H(u_w) = 2\beta W(u_f, \beta) \tag{77}$$

The substitution of these two last expressions into the main expression of the volumetric approach (Equation 64) gives the type curve expression for a single horizontally fractured aquifers as

$$u_f = \left[\frac{e^{2\beta W(u_f,\beta)}}{2} + W(u_f,\beta)\left(\frac{1}{S} - 1\right)\frac{1}{2\beta}\right]^{-1} \tag{78}$$

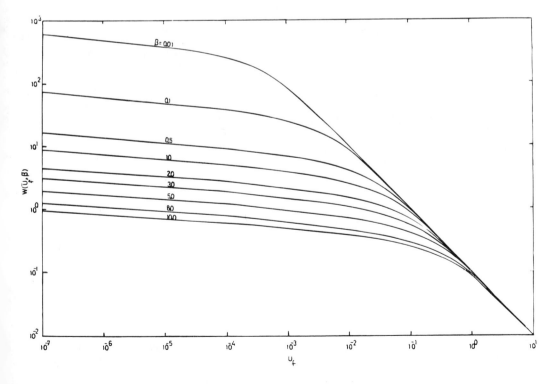

Figure 32 Type curve set for $S = 10^{-1}$.

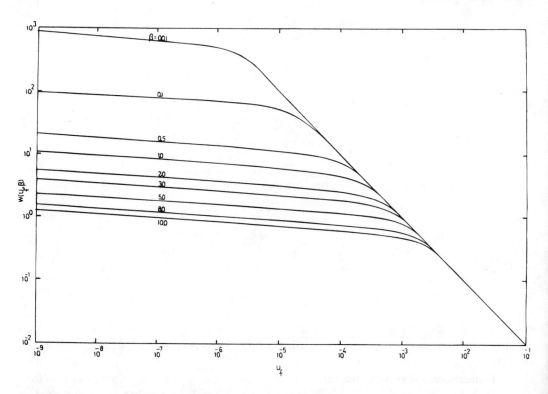

Figure 33 Type curve set for $S = 10^{-3}$.

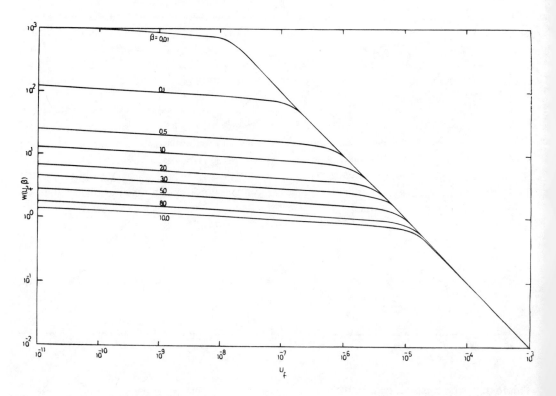

Figure 34 Type curve set for $S = 10^{-5}$.

1. Type Curve Discussion

The plot of the dimensionless time factor vs. the well function for a set of β values and a fixed storage coefficient gives dimensionless standard curves which are referred to as the type curves. Three sets of such type curves are presented in Figures 32 to 34. In general, each type curve has three distinctive portions:

1. The initial straight line portion for small times ($u_f \to \infty$) indicates the effect of the well storage contribution only on the pump discharge. This portion is completely independent from the aquifer properties and serves only as a means of estimating the pump discharge through a simple procedure. Generally, irrespective of any aquifer and flow type, this initial portion exists, provided that the well bore storage is effective, i.e., the well diameter is rather large. However, in the case of horizontally fractured aquifers its duration decreases with increase in the fracture aperture factor, β. In other words, increase in the fracture aperture and/or fracture number causes decrease in this duration.
2. The initial straight line portion terminates with the start of the curved portion during the moderate times which will be referred to as the intermediate portion. The effective duration of this intermediate portion depends on the S value. The smaller the S value the longer is the duration. During this portion the well storage loses its effectiveness gradually at the cost of fracture contribution to the pump discharge.
3. The initial portion for large times ($u_f \to \infty$) does not have any well storage contribution and the quasi-steady-state flow prevails; as a result, the curvature in this portion is rather small. Furthermore, a decrease in the number of fractures or in the total fracture aperture or in both leads to a reduction in the aquifer domain. Physically, this corresponds to a rather quick aquifer response which affects the well storage contribution.

So far as the relative roughness is concerned its increase gives rise to a decrease in the well function (see Equation 75). This leads to the physical implication that, as the fracture surface becomes rougher, the drawdown increases due to the flow-retarding effects of the irregularities on the fracture wall. On the other hand, the proportionality relationship between u_f and r implies that any increase in r decreases the possibility of prompt aquifer response to the pumping.

Figure 35 shows that irrespective of the storage coefficient value the late time behavior of the aquifer appears along the same curve, depending on the specific value of the dimensionless fracture aperture, β. This situation confirms the earlier-mentioned fact that the late portions of the type curves are reflections of the fracture parameters such as the fracture number, aperture, and the relative roughness. Furthermore, these mergers for late time ($u_f \to 0$) imply that many type curves match each other on a double-logarithmic paper. For the last portion of each curve Equation 78 will have its asymptotic value for very small u_f values as

$$u_f = 2\beta exp[-2\beta W(u_f,\beta)] \tag{79}$$

which indicates that only β plays a role in forming these portions. However, by taking the natural logarithm of both sides one can obtain

$$W(u_f, \beta) = Ln\left(\frac{2\beta}{u_f}\right)^{1/2} \tag{80}$$

This expression gives a set of straight lines on a semilogarithmic paper. Each one of the straight lines is labeled by a specific β value as shown in Figure 36.

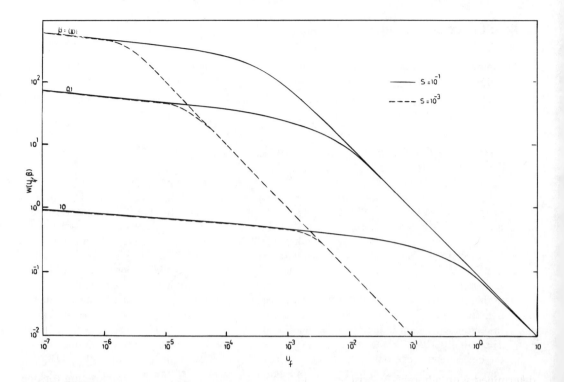

Figure 35 Type curves for various values.

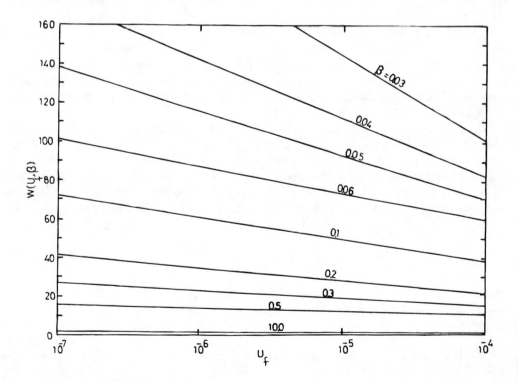

Figure 36 Type straight lines.

2. Application

For the use of described methodology in field applications the following steps are necessary:

1. Plot the field data on a double-logarithmic paper as drawdown on the vertical axis vs. time. This plot is referred to as the field data sheet which should have the same scale as for the type curve sheets.
2. Overlay the field data sheet on the type curve sheet and search for the best matching type curve by shifting the two sheets relatively to each other such that drawdown (1/time) and well function (dimensionless time factor) axis are parallel to each other at all times.
3. From the best matching type curve labels, i.e., S and β, values are recorded.
4. An arbitrary match point is chosen on the common of the two sheets. The coordinates $(u_f)_M$; $[W(u_f,\beta)]_M$ and $(1/t)_M$; and $(S_w)_M$ are recorded from the type curve and field data sheets, respectively.
5. The substitution of relevant values into Equations 73 and 75 yields the estimates of fracture aperture and relative roughness, respectively.

The methodology developed herein for the multihorizontal fracture wells is applied to the field data presented by Gale (1977). The study area is approximately 20 miles southwest of the city of Halifax, Nova Scotia, in the southwestern part of Halifax county near the village of Sambro. This area includes a good combination of well-exposed bedrock and relatively simple fracture geometry. The fracture occurrences in the wells and their apertures were determined with a borehole periscope in addition to the drilling core. The fracture log for one of the main wells is given in Figure 37. It is obvious from this log that some of the fractures are filled with material so that they are tight; others are open but have small apertures. However, the main large and well-developed horizontal fracture appears at a depth of 9.36 ft ($\cong$2.85 m) with the fracture aperture equal to 0.1 ft ($\cong$3 cm). Gale calculated the aperture for the same fracture from injection test results as 0.0037 ft (0.11 cm). There is a considerable difference between this fracture and the one measured directly in the field. The next nearest fracture aperture is 0.01 ft (0.3 cm), the ratio of the main fracture aperture to this being 10. Similar pattern of fractures were observed in six other wells that were distributed around the main well within about a 25-m vicinity. From the fracture logs it was apparent that most of the fractures are continuous between the wells and horizontal or near horizontal planes. This fracture system with borehole can be idealized as a system of multi-horizontal well.

The aquifer test data from this well borehole are presented in Figure 38. The pump discharge has been kept constant at 1.43×10^{-3} cfs (4.04×10^{-5} m³/s). The data were recorded in the main well of 12-in. ($\geqslant$30 cm) diameter and the initial water level was 8 ft (2.44 m) below the ground surface. For the first 250 min the test resulted in the maximum drawdown of about 1 ft (0.3 m). Subsequently, there appeared a sharp rise in the drawdown which is due to the dewatering of large fracture. This large fracture acts as a reservoir, resulting in small drawdowns during the early part of the aquifer test. When it had been dewatered, the rate of drawdown increased rapidly. Gale concluded that, when the water level falls below the large fracture at 9.36 ft ($\cong$2.85 m) level, the discharge from this fracture is independent of the drawdown in the pumping well. Therefore, it is extremely important to know the points along the well borehole at which the main water-conducting fractures intersect the borehole. He noticed that these data were not amenable to standard aquifer test analysis. Obviously, such a pattern of time drawdown cannot be explained by any double-porosity medium fracture models as mentioned earlier. Furthermore, continuation of the sudden rise with significant curvature indicates that there is no other major fracture to influence the drawdown in the well. Otherwise, after the breaking point, a pattern should emerge similar to that shown prior to the breaking point (Rissler, 1978).

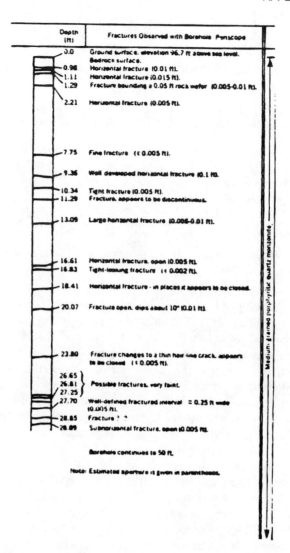

Figure 37 Well bore fracture log (after Gale, 1977).

However, herein after many graphical trials the field data for the first 250 min have been matched with type curve from the set of $S = \Sigma S_i = 1Q^{-3}$ in Figure 38 with $\beta = 1$. The match point coordinates on the field data sheet are chosen arbitrarily as $(1/t)_M = 10^{-1}$ 1/s and $(S_w)_M = 10^0$ ft (0.30 m) whereas on the type curve sheet the corresponding coordinates are $(u_f)_M = 4.7 \times 10^{-4}$ and $[W(u_f,\beta)]_M = 4.1 \times 10^0$. With these numerical values at hand, it is possible to calculate the relative roughness from Equation 75, provided that the kinetic viscosity of water is known, under normal conditions at 62°F (20°C), $= 1.07 \times 10^{-1}$ ft²/s ($\cong 10^{-2}$ m²/s) (Brown et al., 1960). Hence, from Equation 75 after some algebra one can obtain

$$r = \left\{ \frac{1}{8.8} \left[\frac{\nu \pi s_w}{QW(u_f, \beta)} - 1 \right] \right\}^{2/3} \tag{81}$$

Substitution of the aforementioned relevant numerical values into this equation yields $r = 3.32$ which is larger than 0.032, implying that the fracture is very rough. On the other hand,

FRACTURED MEDIUM AQUIFER TESTS

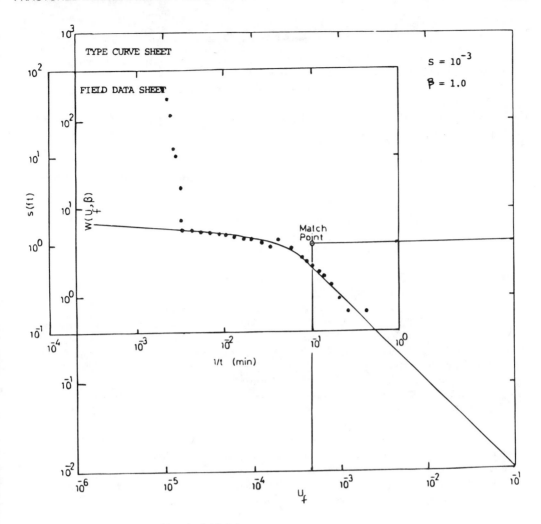

Figure 38 Type curve matching to field data.

Equation 73 gives that $(2a)^2 = 6.1 \times 10^{-5}$ m^3 or $2a = 0.039$ m (0.13 ft). This fracture aperture estimate is very close to the field measurement (0.10 ft) that shows the effectiveness of the presented methodology. It is not necessary to know the location of the fractures to apply the method. The following conclusions can be drawn from this study.

1. Fluctuations of groundwater level in the multi-aquifer wells can be predicted by using the volumetric approach.
2. Description of horizontally fractured aquifer test data requires three physical dimensionless parameters. These are the dimensionless time and fracture aperture factors in addition to the horizontally fractured aquifer well function.
3. The necessary quantities in describing the groundwater flow through horizontal fractures are their number along the well borehole and total and individual aperture as well as the effective roughness.
4. The classical type curve-matching technique can be applied with the type curves in this paper to determine the aquifer parameter estimates which are, in the case of horizontally fractured aquifers, the storage coefficient, fracture aperture, and roughness.

REFERENCES

Barenblatt, G. E., Zheltov, I. P., and Kochina, I. N., 1960. Basic concepts in the theory of homogeneous liquids in fractured rocks, *J. Appl. Math. Mech.*, 24, 1286.

Boulton, N. S., 1963. Analysis of data from nonequilibrium pumping tests allowing for delayed yield from storage, *Proc. Inst. Civil Eng.*, 26,469.

Boulton, N. S. and Streltsova, T. D., 1977a. Unsteady flow to a pumped well in a fissured water bearing formation, *J. Hydrol.*, 35, 257.

Boulton, N. S. and Streltsova, T. D., 1977b. Unsteady flow to a pumped well in a two-layered water bearing formation, *J. Hydrol.*, 35, 257.

Bourdet, D. and Gringarten, A. C., 1980. Determinations of Fissure Volume and Block Size in Fractured Reservoirs by Type Curve Analysis, Presented at the SPE Annual Technical Conference and Exhibition, 21.

Brown, R., Steward, W. E., and Lightfoot, E. N., 1960. *Transport Phenomenon*, John Wiley & Sons, New York.

de Swan, O. A., 1976. Analytical solution for determining naturally fractured reservoir properties by well testing, *Soc. Pet. Eng. J.*, 117.

Ershaghi, I. and Aflani, R., 1985. Problems in characterization of naturally fractured reservoirs from well test data, *Soc. Pet. Eng. J.*, 405.

Freeze, R. A. and Cherry, J. A., *Groundwater*, Prentice-Hall, Englewood Cliffs, NJ.

Gale, J. E., 1977. A Numerical Field and Laboratory Study of Flow in Rocks with Deformable Fractures, Scientific Series, Environment Canada, Water Resources Branch.

Gringarten, A. C. and Witherspoon, P. A., 1972. A method of analyzing pumped test data from fractured aquifers, Proc. Symp. Percolation through Fissured Rocks, Sept. 18–19, 1971, Stuttgart, Inst. Soc. Rock Mech. Ins. Assoc. Engrg. Geol., T3-B-1.

Gringarten, A. C. and Ramey, H. J., 1974. The use of source and Green's functions in solving unsteady flow problems in reservoirs, *Soc. Pet. Eng. J.*, 285.

Gringarten, A. C., Ramey, R. J., and Raghavan, B., 1974. Unsteady flow pressure distributions created by well with a single infinite conductivity vertical fracture, *Soc. Pet. Eng. J.*, 257, 347.

Gringarten, A. C., 1978. A study by the finite element method of the influence of fractures in confined aquifers, *Soc. Pet. Eng. J.*, 183.

Gringarten, A. C., 1984. Interpreting tests in fissured and multilayered reservoirs with double porosity behavior: theory and practice, *Soc. Pet. Eng. J.*, 584.

Jenkins, D. N. and Prentice, J. K., 1982. Theory for aquifer test analysis in fractured rock under linear (nonlinear) flow conditions, *Ground Water*, 20(1), 12.

Kazemi, H., 1969. Pressure transient analysis of naturally fractured reservoirs with uniform fracture distribution, *Soc. Pet. Eng. J.*, 246, 451.

Kazemi, H., Seth, M. S., and Thomas, G. W., 1969. The interpretation of interference tests in naturally fractured reservoirs with uniform fracture distribution, *Soc. Pet. Eng. J.*, 246, 463.

Kohut, A. P., Foweraker, J. A., and Johnson, E. H., 1983. Pumping effects on wells in fractured granite terrain, *Ground Water*, 21(5), 564.

LeGrand, H. E., 1949. Sheet structure, a major factor in the occurrence of groundwater in the granites of Georgia, *Econ. Geol. Bull. Soc. Econ. Geol.*, 44, 110.

Louis, C., 1969. A Study of Groundwater Flow in Jointed Rock and its Influence on the Stability of Rock Mass, Rock. Mech. Rep., 10, Imperial College of Science, Technology and Medicine, London.

Movar, M. J. and Cinco, H., 1977. Transient Pressure Behavior of Naturally Fractured Reservoirs, Paper presented at the SPE California Regional Meeting, Venture, CA.

Najurieta, H. L., 1980. A theory for pressure analysis in naturally fractured reservoirs, *J. Pet. Tech.*, 1241.

Odeh, A. S., 1965. Unsteady flow behavior of naturally fractured reservoirs, *Soc. Pet. Eng. J.*, 234, 60.

Rissler, P., 1978. Determination of the Water Deformability of Jointer Rocks, Int. Foun. Eng. Soil Mech. Rock Mech. and Water Ways Const., University of Aahen, Aahen, Germany.

Rushton, K. R. and Holt, S. M., 1981. Estimating aquifer parameters for large diameter wells, *Ground Water*, 19(5), 505.

Sagar, B. and Runchal, A., 1982. Permeability of fractured rock. Effect of fracture size and data uncertainties, *Water Resour. Res.,* 18(2), 266.

Serra, K. V., Reynolds, A. C., and Raghavan, B., 1983. New pressure transient analysis methods naturally fractured reservoir, *J. Pet. Tech.,* December issue, 2271.

Smith, E. D. and Vaughan, N. D., 1985. Aquifer test analysis in non-radial flow regimes. A case study, *Ground Water,* 23(2), 167.

Snow, D. T., 1965. A Parallel Plate Model of Fractured Permeability Media, Ph.D. thesis, University of California, Berkeley.

Snow, D. T., 1969. Anisotropic permeability of fractured media, *Water Resour. Res.,* 23(6), 1273.

Streltsova, T. D., 1976. Hydrodynamics of groundwater flow in a fractured formation, *Water Resour. Res.,* 42, 405.

Streltsova, T. D., 1983. Well pressure behavior of a naturally fractured reservoir, *Soc. Pet. Eng. J.,* September issue, 769.

Şen, Z., 1986. Aquifer test analysis in fractured rocks with linear flow pattern, *Ground Water,* 24(1), 72.

Şen, Z., 1988. Fractured media type curves by double-porosity model of naturally fractured rock aquifers, *Bull. Tech. Univ. Istanbul,* 41(1), 103.

Şen, Z., 1989. Volumetric approach in multi-aquifer and horizontal fracture, *ASCE Hydraul. Eng.,* 115(12), 1646.

Şen, Z., 1992. Unsteady ground-water flow toward extended wells, *Ground Water,* 30(1), 61.

Witherspoon, P. A., Wang, J. S. Y., Iwai, K., and Gale, J. E., 1980. Validity of cubic law for fluid flow in a deformable rock fracture, *Water Resourc. Res.,* 16(6), 1016.

Warren, J. E. and Root, P. J., 1963. Behavior of naturally fractured reservoirs, *Soc. Pet. Eng. J.,* 228, 245.

CHAPTER 11

Well Test

I. GENERAL

A well test is a specific type of pumping test designed primarily to evaluate well characteristics. Successful completion of any groundwater resources system is possible only after an effective evaluation of aquifer and well characteristics in addition to the reliable predictions of the groundwater variables. Among such variables are the groundwater levels, drawdowns, and discharge which vary with time and space and they are interdependent among themselves. For instance, drawdown in a well is directly proportional with the discharge.

In the previous chapters various methods were developed and applied for an aquifer test leading to numerical calculations of the aquifer parameters only. These parameters are representative of the undisturbed and saturated portions of the groundwater reservoirs, i.e., the zones away from the well location. However, during the drilling of a well and its completion with screens, casing, and gravel pack a certain zone around the well is disturbed and therefore the parameters found from the aquifer tests can no longer represent the well vicinity. These man-made products around the well affect the groundwater movement in a pattern different from the undisturbed aquifers. Generally, each of these products causes additional energy losses during a water particle movement before it reaches the well. It is expected that several discontinuities in the piezometric level within the well vicinity occur as shown in Figure 1.

The total energy loss, s_w within the aquifer for any water particle reaching the well has different contributions from the undisturbed aquifer zone, s_a, the gravel pack, s_g, and casing, s_s,

$$s_w = s_a + s_s + s_g \tag{1}$$

It is possible to consider the total energy loss, i.e., drawdown in the well at any time as composed of the aquifer loss, s_a, and well losses, $(s_g + s_s + \ldots)$. Furthermore, the aquifer and well losses are due to the laminar and turbulent flows, respectively. It is well known in pipe hydraulics that the laminar flow losses are directly proportional with the discharge, whereas the turbulent losses are with the square of the discharge. Considering the statement Jacob (1946) suggested theoretically that the drawdown in a well is composed of two parts as

$$s_w = AQ + CQ^2 \tag{2}$$

in which A and C are the undisturbed aquifer loss and well loss coefficient, respectively. In fact, Equation 2 is written for the steady-state situation but it is not always possible to wait until the steady state is reached in the field. Therefore, in general, for unsteady-state flows

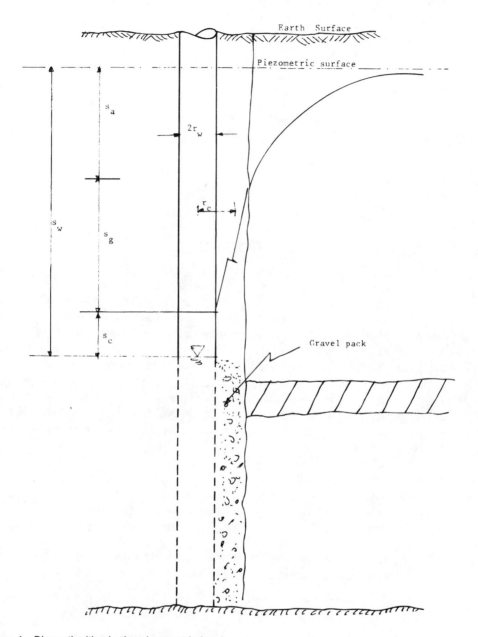

Figure 1 Discontinuities in the piezometric level.

Equation 2 will take the following form:

$$s_w(t) = A(t)Q + CQ^2 \qquad (3)$$

in which $s_w(t)$ is the total drawdown in the well and $A(t)$ is the aquifer loss coefficient both dependent on time, t. In all the theoretical aquifer test analysis the well losses were ignored or assumed as zero. For instance, in the Papadopulos and Cooper (1967) method from Equation 25 in Chapter 9 the well drawdown can be written as

$$s_w(t) = \frac{Q}{4\pi T} W(u_w, \beta) \qquad (4)$$

which means to say that the aquifer loss is

$$A(t) = \frac{W(u_w, \beta)}{4\pi T} \tag{5}$$

Conventional aquifer tests do not yield any useful information about the well losses. It is, therefore, necessary to devise another method specifically for determining the well loss coefficient. In practice, Equation 13 furnishes information about the following important questions:

1. What is the well radius that will give rise to minimum well loss? This radius is referred to as the effective well radius by Jacob. It is the distance measured radially from the well axis at which the theoretical drawdown equals the actual drawdown just outside the well screen. It reflects the effectiveness of well development on increasing the hydraulic conductivity in the immediate vicinity of well. This diameter is directly proportional with the transmissivity but inversely related to the storage coefficient. Besides, there is an inverse relationship between the well loss and diameter.
2. What is the efficiency of a complete well, that is, what portion of the total well drawdown is due to the linear aquifer flow and what percentage is a result of different factors? The smaller the percentage, the more efficient is the well. The well efficiency, η, is defined usually as the ratio of the aquifer loss to the total drawdown as

$$\eta = \frac{A(t)}{A(t) + CQ} \times 100 \tag{6}$$

Most wells exhibit turbulent losses through and near the well vicinity including screen and gravel pack regions resulting in less than perfect efficiencies. The ideal well with 100% efficiency rarely exists in practice with the exception of uncased wells in fractured rock aquifers or cavernous limestones. Badly designed wells will have rather small efficiency percentages. The well losses comprise the turbulent losses about the well screen and frictional losses in the casing. One can seek to minimize the well losses so as to increase the efficiency but the length of the casing is governed by the aquifer depth. As a result the well efficiencies quoted merely as percentages are therefore of no value. However, comparing wells of similar depths, the well efficiency gives a measure of reliability.
3. How can one control the future performance of a well? The operation of the well for water supply purpose will give rise to some aging problems such as corrosion, encrustation, clogging, etc., which will affect the total drawdown in the well even if the discharge is kept constant. Right after the well completion and development different drawdown measurements in the well for different discharges give a discharge-drawdown curve as shown in Figure 2.

With aging the same discharge will cause greater drawdowns (Figure 2a) than the original curve (Figure 2b). If this drawdown is more than, say, 10%, of the initial drawdown, then the well must be cleaned or rehabilitated. The last point shows the need for a special test in the main well with the aim of obtaining a discharge-drawdown curve as in Figure 2. The curves in this figure are represented in general by Equation 2. In practice, however, it is more convenient to transform Equation 2 into a linear form by dividing both sides with Q:

$$\frac{s_w}{Q} = A(t) + CQ \tag{7}$$

in which s_w/Q is named as the specific drawdown which is the reverse of the specific capacity, Q/s_w. On an arithmetic paper Equation 7 yields to a straight line as shown in Figure 3.

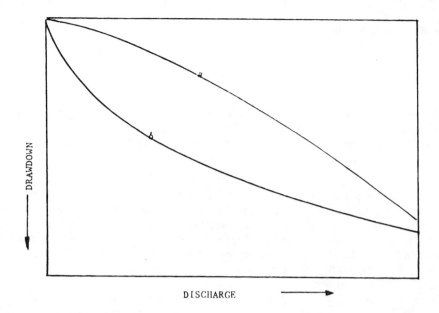

Figure 2 Discharge-drawdown curve.

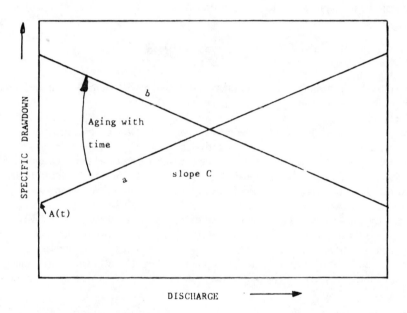

Figure 3 Specific drawdown-discharge relation.

In an ideal situation with no well loss, the relationship appears as a horizontal line parallel to the Q axis at a distance $s_w/Q = A(t)$. Otherwise, it is a straight line with intercept, $A(t)$, and slope equal to C.

Field measurements may reveal some systematic deviations from the ideal situation depending on the aquifer or gravel pack properties as shown schematically in Figure 4. Figure 4a shows characteristic specific capacity-discharge relationship for fractured groundwater reservoir. The initial portion of the curve may have a straight line or upward bending portion that is followed by a sudden uprising. There may be two explanations for this phenomenon. First, after a threshold value of specific discharge, the flow becomes turbulent and, therefore,

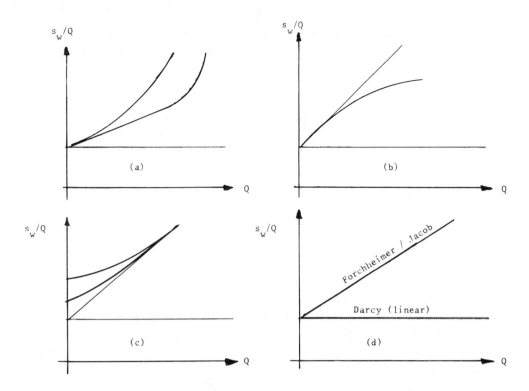

Figure 4 Different specific yield drawdown curves.

additional energy losses due to turbulence take place. Second, increase in drawdown commands expansion of water but compaction of fracture aperture. Because the hydraulic resistance of a fracture is inversely proportional with the cube of the aperture, even a small compaction forms a significant increase in the energy loss, i.e., hydraulic resistance, and hence the specific capacity discharge curve bends upward. This is particularly true in the vicinity of the well as the large drawdowns, specific discharges, and hydraulic gradients are extremely high.

Downward bending in Figure 4b is a result of significant formation heterogeneities in the well vicinity. For instance, if a rather coarse gravel pack is placed on a short distance around the well, then the well losses become important even for moderate Q values. As a result, coarse grains can enter the well with little resistance and hence increase in the discharge in large values does not give very significant changes in the drawdown.

For some cases the specific capacity discharge curve approaches the Jacob line asymptotically as in Figure 4c. The former case means that the hydraulic resistance, i.e., energy loss, decreases with increase in small discharge values. For Darcian (linear) flow the resistance is constant and therefore in Figure 4d a horizontal line represents this situation. However, in the Forchheimer flow the rate of resistance is constant.

Although the Jacob method has a sound physical basis, provided that the medium is porous, homogeneous and the flow is Darcian, many field data fail to fit the formulation in Equation 2. To account for some of these deviations Rorabough (1953) proposed an alternative formula as

$$s_w = A(t)Q + CQ^n \qquad (8)$$

in which n is an exponent often greater than 2 with an average value of 2.5. His method usually provides a value for C smaller than the value calculated from the same data by Jacob's method. Rorabough's formula has an upward curvature and hence is suitable to account for the characteristic specific yield-discharge relationship as in Figure 4a and c. Lennox (1966) applied the Rorabough approach in analyzing 18 wells in Canada in rather low transmissivity aquifers. He observed that this approach seems better than Jacob's. Out of 18 wells tested only 7 showed turbulent flow throughout the range of pumping rates used.

II. STEP DRAWDOWN TEST ANALYSIS

To determine the characteristic shape of specific drawdown-discharge relationship a set of pumping tests, each with different discharge, must be conducted in the field. Although it is possible to conduct independent tests in the same well with different discharges yielding different drawdown values, it is rather cumbersome and requires a longer time which is not a practical solution. However, it is preferable to conduct a continuous pumping test consisting of discharge increments for specific time duration. To save time, step drawdown tests are carried out with no stop between successive steps. As a result the measured drawdown in the field during each step contains the residual drawdowns of the preceding step. As suggested by Jacob (1946) at least three epochs should be included in such a test with constant discharge during each epoch as schematized in Figure 5.

The identification of parameters, namely, $A(t)$ and C from a given pairs of observations sequences on Q and corresponding s_w values, is possible using different methods.

A. JACOB METHOD

This is a graphical method that helps to find the well loss coefficient from a general expression and data concerning two successive epochs, e.g., $(i - 1)$ and ith epochs as

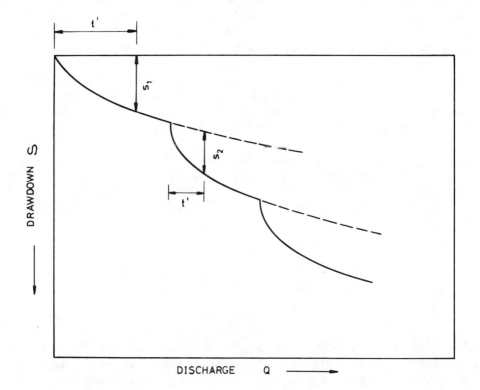

Figure 5 Step drawdown procedure.

$$C_i = \frac{\dfrac{\Delta s_{wi}}{\Delta Q_i} - \dfrac{\Delta s_{w(i-1)}}{\Delta Q_{(i-1)}}}{\Delta Q_i + \Delta Q_{(i-1)}} \quad (i = 1, 2, \ldots, j) \tag{9}$$

in which j is the epoch number and other terms are self explanatory from Figure 5 and accordingly Table 1 is prepared for the systematic calculation of necessary steps.

The solution of Equation 2 needs a graphical representation of time-drawdown data on a semi-logarithmic paper for genuine drawdown increments in each step. The drawdown curves of each step are extended smoothly until the end of the succeeding step ends. The incremental drawdown, Δs_i, is read and written on Table 1 into column 4. The remaining columns in the table are filled accordingly which leads to individual C estimations for each step and then to an average value as a representative of the well loss factor in that site.

As stated by Clark (1977) the Jacob method of analysis depends on the measurement of increments of drawdown and discharge rate and both of these measurements are open to error. For instance, a faulty reading on one step will affect the well loss calculations for that step and the one following and commonly leads to significant errors in the calculated average value for C. One of the drawbacks in this method is that an anamolously high incremental specific drawdown leads to a negative value for the well loss coefficient.

Example 1—A step drawdown test data conducted by Bierschenk (1964) with five steps are presented in Figure 6. The necessary numerical values for the test are given in Table 2. All the necessary calculations are performed and included in the same table.

The average value of well loss coefficient is 0.000265 (gpm)²/ft².

B. BIERSCHENK METHOD

The local idea of this method is to fit a straight line to the specific-drawdown and -discharge scatter diagram on an arithmetic graph paper. The arithmetic plot of s_w/Q vs. Q should lie around a straight line with intercept on the s_w/Q axis equal to the aquifer loss factor, A, and the slope equal to C. The best-fitting straight line is determined either by eye as is the common practice or A and C can also be evaluated in a systematic manner statistically by fitting the best straight line to the points on the basis of the least-squares technique. The necessary calculations for this technique are shown in Table 3.

Consideration of classical least-squares technique procedure from any textbook on statistics with the notations at hand leads to well and aquifer loss coefficient estimations as

$$C = \frac{s_T - \dfrac{1}{n}(s/Q)_T}{(Q^2)_T - \dfrac{1}{n}Q_T^2} \tag{10}$$

$$A = \frac{s_T - C(Q^2)_T}{Q_T} \tag{11}$$

Table 1 Jacob Calculations of C Value

Step no. (1)	Q_i (m³/s) (2)	ΔQ_i (m³/s) (3)	Δs_{wi} (m) (4)	$\Delta s_{wi}/\Delta Q_i$ (s/m²) (5)	C_i (from Equation 9) (s²/m⁵) (6)
0	0	0	0	0	
1	Q_1	Q_1	s_{w1}	s_{w1}/Q_1	C_1
2	Q_2	$Q_2' = Q_2 - Q_1$	s_{w2}	s_{w2}/Q_2'	C_2
3	Q_3	$Q_3' = Q_3 - Q_2$	s_{w3}	s_{w3}/Q_3'	C_3
4	Q_4	$Q_4' = Q_4 - Q_3$	s_{w4}	s_{w4}/Q_4'	C_4
5	Q_5	$Q_5' = Q_5 - Q_4$	s_{w5}	s_{w5}/Q_5'	C_5

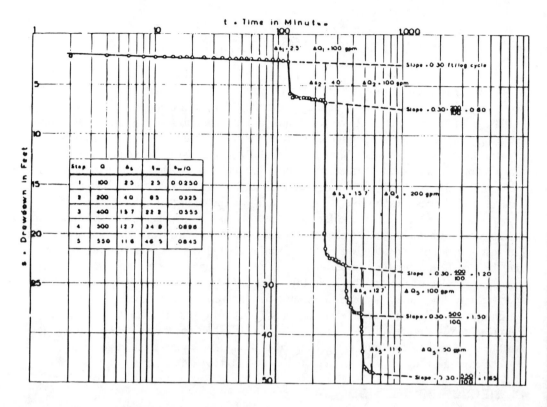

Figure 6 Step drawdown field data (from Clark, 1977).

Table 2 Step-Drawdown Data

Step no.	Q_i (gpm)	ΔQ_i (gpm)	s_{wi} (ft)	Δs_{wi} (ft)	$\Delta s_{wi}/\Delta Q_i$ (ft)	$C_i (\times 10^{-5})$ (min²/ft⁵)	η (%)	CQ^2 (ft)	$s_w - CQ$ (ft)
1	100	100	2.5	2.5	0.025	—	91.6	0.212	2.29
2	200	100	6.5	4.0	0.040	7.500	86.9	0.858	5.65
3	400	200	22.2	15.7	0.0785	0.130	84.7	3.390	18.81
4	500	100	34.9	12.7	0.1270	0.160	84.8	5.300	29.60
5	550	50	46.5	11.6	0.2320	0.700	86.2	6.410	40.09

Table 3 Bierschenk Calculations

Step no. (1)	Discharge Q_i (2)	Drawdown increment Δs_{wi} (3)	Cumulative drawdown (4)	Specific drawdown (5)	$Q_i^2 Q_i s_{pi}$ (6)
1	Q_1	Δs_{w1}	$s_1 = \Delta s_{w1}$	s_1/Q_1	Q^2_{12}
2	Q_2	Δs_{w2}	$s_2 = s_1 + \Delta s_{w2}$	s_2/Q_2	Q^2_{22}
3	Q_3	Δs_{w3}	$s_3 = s_2 + \Delta s_{w3}$	s_3/Q_3	Q^2_{32}
4	Q_4	Δs_{w4}	$s_4 = s_3 + \Delta s_{w4}$	s_4/Q_4	Q^2_{42}
5	Q_5	Δs_{w5}	$s_5 = s_4 + \Delta s_{w5}$	s_5/Q_5	Q^2_{52}
Total	Q_T		s_T	$(s/Q)_T$	$(Q^2)_T$

respectively, This method is more reliable than the Jacob way of calculations because all the data are fitted to a smooth line rather than the individual step calculations.

Example 2—The scatter diagram for the data in Table 2 is given in Figure 7. The calculations are given in Table 4.

The best fitting straight line seen with the naked eye has the well loss coefficient as the slope, $C = 1.2 \times 10^{-4}$ ft/(gpm)² and the aquifer loss coefficient equal to the intercept on the

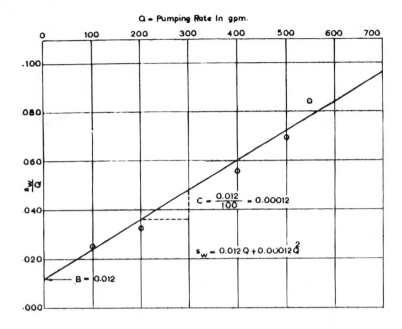

Figure 7 Specific drawdown graph.

Table 4 Step Drawdown Test Calculations

Step no.	Q (gpm)	Δs_wi (ft)	s_wi = ΣΔs_w (ft)	ΣΔs_w/Q (min/ft²)	Q_i² (gpm)²
1	100	2.5	2.5	0.0250	10000
2	200	4.0	6.5	0.0325	40000
3	400	15.7	22.2	0.0555	160000
4	500	12.7	34.9	0.0698	250000
5	550	11.6	46.5	0.0845	302500
Total	1750		112.6	0.2673	762500

vertical axis as $A = 1.2 \times 10^{-2}$ ft/gpm. Consequently, the well drawdown discharge relationship for the well is

$$s_w = 0.012Q + 0.00012Q^2$$

On the other hand, the necessary preliminary calculations for the application of the least-squares technique are presented in Table 4 for the same data. Substitution of totals from the last row in this table into Equations 10 and 11 yields $C = 1.27 \times 10^{-4}$ ft/(gpm)² and $A = 9 \times 10^{-4}$ ft/gpm.

It is worthy to notice that the Jacob method C value is more than the Bierschenk procedure value. The two methods are expanded to yield consistent results, provided that the points of scatter on s_w/Q vs. Q diagram follows closely a straight line. However, for the sample data at hand, the slope between the two last points on the scatter diagram in Figure 7 shows significant deviation from the overall slope of the straight line ignoring the last point in the Jacob calculation that the average C value becomes $(0.750 + 12.80 + 16.17)/3 = 1.22 \times 10^{-4}$ ft/(gpm)².

Example 3—Another example to illustrate the Bierschenk method is provided from step-drawdown test data performed in the Umm Er Radhuma limestone aquifer in the eastern part of the Kingdom of Saudi Arabia. The data for four steps are presented in Table 5. Figure 8 shows the semilogarithmic time-drawdown plot and for each step during a fixed time interval, say, $\Delta t = 100$ min, the incremental drawdowns are determined. The specific drawdown calculations together with the necessary quantities are shown in Table 6.

Plotting the specific drawdown-discharge data on arithmetic paper gives a straight line with slope $C = 4.5$ min^2/m^5 and the intercept point on the vertical axis ($Q = 0$) as $A = 7.0$ min/m^2 (see Figure 9). Hence, the drawdown equation for the Umm Er Radhuma aquifer well location becomes

$$s_w = 7.0Q + 20.25Q^2$$

The discharge must be in cubic meters per minute to use this equation.

C. EDEN-HAZEL METHOD

Neither the Jacob nor the Bienschenk method takes into account the time variability of the aquifer loss factor. These methods yield satisfactory results if in each step the steady-state

Table 5 Umm Er Radhuma Aquifer Step-Drawdown Test Results

Step	1	2	3	4
Discharge (m³/min)	0.342	0.468	0.552	0.714
Time from beginning of the test (min)		Drawdown (m)		
0.5	1.49	3.78	5.22	7.20
1.0	1.77	3.84	5.28	7.21
1.5	1.90	3.90	5.32	7.28
2.0	1.98	3.93	5.35	7.34
2.5	2.03	3.95	5.37	7.37
3.0	2.05	3.96	5.40	7.41
3.5	2.09	4.00	5.41	7.44
4.0	2.13	4.09	5.43	7.48
5.0	2.20	4.06	5.46	7.53
6.0	2.26	4.09	5.47	7.58
7.0	2.33	4.11	5.50	7.61
8.0	2.35	4.12	5.51	7.66
9.0	2.40	4.15	5.53	7.69
10.0	2.41	4.17	5.55	7.72
11.0	2.46	4.18	5.56	7.74
12.0	2.48	4.20	5.58	7.77
14.0	2.50	4.23	5.59	7.80
16.0	2.52	4.26	5.62	7.84
18.0	2.57	4.27	5.63	7.87
20.0	2.60	4.29	5.65	7.90
25.0	2.66	4.33	5.69	7.97
30.0	2.70	4.35	5.71	8.01
35.0	2.76	4.40	5.73	8.06
40.0	2.78	4.41	5.74	8.07
45.0	2.81	4.42	5.75	8.11
50.0	2.85	4.44	5.73	8.14
55.0	2.88	4.45	5.73	8.16
60.0	2.89	4.46	5.75	8.18
70.0	2.91	4.50	5.77	8.20
80.0	2.95	4.49	5.78	8.24
90.0	2.98	4.53	5.78	8.26
100.0	2.99	4.55	5.80	8.29
110.0	3.02	4.57	5.81	8.20
120.0	3.02	4.58	5.81	8.32
140.0	3.02	4.59	5.84	8.41
160.0	3.03	4.61	5.84	8.45
180.0	3.04	4.59	5.84	8.47

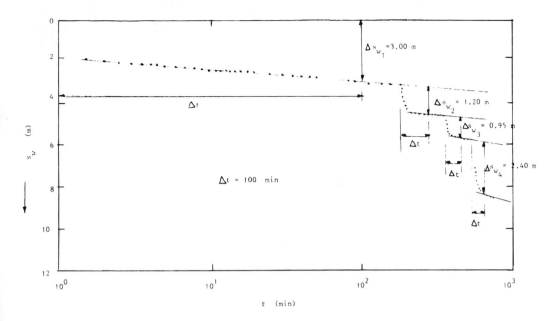

Figure 8 Semilogarithmic time-drawdown plot.

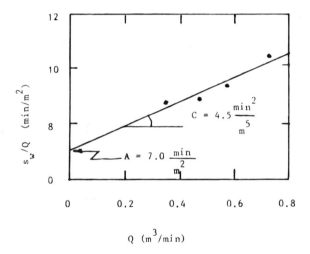

Figure 9 Specific drawdown plot.

flow is reached which requires a long time to complete the step-drawdown test in the field. On the other hand, Hazel (1973) suggested a method that considers the time variability of $A(t)$. In any step-drawdown test, the data for the first step are the true drawdown for the specified discharge in this step. However, the true drawdown curve for, say, the second step is calculated by finding the incremental drawdown between steps 1 and 2. The incremental drawdown in the second step at any point is measured together with the time from the start of step 2. The increment in drawdown during this step is then added to the drawdown in the first step at a time equal to the time measured from the start of step 2. For instance, the incremental drawdown after 30 min from the start of second step is added to the drawdown 30 min from the start of the first step. Likewise, each data point in the second or any other step is transferred in this way to reconstruct the whole drawdown curve at the corresponding

Table 6 Specific Drawdown Calculations for Umm Er Radhuma

Step no.	Drawdown increment (m)	Total drawdown (m)	Discharge (m³/min)	Specific drawdown (min/m²)
1	3.00	3.00	0.342	8.77
2	1.20	4.20	0.468	8.97
3	0.95	5.15	0.552	9.33
4	2.40	7.55	0.714	10.57

discharge rate of that step. The resultant drawdown time plots are shown schematically in Figure 10.

This figure provides the complete drawdown variation with time for individual steps. The data points for each step are matched with the best straight lines that are extended down to have intercepts on a drawdown axis as shown in Figure 10. It is obvious that the drawdown changes rather slightly during each step and therefore the aquifer loss factor should change accordingly with time.

The mathematical expression for these drawdown changes within each step has been modified from Jacob's original work by Sternberg (1968) as

$$s_w(t) = (a + b\log t)Q + CQ^2 \tag{12}$$

in which $s_w(t)$ is the total drawdown in the well at time t; a and b are coefficients defined as

$$a = \frac{2.3}{4\pi T} \log \frac{2.25T}{r_w^2 S} \tag{13}$$

and

$$b = \frac{2.3}{4\pi T} \tag{14}$$

respectively. In fact, Hazel's method uses Equation 12 and its combination with Figure 10 to arrive at the parameters estimation as follows:

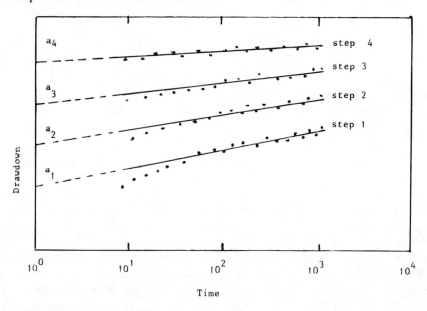

Figure 10 Time-drawdown plot.

1. The slope of each straight line constructed in Figure 10 is equal to the increment, Δs_{wi}, of drawdown over one log cycle. Hence, $b_i = \Delta s_{wi}/Q_i$. Overall representative b value is obtained by averaging b_is for each step.
2. Intercept of each straight line with the drawdown axis corresponding to time as $t = 1$ min in Figure 10 yields a_i value for this step, because $b \log 1 = 0$.
3. By knowing a_is and b_is one can calculate $a_i + b_i \log t$ for a given t value which is preferably taken as the longest time possible in the step drawdown test.
4. Equation 12 can be rearranged for each step as

$$\frac{s_{wi}(t)}{Q_i} - a_i = b_i \log t + CQ_i \qquad (15)$$

and the plot of $s_{wi}(t)/Q_i - a_i$ vs. Q_i leads to a straight line the slope of which gives C value and the intercept leads to over all aquifer loss factor $A(t)$.

A step-drawdown test was conducted in the Umm Er Radhuma limestone aquifer (Table 6). Figure 11 shows time-drawdown plot on a semilogarithmic graph paper. The test specifications and necessary calculations are given in Table 7.

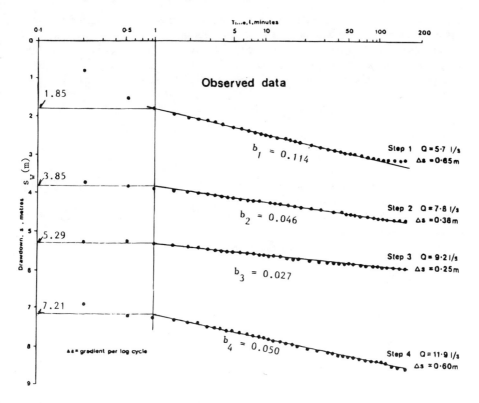

Figure 11 Step drawdown field data for Umm Er Radhuma aquifer.

Table 7 Step-Drawdown Test Calculations

Step no.	Q_i (m³/min)	s_{wi} (m)	s_{wi}/Q_i (min/m²)	Δs_{wi} (m)	a_i (m)	$b_i = \Delta s_{wi}/Q_i$ (min/m²)	$s_{wi}/Q_i - a_i$ (m)
1	0.342	1.48	4.330	0.65	1.85	1.900	1.480
2	0.468	3.21	6.859	0.38	3.85	0.812	3.010
3	0.552	4.78	8.659	0.25	5.29	0.453	3.370
4	0.714	5.40	7.563	0.60	7.21	0.840	0.313

On the other hand, in light of Equation 12 it is possible to state that, in general, a step starting at time t_i with a discharge rate increment of $\Delta Q_i = Q_i - Q_{i-1}$, the head loss due to this increment at any time t after the start of the first step, can be written as

$$s_{wi}(t) = \Delta Q_i[a + b\log(t - t_i)] + C_i Q_i^2 \tag{16}$$

or as the total drawdown, $s_w(t)$ due to all the steps is

$$s_w(t) = aQ_n + bH_j + CQ_n^2 \tag{17}$$

in which the subscript j indicates the last step and

$$H_j = \sum_{i=1}^{j} \Delta Q_i \log(t - t_i) \tag{18}$$

For the last step $j = n$ and $H_j = H_n$. Eden and Hazel (1973) presented a step-drawdown test evaluation method for finding aquifer and well losses through the following procedure:

1. In any step considered, H_j is calculated for each drawdown observation using the corresponding discharge increment rate ΔQ_i, and the time. The plot of H_j vs. drawdowns on an arithmetic graph paper exhibits a set of parallel straight lines for each discharge value. The slopes of these lines give a set of b values (see Figure 11).
2. The intercept A_j of these lines on the drawdown axis corresponding to $H_n = 0$ gives the drawdowns at unit time.
3. These drawdown values at unit time are used with the irrespective rates to constitute a specific drawdown-discharge graph as in the Hazel method.
4. The slope of the best-fitting straight line will be one and the intercept on the specific drawdown axis will be a.

Unfortunately, the calculations involved are tedious by hand but a computer program has been written by Holloway (1972).

Example 4—Application of this method will be illustrated with the Umm Er Radhuma data in Table 5. Using Equation 18 values of H_j are calculated. The necessary calculations for each step with numerical examples are as follows:

Step 1

$$H_1 = 0.342 \log t$$

For $t = 60$ min $H_1 = 0.60$ (m³/min)log(min)

Step 2

$$H_2 = H_1 + (0.468 - 0.342)\log(t - 180)$$
$$= H_1 + 0.126 \log(t - 180)$$

For $t = 250$ min $H_2 = 1.04$ (m³/min)log(min)

Step 3

$$H_3 = H_2 + (0.552 - 0.468)\log(t - 360)$$
$$= H_2 + 0.084 \log(t - 360)$$

For $t = 510$ min $H_3 = 1.43$ (m³/min)log(min)

Step 4

$$H_4 = H_3 + (0.714 - 0.552)\log(t - 540)$$
$$= H_3 + 0.126 \log(t - 540)$$

For $t = 650$ min $H_4 = 1.84$ (m³/min)log(min)

Similarly by using the same formulations for H_1, H_2, H_3, and H_4 necessary calculations should be carried out for each time during every step given in Table 5. Consequently, the arithmetic plot of drawdowns vs. H_n appears as in Figure 12. The slope of these parallel straight lines is $b = 1.2$ m³/min and its substitution into Equation 14 gives the transmissivity value $T = 0.152$ m²/min $= 219$ m²/d.

The plot of specific intercept values defined as A_n/Q_n vs. discharge values gives a straight line with a slope equal to the well loss coefficient. Such a plot is shown in Figure 13. The slope value is $C = 5.3$ min²/m⁵ and the intersection of the straight line at the ordinate where $Q = 0$ yields $a = 4.55$ min/m². Finally, the drawdown equation for the well after pumping at a constant discharge Q becomes

$$s_w = (4.55 + 1.2 \log t)Q + 5.3Q^2$$

D. RORABAUGH METHOD

Rorabaugh (1953) noted that the treatment of discharge as a second-order variable in the well loss term of the Jacob equation was overrestrictive and he suggested a more general form as mentioned earlier in this chapter (see Equation 8). Hence, instead of two parameters there are now three unknown parameters including the exponent value, n, which is to be determined from the field data. None of the previously mentioned methods are capable of

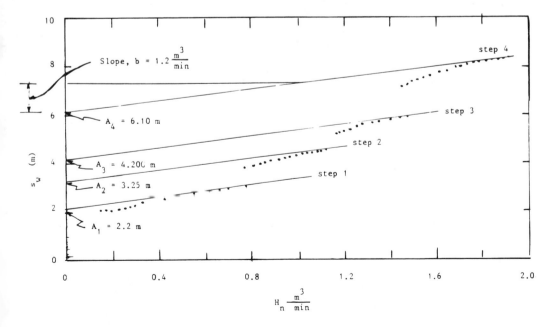

Figure 12 Drawdown vs. H_n plot.

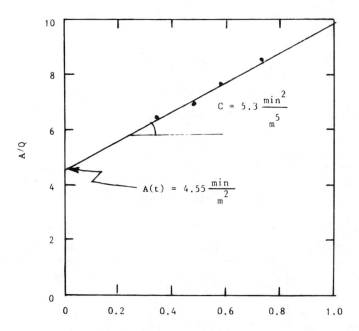

Figure 13 Hazel analysis specific drawdown-discharge plot.

yielding the three estimates. Similar to the Jacob method Rorabaugh transformed in two steps Equation 8 into a linear form first by dividing by Q

$$\frac{s_w}{Q} = A + CQ^{n-1} \tag{19}$$

and then by a logarithmic transformation as

$$\log\left(\frac{s_w}{Q} - A\right) = \log C + (n-1)\log Q \tag{20}$$

This expression appears as a set of curves on a double-logarithmic plot $(s_w/Q - A)$ vs. Q as shown in Figure 14.

Among these plots there is one that gives a straight line for a certain value of A. Because A is unknown a series of trial values are assumed for it, and the true value is the one to give the closest straight line plot. The slope of this line is then equal to $(n - 1)$ and the intercept on $s_w/Q - A$ axis at unit discharge rate will equal C.

Example 5—To demonstrate this method specific drawdown-discharge data in Table 8 are adopted from a step-drawdown test analysis carried out in the thick Quaternary deposit within the western part of the Kingdom of Saudi Arabia.

In the same table values of $s_{wi}/Q_i - A$ are provided for $A = 0.3$ m and $A = 0.6$ m. Figure 15 shows a double-logarithmic plot of $s_{wi}/Q_i - A$ vs. Q_i for these values in addition to $A = 0$. It is obvious that the plotted points fall on a straight line for $A = 0.6$ min/m². The slope of this line is

$$\Delta s = \frac{\log 10^1 - \log 10^0}{\log 3.95 - \log 2.3 \times 10^{-2}} = 1.447$$

Theoretically, the slope is equal to $(n - 1)$ and therefore $n = 1.447$. The intercept of the line

WELL TEST

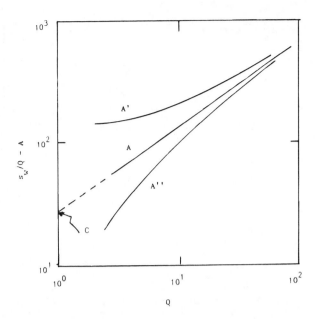

Figure 14 Rorabough method.

Table 8 Rorabough Calculations

Step no.	Q_i (m³/min)	s_{wi} (m)	s_{wi}/Q_i (min/m²)	$s_{wi}/Q_i - A$ (min/m²) $A = 0.3$	$A = 0.6$
1	2.00	1.22	0.61	0.31	0.01
2	2.75	2.17	0.79	0.49	0.19
3	4.50	5.17	1.15	0.85	0.55
4	7.00	15.05	2.15	1.85	1.55
5	10.00	45.30	4.53	4.25	3.93

corresponding to log $Q = 0$ axis, i.e., $Q = 1$ m³/min appears as $C = 2.3 \times 10^{-2}$. The well drawdown equation finally is

$$s_w = 0.6Q + 0.023Q^{1.447}$$

E. SHEAHAN TYPE-CURVE METHOD

Practical use of Rorabaugh procedure has been impelled by the need to resort to graphical means to solve the resulting system of nonlinear equations. Sheahan (1971) presented a set of type curves that simplified and quickened the solution of the Rorabaugh equation for drawdown in a pumping well by eliminating the trial-and-error computation. The type curve set is reproduced in Figure 16. The type curve matching to the specific discharge drawdown data and thereof parameter estimations are obtained according to Sheahan as follows:

1. The field data on drawdowns and discharges in each step are converted first into specific drawdowns, s_w/Q, for a selected time of step pumping similar to the Jacob method.
2. Specific drawdowns are plotted against the discharges on double-logarithmic paper of the same scale as the type curves (see Figure 6).
3. The type curve sheet is then placed over the plotted data such that the s_w/Q and Q scales of the two sheets are parallel to each other.

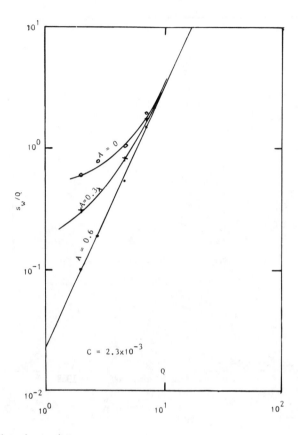

Figure 15 Specific drawdown plot.

4. While keeping the axis parallel, the type curves are moved until the best match is obtained between the plotted points and one of the type curves or an interpolated curve.
5. At the intersection of the "index line" and the curve of the best fit the values of $(s_w/Q)_R$ and Q_R are read from the plotted data sheet.
6. The exponent, n, is read from the best matching-type curve.
7. The values of B and C are computed as follows:

$$A = \left(\frac{s_w}{Q}\right)_{R/2} \tag{21}$$

and

$$C = \frac{A}{Q_R^{n-1}} \tag{22}$$

To define field curve distinctively more than three data points are preferable.

Example 6—The step drawdown test results from a field study are presented in Table 9. The plot of data from this table as s_w/Q vs. Q on log-log paper and after the matching procedure the best Sheahan's type curve appears for $n = 2.8$ as shown in Figure 17. At the intersection of the under line and the curve of best fit the values of $s_w/Q = 0.034$ and $Q = 900$ are read from the field data sheet. From Equations 21 and 22 the values of A and C are

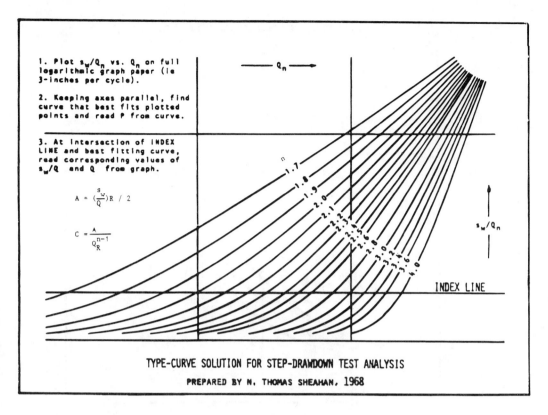

Figure 16 Step-drawdown test type curves (after Sheahan, 1968).

Table 9 Step-Drawdown Test Data

Step no.	Drawdown (m)	Discharge (m^3/min)	Specific drawdown (m^2/min)
1	8.6	400	0.0215
2	20.0	700	0.0286
3	56.5	1200	0.0470
4	141.0	1800	0.0780

0.017 min/m^2 and 8.2×10^{-8} (min)^2/m^5, respectively. Hence, the drawdown equation can be written as $s_w = 0.017Q + 8.2 \times 10^{-8} Q^{2.8}$.

F. MILLER-WEBER MODEL

This method is a numerical calculation technique for determining the unknown parameters in Rorabaugh's equation without resorting to any graphical means or extensive computer facilities. In fact, it is readily usable in the field. Miller and Weber (1983) assumed that the loss coefficients are independent of the discharge rate and three equations are necessary in solving three unknowns which are A, C, and n. Hence,

$$s_1 = AQ_1 + CQ_1^n \tag{23}$$

$$s_2 = AQ_2 + CQ_2^n \tag{24}$$

$$s_3 = AQ_3 + CQ_1^n \tag{25}$$

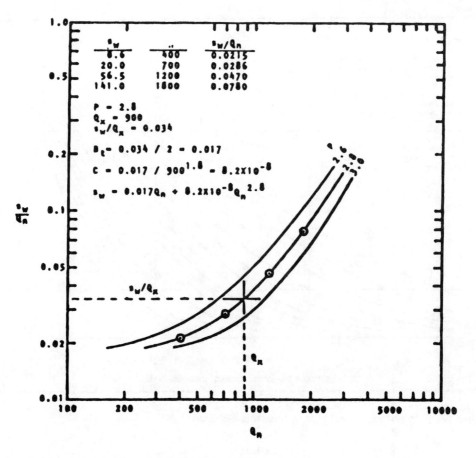

Figure 17 Type curve matching (after Sheahan, 1968).

where $Q_1 < Q_2 < Q_3$ and $s_1 < s_2 < s_3$. Elimination of A between Equations 23 and 24 and then between Equations 24 and 25 results in

$$s_1 - \frac{Q_1}{Q_2} s_2 = C\left(Q_1^n - \frac{Q_1}{Q_2} Q_2^n\right) \tag{26}$$

and

$$s_3 - \frac{Q_3}{Q_2} s_2 = C\left(Q_3^n - \frac{Q_3}{Q_2} Q_2^n\right) \tag{27}$$

respectively. Furthermore, elimination of C between Equations 26 and 27 leads after some definitions and simple algebra to

$$n = \frac{\log[\epsilon + (Q_3/Q_2) - (K_2/K_1)(Q_1/Q_2)]}{\log(Q_3/Q_2)} \tag{28}$$

in which

$$K_1 = s_1 - \frac{Q_1}{Q_2} s_2 \tag{29}$$

$$K_2 = s_3 - \frac{Q_3}{Q_2} s_2 \tag{30}$$

and

$$\epsilon = \frac{K_2}{K_1}\left(\frac{Q_1}{Q_2}\right)^n \tag{31}$$

Because always $Q_2 > Q_1$ for large values of n ($n > 5$), ϵ becomes negligible. However, this is a very rare occurrence in practice. For a practically valid range ($2 < n < 3$), the substitution of ϵ in Equation 28 becomes rather significant. In such a situation a trial and error method is used in an iterative way with Equations 28 and 31. First, an n value is guessed and ϵ is calculated from Equation 31. Its substitution into Equation 28 yields a new estimate for n. If this estimate is close enough to the guessed value it is regarded as the desired n value. Otherwise, iteration is continued for the accomplishment of the equality in Equation 28.

Example 7—For the application of this method the data given by Miller and Weber (1983) as presented in Figure 18 will be used. Because three equations are needed to solve for the three unknowns, three representative points are chosen from Figure 18 to give the step-drawdown data presented in Table 10.

Utilizing relationships given in Equations 28 and 31, the system may be solved for the conditions given in Table 10 to develop the information presented in Table 11. Noticing that n converges on a value of 2.80 allows for the rapid calculation of $C = 1 \times 10^{-6}$ (min)^2/m^5 and $A = 1 \times 10^{-1}$ min/m^2. Substitution of these coefficients into the general equations gives the relationship

$$s_w = 0.1Q + 10^{-6}Q^{2.8}$$

This relationship is represented by the solid line drawn through the data points as in Figure 18.

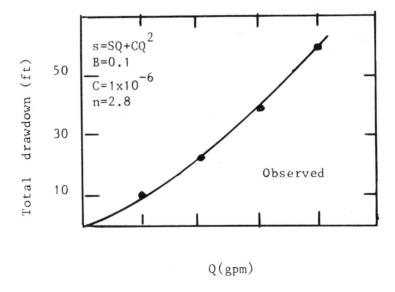

Figure 18 Drawdown discharge plot.

Table 10 Step-Drawdown Data

I	Q_i(m³/min)	s_i(m)
1	1.00	10.40
2	2.00	22.77
3	3.00	38.63

Table 11 Intermediate Equation Solution Results

I	ϵ_i	n_i
1	0	3.27
2	−0.469	2.94
3	−0.589	2.85
4	−0.627	2.82
5	−0.640	2.81
6	−0.645	2.80
7	−0.645	2.80

The alternative solution approach is to predetermine the step-drawdown tests that will be run for Q_1/Q_2 and Q_3/Q_2 values of 0.5 and 1.5, respectively, or that sufficiently close operational values will be obtained to enable valid interpolation of the pumping test data to these values. It may be demonstrated that for data obtained at these pumping ratio values the relationship between n and $\bar{n}_\epsilon = 0$ for the example from Equation 28 yields 3.27. Reference to Figure 19 then gives a corresponding value of n of 2.80, the same value as obtained by the first method. Calculations for the remaining coefficients are then made in a manner similar to that employed in the first method.

III. PRACTICAL IMPLICATIONS OF WELL LOSS FACTOR

The well loss factor, C, helps to make various interpretations about the well development and performance. Much of these interpretations have been noted by Walton (1962). The main points are as follows:

1. Knowing the well loss factor values the drawdown measurement are corrected as

$$s_{wc} = s_w - CQ^2 = AQ \tag{32}$$

where s_{wc} is the corrected drawdown. Hence, the well loss is eliminated from the drawdowns

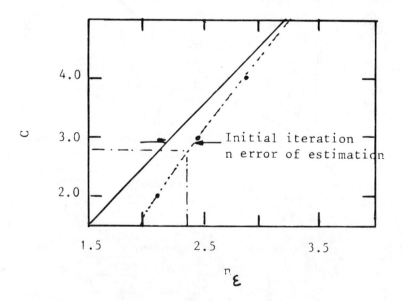

Figure 19 Calculation of n.

and therefore aquifer test analysis such as the Papadopulos and Cooper (1967) method can be applied safely.
2. The value of C for a properly designed and developed well is generally less than 5 sec^2/ft^5 (1490 s^2/m^5).
3. Values of C within the range 5 (1490) to 10 (2980) s^2/ft^5 (s^2/m^5) indicate mild deterioration and clogging is severe when C is greater than 10 (2980) s^2/ft^5 (s^2/m^5).
4. It is difficult and sometimes impossible to restore the original capacity if C is greater than 40 (11920) s^2/ft^5 (s^2/m^5).
5. The magnitude of C is influenced by the specific capacity and pumping rate, i.e., the higher the specific capacity and the pumping rate, the lower is the C value.
6. All the models for step drawdown test data evaluations are for confined aquifers and hence their use in the case of unconfined aquifers is allowable if the drawdown is small at the maximum pumping rate in the last step.

IV. SPECIFIC CAPACITY OF AN AQUIFER

The specific capacity is defined as the ratio of discharge to the drawdown in a well. It is discharge per unit drawdown and usually expressed in (lt/min)/m. It provides a measure for well behavior and productivity. Wells are productive provided that small drawdowns are coupled with high discharges. The larger the specific capacity the better the well. Table 12 gives limits for well classification from the practical point of view.

The specific capacity value of a well is not constant but dependent on time. A high specific capacity indicates efficient well design and construction. A decline in its value may indicate failure of well screen due to clogging or decline in S and T values and hence a fall in piezometric level. In practice, most often the well discharge is kept constant and, therefore, the drawdown increases with time only. Comparison of two or more wells from the productivity point of view is possible provided that the same time duration is considered. The best, however, is to consider the steady-state or quasi-steady-state drawdowns. Such a comparison is valid only when the well diameters are equal to each other. Otherwise, as explained in Chapter 6, under the same conditions the small diameter causes more drawdown which introduces a source of bias in this comparison.

Confined aquifer steady-state flow specific capacity can be written explicitly from Equation 10 in Chapter 8 by considering the maximum drawdown at the well surface, i.e., $r = r_w$ and $s(r) = s_w$ and

$$\frac{Q}{s_w} = \frac{2\pi T}{2.3 \log(R_0/r_w)}$$

The accuracy of the calculations depends greatly on the precision in measuring the radius of influence, R_0, or on the ratio R_0/r_w. However, because of the logarithmic operation on this ratio its effect is reduced considerably. For instance, if the two values of this ratio are 50 and 500, respectively, their logarithmic values are 1.2 and 2.69. It is obvious that, although the ratio between the values is 10-fold, similar ratio in logarithmic scale is about 1.5.

Table 12 Well Productivity [(lt/min)/m]

Highly productive	$Q/s_w > 300$
Moderate productive	$300 > Q/s_w > 30$
Low productive	$30 > Q/s_w > 3$
Very low productive	$3 > Q/s_w > 0.3$
Negligibly productive	$Q/s_w < 0.3$

In a confined aquifer by assuming $R_0 = 3000$ m and $r_w = 0.20$ cm the last equation yields

$$\frac{Q}{s_w} = \frac{T}{1.6} \tag{33}$$

However, the radius of influence is practically smaller in the unconfined aquifer and hence by adopting $R_0 = 300$ m and $r_w = 0.15$ cm the same equation gives

$$\frac{Q}{s_w} = \frac{T}{1.2} \tag{34}$$

Similar results were obtained initially by Logan (1964). It should be borne in mind that these are only rough approximations which may include up to 40% error.

V. RECOVERY TEST ANALYSIS

After water abstraction the shut down of pumping causes rise in the piezometric level which is referred to as the recovery test. It is not a main test by itself but provides supplementary information after any aquifer or well test completion. Records of the piezometric levels both in the main and observation wells during the recovery periods furnish useful quantitative and qualitative information about the aquifer and/or the well characteristics. The quantitative interpretation of the recovery test data is used very often in determining the hydraulic behavior of the oil and gas deposits; however, the recovery data are rarely used in the hydrogeology literature. Usually the data are presented in the graph form without quantitative interpretation. This is because there are few techniques developed to deal with the recovery period.

At the time, t_0, of pump shut down there is the maximum drawdown, $s_M(r,t_0)$ at any radial distance. With the rise in water level the maximum drawdown is divided into two complementary parts: (1) the recovery drawdown, $s_{rec}(r,t)$, that is expressed as the difference between the water level measured at a time, t', and the level at the time of pump stop and (2) the residual drawdown, $s_{res}(r,t)$, expressed as the difference between the static water level before the water abstraction and the water level at current time during recovery. Figure 20 shows the three drawdowns and they are related to each other as

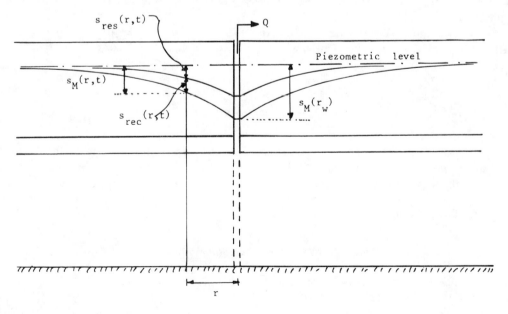

Figure 20 Recovery test setup.

WELL TEST

$$s_M(r,t) = s_{res}(r,t) + s_{rec}(r,t)$$

In practice, the recovery test data are treated on the basis of residual drawdowns. Recovery tests cost very little compared to the pumping test and they provide independent transmissivity estimation. Although the residual drawdown data are more reliable than aquifer test data because recovery occurs at a constant rate, unfortunately they do not yield storage coefficient estimations. Figure 21 shows the change in water level with time during and after an aquifer test. This method permits the calculation of the transmissivity only and thus gives an opportunity to check the results obtained from the preceding aquifer or well tests.

A. THEIS RECOVERY METHOD

The recovery of water level in a well after the pump is shut off can be simulated by continuing pumping as before, and recharging by an imaginary recharge well at the same rate during the recovery period. The assumptions listed for the Jacob straight line method are valid, apart from the fact that during the recovery the discharge is perfectly constant and equal to zero. This is equivalent to saying that there exists no discharge fluctuation during this test which might lead to uncontrollable errors in the parameter determinations. Cooper and Jacob (1946) assumed that the same pumping discharge continues during the recovery period in addition to injection of water into the well with the same discharge starting from the time of pump shut down. Hence, the residual drawdown, $s_{res}(r,t)$, during the recovery period for large times according to the Jacob method is given by

$$s_{res}(r,t) = \frac{2.3Q}{4\pi T}\left(\log \frac{4Tt}{r^2 S} - \log \frac{4Tt'}{r^2 S'}\right) \quad (35)$$

where S' is the storage coefficient during recovery; t' is the time in days since pumping

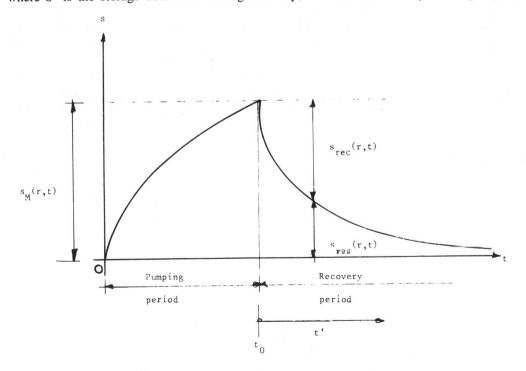

Figure 21 Recovery test plot.

stopped; and other terms are as explained before S and S' are assumed to be the same and therefore Equation 35 can be written succinctly as

$$s_{res}(r,t) = \frac{2.3Q}{4\pi T} \log \frac{t}{t'} \tag{36}$$

The plot of $s_{res}(r,t)$ data from the main or an observation well vs. the t/t' ratio on a semilogarithmic paper appears as a straight line the slope, Δs_r, is the amount of drawdown increment for any full-logarithmic cycle and it helps to determine the transmissivity from Equation 35 as

$$T = \frac{2.3Q}{4\pi \Delta s_r} \tag{37}$$

which is the same as the time-drawdown straight line method already given in Equation 13 in Chapter 9.

If the storage coefficients of pumping and recovery periods are constant but not equal to each other, then the straight line intercepts the time axis at a point $(t/t')_0$ where $s_{res}(r,t) = 0$ and the substitution of this intercept coordinates into Equation 37 leads to

$$\frac{S}{S'} = \left(\frac{t}{t'}\right)_0 \tag{38}$$

Hence, the ratio of two storativities is equal to the time intercept $(t/t')_0$. However, it is not possible to obtain absolute values of S or S' but the relative change in S only.

Apart from the aforementioned quantitative information, it is possible to obtain additional qualitative knowledge especially from the initial and final portions of the recovery plot. If the whole assumptions of the Theis method are satisfied within the aquifer, then the plot is a straight line. The geometric shape as well as the heterogeneous behavior of the aquifer influence the graph in its final portion. If the aquifer is bounded by an impermeable barrier the observed residual drawdown pattern will deviate upward from the infinite reservoir represented by the straight line. The same effect is felt if the transmissivity decreases due to either hydraulic conductivity and/or aquifer thickness reductions. On the contrary, the observed drawdown pattern deviates downward from the ideal straight line if there is a recharge source or transmissivity increase. Furthermore, at the initial times the drawdown deviation always occurs due to the well storage effect. The larger the duration the larger is the well diameter or vice versa.

Example 8—A 50-cm well is pumped at a constant rate $Q = 100$ lt/min for 18 h and at the time of the drawdown in the well reached 1.90 m the pump was stopped. Table 13 presents the residual drawdown during recovery for 60 min. Determine the aquifer transmissivity. The values of $s_{res}(r,t)$ vs. t/t' are plotted on a semilogarithmic paper as in Figure 22. The residual drawdown $\Delta s_{res}(r,t)$ per logarithmic cycle of time t/t' is measured and T is determined from Equation 37 as

$$T = \frac{2.3 \times 1000 \times 10^{-3}}{4 \times 3.14 \times 0.25} = 0.732 \text{ m}^2/\text{min} = 1054 \text{ m}^2/\text{day}$$

VI. NONPUMPING AQUIFER TESTS

In all the methods discussed so far the main instrument to excite the piezometric level in the aquifer was pumping which may not be suitable or available in some of the field studies because of some geological or economical reasons:

WELL TEST

Table 13 Residual Drawdown Data

Time since pump stop (min)	Residual drawdown $s_{res}(r,t)$ (m)	Time ratio $\dfrac{t}{t'} = \dfrac{18 \times 60 + t'}{t'}$
1	0.875	1080
2	0.735	541
3	0.690	361
4	0.662	271
5	0.640	217
6	0.625	181
8	0.590	136
10	0.570	109
12	0.556	91
16	0.536	68
20	0.498	55
30	0.458	37
40	0.423	28
45	0.410	25
60	0.382	19

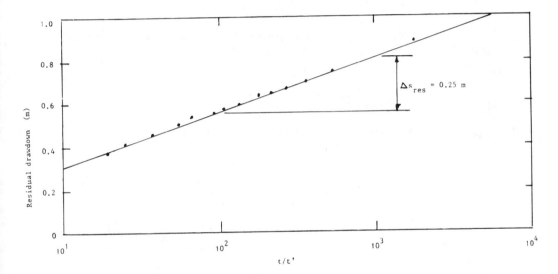

Figure 22 Recovery test field data.

1. In the geologic formations with very low hydraulic conductivity the use of a pump may not be justified, especially in small-diameter deep wells.
2. If the piezometric level is at a great distance from the earth surface the construction of pipes or experimental pipes could be too expensive.
3. Mostly in arid zones of developing countries many wells operate without pumps but with buckets or other containers.
4. Not the whole of the aquifer but a part of it along the saturation thickness needs to be defined hydraulically due to some local structures such as fractures, joints, cavities, or heterogeneous layers.
5. The aquifer test provides data for parameter estimates under the influence of the depression cone geological composition around the well vicinity. At times there may be preferential directions and the parameters have to be evaluated for this portion of the aquifer only.
6. During the drilling phase some preliminary information about the aquifer parameters may be needed. Hence, without the pumping the goal must be achieved.

To accommodate one or more of the above points different tests are developed for practical purposes.

A. BAILER TEST

This method is used especially if there are difficulties of transporting water from one source for exciting the groundwater level in the well. The main principle of the method is to bail water from the well storage by a bailer once or intermittently several times to lower the groundwater level from its original position. Consequently, due to the hydraulic head difference between the heads in the well and the aquifer the groundwater will flow toward the well. It is similar in principle to the recovery test, the only difference being in its intermittent water withdrawals.

The basic formulations of the problem is given by Skibitzke (1958) for a single instantaneous discharge of bail under the assumptions

1. That the aquifer is homogeneous, isotropic, and infinite in areal extent.
2. The coefficients of storativity and transmissivity do not change with time or space.
3. That the well penetrates the entire aquifer thickness.
4. That the water is removed instantaneously causing water level decline.

The residual drawdown, $s_{res}(t)$, is related to the aquifer properties and the volume of water bailed, V, during time t as

$$s_{res}(t) = \frac{V}{4\pi t T} e^{-\frac{r_w^2 S}{4tT}} \tag{39}$$

in which r_w is the well radius. The corresponding sequences of residual drawdown and time help to identify the aquifer parameters from Equation 39. To arrive at a simple graphical procedure this equation can be rewritten in logarithmic form as

$$\log[s_{res}(t)t] = \log \frac{V}{4\pi T} - \frac{r_w^2 S}{1.73T} \frac{1}{t} \tag{40}$$

On a semilogarithmic paper this expression yields to a straight line as shown in Figure 23

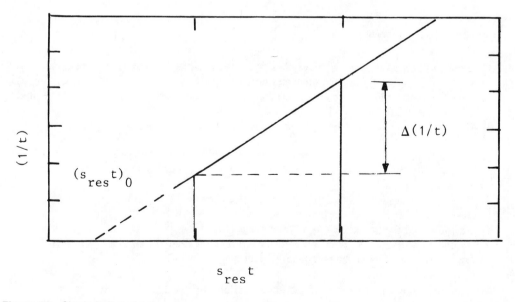

Figure 23 Single bail method.

where $1/t$ is plotted against $\log[s_{res}(t)t]$. The slope, $\Delta(1/t)$, and the intercept, $(s_{res}t)_0$, on the logarithmic axis are related to Equation 40 as

$$\frac{T}{S} = \frac{r_w^2}{1.73\Delta(1/t)} \tag{41}$$

and

$$s_{res}(t) = \frac{V}{4\pi t T} \tag{42}$$

Hence, arithmetic plot of $s_{res}(t)$ vs. $1/t$ appears as a straight line passing from the origin and its slope, $\Delta(1/t)$, gives the estimation of the transmissivity as

$$T = \frac{V}{4\pi \Delta(1/t)} \tag{43}$$

The restrictive assumption for the above derivation is that the bail discharge is instantaneous. However, in practice, repeated bailing of a well takes place with intermittent time. The formula for the recovery of the water level in confined aquifers, in general, is given by Skibitzke (1958) for large times as

$$s_{res}(t) = \frac{1}{4\pi T} \sum_{i=1}^{n} \frac{V_i}{t_i} \tag{44}$$

where n is the number of bails and V_i and t_i are the volume and elapsed time between the instant at which the ith bailer of water was removed and the instant at which the observation of residual drawdown was made. Provided that the volume in each bail is the same, Equation 44 simplifies further to

$$s_{res}(t) = \frac{V}{4\pi T} \sum_{i=1}^{n} \frac{1}{t_i} \tag{45}$$

Similar to the earlier mentioned technique the plot of $s_{res}(t)$ vs. $\Sigma(1/t_i)$ gives a straight line and its slope helps to estimate the aquifer transmissivity from Equation 45. Sufficiently large time must elapse since bailing stopped for reliable estimates.

Skibitzke remarked that the use of the bailing method will require a great deal of computation, particularly if the number of bailing cycles is large. However, Lennox and Jones (1964) suggested a considerable simplification if the time intervals between successive bailer cycles were made equal. In this case, the time at which the drawdown observation is made can be chosen so that

$$t_i = (p + n - 1)t \tag{46}$$

where t is the time interval between successive bailer cycles and p is the number of bailer cycles between the time of removal of the last bailer and the time of observation. With this concept Equation 45 can be written succinctly as

$$s_{res}(t) = \frac{V}{4\pi T t} [S_q - S_{p-1}] \tag{47}$$

where $q = (p + n - 1)$ is the number of bailer cycles between the time of removal of the first bailer and the time of observation and s_q is the sum of the reciprocal of the first q natural numbers, and s_{p-1} has similar meaning. The transmissivity can be determined from Equation 47 as

$$T = \frac{V}{4\pi t s_r} [S_q - S_{p-1}] \qquad (48)$$

Under these conditions all that is needed, even if the number of bailer cycles is large, is a table of the sums of s_j of the reciprocals of the natural numbers (see Table 14)

VII. SLUG TEST

The change of groundwater level is originated by adding a column of water to the well. Ferris et al. (1962) introduced the usefulness of the instantaneous line source slug test for determining aquifer transmissivity. A known volume of water is injected suddenly into the well and the drawdown measurements are recorded with time until the water level rise in the well returns to its original position. Carrying out pumping test becomes problem in hard rocks due to high drawdown and small well bore diameters. In such situations, a preliminary evaluation of yield may be achieved by slug tests. Unlike the pumping tests, the volume of water injected into the well is rather small and consequently it does not make significant effects on the piezometric level. Hence, the storativity of aquifer plays a negligible role. It is for this reason that slug tests are performed for determining the aquifer transmissivity only. Slug tests are useful for tight formations such as compacted clay, rocks in which the fissures are partially closed with filling material, or matrix blocks with weak features. Currently low-permeability rock environments are sought for land disposal of hazardous wastes which require determination of hydraulic conductivity. For this purpose the slug tests are one of the primary methods used in routine analysis of properties of low-permeability units below hazardous

Table 14 Aquifer Evaluation by the Bail Test Method

j	S_j	j	S_j	j	S_j	j	S_j
1	1.0000	26	3.8545	51	4.5187	76	4.9144
2	1.5000	27	3.8915	52	4.5379	77	4.9274
3	1.8333	28	3.9272	53	4.5568	78	4.9402
4	2.0833	29	3.9617	54	4.5753	79	4.9529
5	2.2833	30	3.9950	55	4.5935	80	4.9654
6	2.4500	31	4.0272	56	4.5114	81	4.9777
7	2.5929	32	4.0585	57	4.6289	82	4.9899
8	2.7179	33	4.0888	58	4.6461	83	5.0019
9	2.8290	34	4.1182	59	4.6630	84	5.0138
10	2.9290	35	4.1468	60	4.6797	85	5.0256
11	3.0199	36	4.1746	61	4.6961	86	5.0372
12	3.1032	37	4.2016	62	4.7122	87	5.0487
13	3.1801	38	4.2279	63	4.7281	88	5.0601
14	3.2515	39	4.2535	64	4.7437	89	5.0713
15	3.3182	40	4.2785	65	4.7591	90	5.0824
16	3.3807	41	4.3029	66	4.7743	91	5.0934
17	3.4395	42	4.3627	67	4.7892	92	5.1043
18	3.4951	43	4.3500	68	4.8039	93	5.1151
19	3.5477	44	4.3727	69	4.8184	94	5.1257
20	3.5977	45	4.3949	70	4.8327	95	5.0256
21	3.6453	46	4.4166	71	4.8468	96	5.1466
22	3.6908	47	4.4379	72	4.8607	97	5.1569
23	3.7343	48	4.4587	73	4.8744	98	5.1671
24	3.7760	49	4.4791	74	4.8879	99	5.1772
25	3.8160	50	4.4991	75	4.9012	100	5.1872

waste sites. Slug test results of hydraulic conductivity represent only the aquifer material close to the well. In the regional groundwater resources especially, evaluations at a large number of slug test points provide hydraulic conductivity estimates more efficiently, economically, and easily than a single long-term pumping test. It should be borne in mind that the slug test hydraulic conductivity is higher than the long-term pumping test results. This is due to the fact that slug tests cover a small region around the well which in most of the cases includes a low-permeability layer on the well wall because of the drilling mud. The slug test is quicker than the pumping test because observation wells and pumping periods are not needed.

A. CONFINED AQUIFER

For the analytical solution of slug test analysis the same assumptions are valid as were for the bailer test. The exact solution for the water level drop in a finite diameter well after an instantaneous surge of water has been given by Cooper et al, (1967) and later extended by Papadopulos et al. (1973). They considered a confined aquifer with fully penetrating well cased at diameter $2r_c$ to the top of a homogeneous isotropic artesian aquifer of uniform thickness as shown in Figure 24. With the injection of a certain volume, V, of water into the well, the static piezometric level rises up to the height, $H_0 = V/\pi r_c^2$, subsequently begins to retract to its initial level after a certain period during which the head changes with time, $H(t)$. Their solution can be expressed succinctly as

$$\frac{H(t)}{H_0} = F(\alpha,\beta) \qquad (49)$$

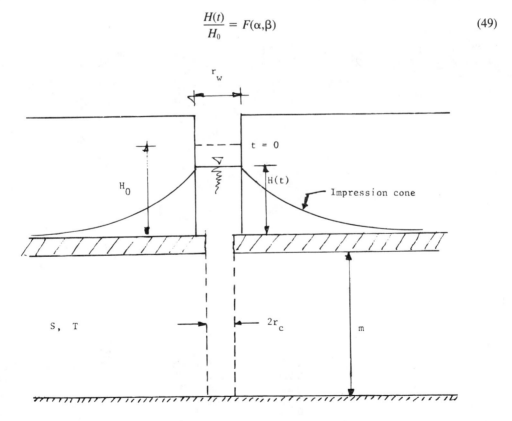

Figure 24 Rised piezometric level in slug test.

where the well size and dimensionless time factors are

$$\alpha = \frac{r_w^2 S}{r_c^2} \tag{50}$$

and

$$\beta = \frac{Tt}{r_c^2} \tag{51}$$

in which r_w is the effective radius of the well; S and T are the storativity and transmissivity, respectively. Finally, the slug test well function, $F(\alpha,\beta)$ is given in Table 15 for different values. In Figure 25 the values of this table are presented as a set of type curves on a semilogarithmic paper. It is notable from this figure that especially for small values of α the type curves are very similar and close to each other almost in a parallel manner. This fact indicates that the determination of storativity by this method has questionable reliability. Even the most precisely collected data and their careful plots on the same scale semilogarithmic paper could easily be matched with more than one type curves. Papadopulos et al. (1973) stated that an analysis in the range $\alpha < 10^{-5}$ indicates that, if the value of α for the matched type curve is within two orders of magnitude of its actual value, the error in determined T would be less than about 30%. This possible error should be kept in mind in using the slug test results. The aforementioned technique is limited to fully penetrating wells in confined aquifers.

B. UNCONFINED AQUIFER

Simple extensions of the slug test to cover unconfined aquifers with partially penetrating wells were presented by Bouwer and Rice (1976). The geometry and symbols of a partially penetrating well in unconfined aquifer are shown in Figure 26. The water level in the well is suddenly raised and subsequently drawdowns are measured. The amount of discharge, Q, in the steady-state case can be calculated by modifying the Thiem equation to

Table 15 Values of $F(\alpha,\beta)$ for a Well of Finite Diameter

	$F(\alpha,\beta)$				
	$\alpha = 10^{-1}$	$\alpha = 10^{-2}$	$\alpha = 10^{-3}$	$\alpha = 10^{-4}$	$\alpha = 10^{-5}$
1.00×10^{-3}	0.9771	0.9920	0.9969	0.9985	0.9992
2.15	0.9658	0.9876	0.9949	0.9974	0.9985
4.64	0.9490	0.9807	0.9914	0.9954	0.9970
1.00×10^{-2}	0.9238	0.9693	0.9853	0.9915	0.9942
2.15	0.8860	0.9505	0.9744	0.9841	0.9888
4.64	0.8293	0.9187	0.9545	0.9701	0.9781
1.00×10^{-1}	0.7460	0.8655	0.9183	0.9434	0.9572
2.15	0.6289	0.7782	0.8538	0.8935	0.9167
4.64	0.4782	0.6436	0.7436	0.8031	0.8410
1.00×10^{0}	0.3117	0.4598	0.5729	0.6520	0.7080
2.15	0.1665	0.2597	0.3543	0.4364	0.5038
4.64	0.07415	0.1086	0.1554	0.2082	0.2620
7.00	0.04625	0.06204	0.08519	0.1161	0.1521
1.00×10^{1}	0.03065	0.03780	0.04821	0.06355	0.08378
1.40	0.02092	0.02414	0.02844	0.03492	0.04426
2.15	0.01297	0.01414	0.01545	0.01723	0.01999
3.00	0.00907	0.00961	0.01016	0.01083	0.01169
4.64	0.00571	0.00591	0.00611	0.00631	0.00655
7.00	0.00372	0.00380	0.00388	0.00396	0.00404
1.00×10^{2}	0.00257	0.00261	0.00265	0.00268	0.00272
2.15	0.00117	0.00118	0.00119	0.00120	0.00129

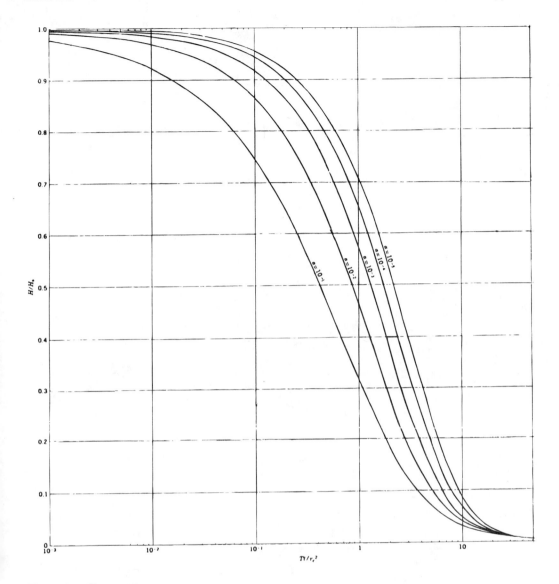

Figure 25 Slug test-type curves.

$$Q = 2\pi KL \frac{y}{\ln\left(\dfrac{R}{r_c}\right)} \tag{52}$$

in which K is the hydraulic conductivity, L is the height of well portion through which water enters, y is the vertical distance between water levels in the well and equilibrium state, R is the radius of influence, and r_w is the well radius plus gravel pack zone. On the other hand, the change of drawdown within the well is

$$\frac{dy}{dt} = -\frac{Q}{\pi r_c^2} \tag{53}$$

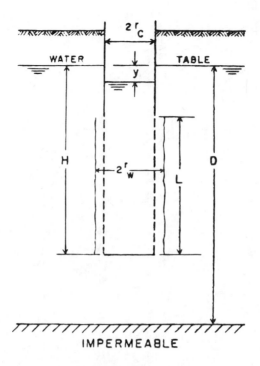

Figure 26 Geometry and symbols of partial penetration.

where πr_c^2 is the cross-sectional area of the casing. Combination of Equations 52 and 53 yields after some algebra to

$$\frac{mr_c^2 \ln(R/r_w)}{2L} \frac{1}{t} \ln\left(\frac{y_0}{y}\right) \tag{54}$$

in which m is the aquifer thickness. This expression yields to a straight line on a semilogarithmic paper as schematically shown in Figure 27. Thus the field data should yield a straight line when they are plotted as Lny vs. t. The variable composite term $(1/t)[Ln(y_0/y)]$ in Equation 54 is then obtained from the best-fitting line in a plot of Lny vs. t. The remaining unknown term $Ln(R/r_w)$ is dependent on well and aquifer configuration, i.e., on H, m, L, and r_w and can be obtained from the procedure given by Bouwer and Rice (1976). The empirical equation is proposed for determining the value of $Ln(R/r_w)$ for partially penetrating wells as

$$\ln\left(\frac{R}{r_w}\right) = \left\{\frac{1.1}{\ln(H/r_w)} + \frac{A + B\ln[(m-H)/r_w]^{-1}}{L/r_w}\right\} \tag{55}$$

and for fully penetrating wells it becomes

$$\ln\left(\frac{R}{r_w}\right) = \left[\frac{1.1}{\ln(H/r_w)} + \frac{C}{L/r_w}\right]^{-1} \tag{56}$$

in which A, B, and C are dimensionless coefficients that are functions of L/r_w and are shown in Figure 28. For any L/r_w values the corresponding A, B, or C can be read from Figure 28 and hence $Ln(R/r_w)$ can be calculated from Equations 55 or 56 accordingly. The substitution of a $Ln(R/r_w)$ value into Equation 54 yields the transmissivity of the aquifer.

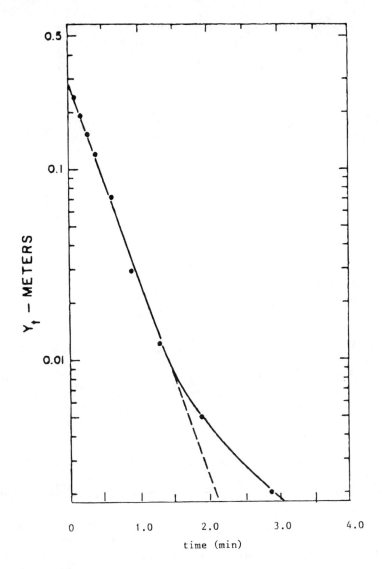

Figure 27 y vs. t in a slug test.

VIII. PACKER TEST

The previous methods of aquifer and well tests give information about the whole thickness of the aquifer. If there are alternate aquifers with different hydraulic conductivities or fractures between an almost impervious matrix block the tests mentioned above yield transmissivity or storativity values that are a global representation of the whole layer. They cannot provide individual values for each layer or fracture. However, in practice, at times it is most desirable to determine the aquifer parameters not for the whole aquifer thickness but a portion of interest from it or parameters of different layers or hydraulic conductivity of individual fractures. The Packer test, which is also referred to as the Lugeon (1933) test, enables the hydraulic conductivity of a selected section of rock to be determined. This selected section can be above or below the water table, in fact, at any desired depth along the well bore hole.

The basic principle of such a test is that water is injected directly into the layer of interest through a perforated tube which is located between two inflated packers or into a hole below a single packer as shown in Figure 29.

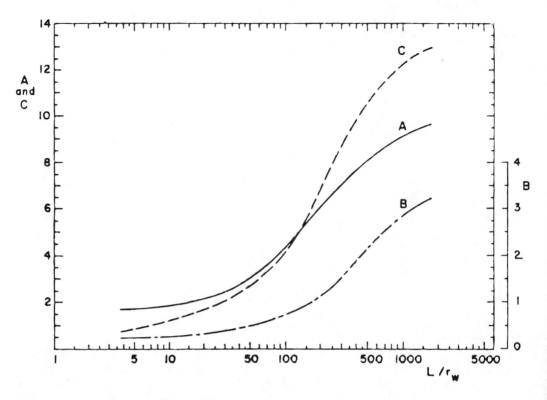

Figure 28 A, B, and C coefficients in relation to L/r_w.

To minimize the end effects the test length should be at least ten times longer than the hole diameter. Water is pumped into test section under pressure and after allowing time for saturation of the groundwater the steady-state flow rate is recorded. The test is carried out at a series of pressures but should not exceed overburden pressure of the surrounding rocks in order not to give way to hydraulic fracturing.

The very high cost of deep drilling generally determines that the test procedures must be confined to the drilled well only. Additional observation wells are not practicable due to the difficulties, time, and cost involved in monitoring a multiple aquifer system. In any groundwater exploration the direct information comes from exploratory bore holes. The cost reduction of the overall project depends on the type of information one can obtain from the drilled well bore hole. Among such information are the aquifer location, alternate sequences of permeability, semipermeable and impermeable layers, and the existence of fractures, solution cavities, fissures, pore channels, and piezometric levels. Conventional aquifer tests do not yield detailed information essentially on the heterogeneous nature of fissures, solution channels, and joint rock.

Hvorsley (1951) provided a formula for analyzing packer test results in the field. He considered a number of conditions at the end of a bore hole. If Q is the injected quantity of water between the two packers that are L apart from each other in a bore hole of radius, r_w, then the test section hydraulic conductivity, K, can be calculated as

$$K = \frac{Q}{2\pi LH} \ln\left[\frac{L}{2r_w} + \sqrt{1 + \left(\frac{L}{r_w}\right)^2}\right] \tag{57}$$

in which H is the increase or decrease in the piezometric head compared with the initial head. However, in practice, it is useful to choose the length of test section more than ten times the

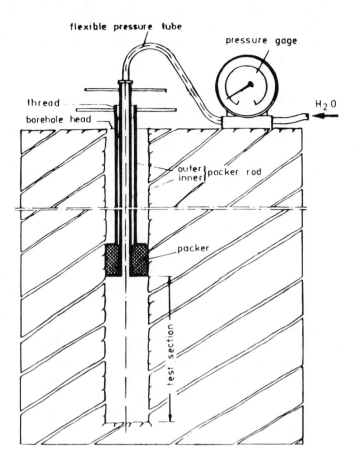

Figure 29 Packer.

diameter of the well bore in which case the square-root term in Equation 57 becomes equal to $L/2r_w$ and the equation itself reduces to

$$K = \frac{Q}{2\pi LH} \ln \frac{L}{r_w} \quad (L \geq 10r_w) \tag{58}$$

or when $10 < L < r_w$ then the approximation becomes

$$K = \frac{Q}{2\pi LH} \sinh^{-1}\left(\frac{L}{2r_w}\right) \tag{59}$$

In the theoretical derivation of Equation 58 Hvorslov adopted an approximation for flow from a line source in an infinite aquifer for which the equipotential surfaces are semi-ellipsoids as shown in Figure 30.

Hvorslev modified his approach for isotropic situation where the vertical, K_z, and horizontal, K_r, hydraulic conductivities are different from each other. By introducing the square-root transformation of the hydraulic conductivity ratios as

$$m = \sqrt{K_r/K_z} \tag{60}$$

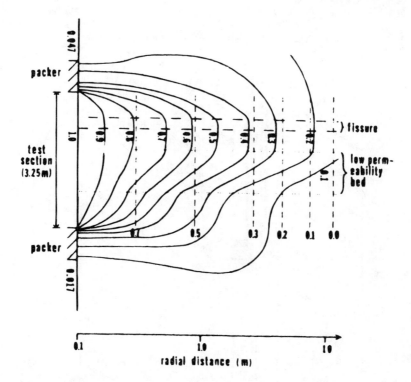

Figure 30 Theoretical and possible field flow nets.

similar to Equation 56 one can write

$$K = \frac{Q}{2\pi LH} \ln\left(\frac{mL}{r_w}\right) \tag{61}$$

In addition to this anisotropy, if there is a single fissure of hydraulic conductivity, K_f, with aperture b, then the modification due to Barker (1982) must be used as

$$K_f = \frac{Q}{2\pi bH} \ln[K_f b/1.78 r_w \sqrt{K_r K_z}] \tag{62}$$

Two important points in using this equation are

1. That the horizontal and vertical hydraulic conductivities in the aquifer should be determined.
2. That on both sides of the equation K_f appears and therefore an iterative method should be employed for finding the K_f value.

The assumptions in deriving Equation 61 are that

1. The fracture and surrounding aquifer are uniform, horizontal, and extensive.
2. The matrix material has horizontal and vertical hydraulic conductivity components.
3. The flow both in the fracture and matrix are Darcian and no vertical flow exists within the fracture.
4. The flow is in steady-state condition.

These assumptions should be borne in mind during the use of aforementioned formula. Bliss and Rushton (1984) have demonstrated by numerical analysis that the steady-state packer tests provide reliable estimations of hydraulic conductivities. These tests are particularly valuable for aquifers with significant hydraulic conductivity variations.

REFERENCES

Barker, J. A., 1982. Laplace transform solution for solute transport infissuredaquifers, *Adv. Water Resour.,* 5, 98.
Bouwer, H. and Rice, R. C., 1976. A slug test for determining hydraulic conductivity of unconfined aquifers with completely or partially penetrating wells, *Water Resour. Res.,* 12, 423.
Bierschenk, W. H., 1964. Determining well efficiency by multiple step-drawdown tests, *Int. Assoc. Sci. Hydrol.,* 64, 1.
Bliss, J. C. and Rushton, K. R., 1984. The reliability of packer tests for estimating the hydraulic conductivity of aquifers, *Q. J. Eng. Geol.,* 17, 81.
Clark, L., 1977. The analysis and planning of step drawdown tests, *Q. J. Eng. Geol.,* 10, 125.
Cooper, H. H. and Jacob, C. E., 1946. A generalized graphical method for evaluating formation constants and summarizing well field history, *Trans. Am. Geophys. Union,* 27, 526.
Cooper, H. H., Bredehoeft, J. D., and Papadopulos, I. S., 1967. Response of a finite-diameter well to an instantaneous charge of water, *Water Resour. Res.,* 3, 263.
Eden, R. N. and Hazel, C. P., 1973. Computer and graphical analysis of variable discharge pumping tests of wells, *Civil Eng. Trans. Inst. Eng. Austr.,* 5.
Ferris, G. D., Knowles, D. B., Brown, R. N., and Stallman, R. W., 1962. Theory of Aquifer Test, U.S. Geol. Survey, Water Supply Paper, 1536-E, 69.
Hazel, C. P., 1973. Lecture Notes: Groundwater Hydraulics, Australian Water Resources Council, Groundwater School, Adelaide.
Holloway, H. G., 1972. Analysis of Pump Test Data, Programme WU71, Systems Branch, Irrigation and Water Supply Commission, Queensland.
Hvorsley, M. J., 1951. Time Lag and Soil Permeability in Groundwater Observations, Bull. 36, Waterways Experiment Stations Corps of Engineers, U.S. Army, Vicksburg, MS.
Jacob, C. E., 1946. Drawdown test to determine effective radius of artesian well, *Proc. ASCE,* 72, 629.
Lennox, D. H. and Jones, J. F., 1964. Transmissibility determination by the bailer-test method, *Groundwater,* 2(1), 38.
Lennox, D. H., 1966. Analysis and application of step drawdown tests, *J. Hydraul.,* HY6, 25.
Logan, J., 1964. Estimating transmissibility from routine production tests of water wells, *Groundwater,* 2(1), 35.
Lugeon, M., 1933. *Barreges et Geologie,* Dunod, Paris.
Miller, C. T., and Weber, W. J., 1983. Rapid solution of the nonlinear step-drawdown equation, *Ground Water,* 21(5), 584.
Papadopulos, I. C. and Cooper, H. H., Jr., 1967. Drawdown in a well of large diameter, *Water Resour. Res.,* 3, 241.
Papadopulos, I. S., Bredehoeft, J. D., and Copper, H. H., 1973. On the analysis of slug test data, *Water Resour. Res.,* 9, 1087.
Rorabaugh, M. I., 1953. Graphical and theoretical analysis of step drawdown tests of artesian wells, *Proc. ASCE,* 79, 1.
Sheahan, N. T., 1971. Type-curve solution of step-drawdown test, *Groundwater,* 9(1), 25.
Skibitzke, H. E., 1958. An equation for potential distribution about a well being bailed, U.S. Geol. Survey Groundwater Note 35.
Sternberg, Y. M., 1968. Simplified solution of variable rate pumping tests, *J. Hydraul.,* HY1, 177.
Walton, W. C., 1962. Selected Analytical Methods of Well and Aquifer Evaluation, Bull. 49, State of Illinois, Deptment of Registration Education, Urbana.

CHAPTER 12

Nonlinear Flow Aquifer Tests

I. GENERAL

It was pointed out in Chapter 5 that the groundwater flow equation is dependent on the medium composition and the hydraulic gradient and that the continuity equation is universal and can be applied in any situation. So far in this book the groundwater flow was assumed to abide with the Darcy law which is valid for laminar (linear) flow types only. However, in practice, there appears many situations where its application is questionable. Nonlinear flow is seldom, if ever, considered in groundwater movement, except for flow in the immediate vicinity of pumped wells. Recently, an interest has arisen for evaluation of the aquifer parameters with nonlinear flow. It is the purpose of this chapter to give available pumping test analysis of nonlinear flow.

Although the inadequacy of Darcy's law at high Reynolds numbers has been widely recognized, the analytical treatments of the nonlinear groundwater flow problems toward wells have been very limited and for the steady-state flows only. Mathematical consequences of nonlinear flow laws are not simple to tackle. In fact, combinations of the continuity and nonlinear flow laws lead to a nonlinear partial differential equation for which the solution is not possible with the conventional Laplace transformations. However, it is not justifiable to neglect deviations from the linear Darcy law because of this mathematical difficulty. In the hydrogeology domain where the flow of groundwater toward a well provides basic data in determining the aquifer parameters, such deviations are frequently encountered in the coarse-grained, fractured, cavernous limestone, and dolomite media or around the well where the hydraulic gradient is relatively higher than that remaining in the aquifer. Even a reasonable attempt at practical solutions and application of their consequences lead to benefits that may not be commensurated with extra difficulty in the analysis.

II. NONLINEAR FLOW

Traditionally aquifer tests are interpreted on the basis of linear flow through the whole aquifer. Nonlinearity in the flow at the pumping wells is attributed solely to "well losses" as explained in the previous chapter. However, laboratory experiments by many investigators and field tests have demonstrated that, when high hydraulic conductivities and/or steep hydraulic gradients are involved, it is necessary to look for nonlinear laws to describe the flow. Otherwise, conclusions drawn from the linear procedures will not correspond to reality. Deviations from the Darcian linear response are expected because of two groups of factors. In the first group there are factors that are related to the flow properties such as the specific discharge and the hydraulic gradient, whereas in the second group are the characteristics of the aquifer material as the effective porosity and the hydraulic conductivity.

Table 1 Critical Specific Discharge Values (m/d)

Porosity (%)	Grain diameter (mm)							
	0.1	0.2	0.5	1	2	4	10	20
25	39.0	19.6	7.8	3.9	1.9	1.0	0.4	0.21
30	30.5	15.2	6.0	3.0	1.5	0.8	0.3	0.15
35	24.3	12.1	4.8	2.4	1.2	0.6	0.24	0.12
40	19.6	9.8	3.9	1.9	1.0	0.48	0.21	0.90

Practically, in order for the groundwater flow to be in the linear form, the specific discharge must not exceed threshold values given in Table 1 for different porosity and grain diameters in a porous medium, (Castany, 1967).

It is rather obvious from this table that decreases in the grain size and/or in the porosity value give rise to large critical specific discharges. On the other hand, theoretically, the Reynolds number, Re, relates the specific discharge, q, and grain size diameter, d, as

$$Re = \frac{qd}{v} \tag{1}$$

in which v is the kinematic viscosity of the fluid. According to various laboratory experiments which are given in Table 1 in Chapter 5, the nonlinear flow occurs at values of Reynolds numbers greater than 1; the classical linear Darcy law cannot be adopted confidently since the flow becomes turbulent, i.e., nonlinear.

On the other hand, under the normal conditions for the linear flow to be valid the hydraulic gradient should be rather small. Excessive increases in the hydraulic gradient values may give rise to the nonlinear flow. It has been observed that in the sand medium hydraulic gradients between 3.3/1000 and 119/1000 lead to linear flow patterns. According to another hypothesis by Meinzer and Fisher (1934), the linear flow is valid even for some rather small hydraulic gradient such as 0.19/1000. Furthermore, Fisher (1935) suggested values such as 0.03 and

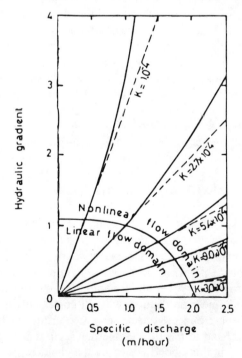

Figure 1 Hydraulic gradient-conductivity-specific discharge relationship.

0.045 per 1000. Sichard (1929) searched for the valid boundaries between the linear and nonlinear flow domain on the hydraulic gradient and conductivity Cartesian coordinate system. In fact, his work was based on the previous study by Prinz (1886), who obtained empirically the relationship between the hydraulic gradient-hydraulic conductivity-specific discharge triple, as shown in Figure 1, where K denotes the hydraulic conductivity. In this figure broken lines correspond to Darcy's linear flow, whereas the solid curves that pass from the origin are the nonlinear flow hydraulic gradient-specific discharge relationship. The connection of deviation points between linear and nonlinear flow relationships defines the boundary that separates the valid and invalid regions of the linear flow. It is clear that the area close to the origin represents the validity domain of the linear flow law. Furthermore, Sichard presented the valid and invalid regions on the hydraulic gradient-hydraulic conductivity and specific discharge-hydraulic conductivity Cartesian coordinate systems, as shown in Figures 2 and 3, respectively. These graphs can be used as charts in deciding whether the groundwater flow has a linear or nonlinear regime at any time, provided that two of the three basic quantities (specific discharge, hydraulic conductivity, and hydraulic gradient) are known.

A. STEADY-STATE FLOW

All the assumptions listed in Chapter 8 are considered valid except that the groundwater movement takes place according to the nonlinear flow law expressed by Forchheimer (1901) as in Equation 17 in Chapter 5. This law relates the hydraulic gradient nonlinearly to the specific yield in differential form as

$$\frac{ds(r)}{dr} = aq(r) + bq^2(r) \tag{2}$$

On the other hand the continuity equation gives

$$Q = 2\pi rmq(r) \tag{3}$$

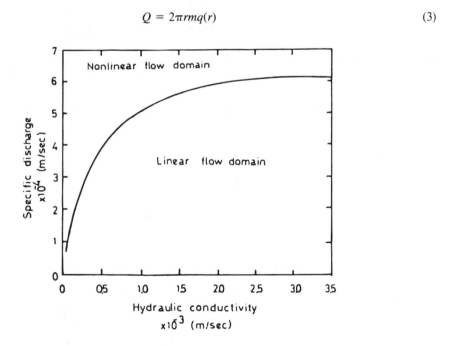

Figure 2 Hydraulic conductivity—specific discharge relationship.

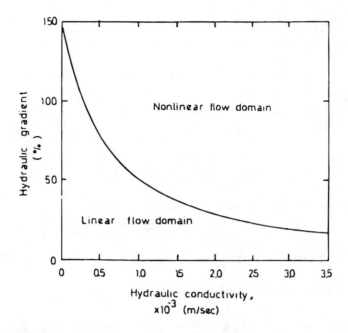

Figure 3 Hydraulic gradient—hydraulic conductivity relationship.

Similar to the linear flow case in Chapter 8, Section IX elimination of $q(r)$ between these two equations results in

$$s(r) - s(r) = \frac{aQ}{2\pi m} Ln(R/r) + b\left(\frac{Q}{2\pi m}\right)^2 \left(\frac{1}{r} - \frac{1}{R}\right) \quad (4)$$

In terms of the radius of influence, R_0, this last expression leads to

$$s(r) = \frac{aQ}{2\pi m} Ln\left(\frac{R_0}{r}\right) + b\left(\frac{Q}{2\pi m}\right)^2 \left(\frac{1}{r} - \frac{1}{R_0}\right) \quad (5)$$

Comparison of this expression with Equation 10 in Chapter 8 indicates the following significant points:

1. Equation 5 reduces to the Darcy flow case for $b = 0$ which implies that $a = 1/K$.
2. At the same distance from the well there is more drawdown in the nonlinear flow case than the linear flow situation. Extra drawdown is represented by the second term on the right-hand side of Equation 5. Furthermore, for large distances the drawdown becomes equal both in linear and nonlinear flow cases.
3. It is obvious that the semilogarithmic plots of Equation 5 do not appear as a straight line but curves and this equation may be rearranged to yield a dimensionless form as

$$\frac{2\pi m}{aQ} s(r) = Ln\left(\frac{R_0}{r}\right) + \frac{bQ}{2\pi m a R_0}\left(\frac{R_0}{r} - 1\right) \quad (6)$$

Dimensionless distance, turbulence, and drawdown factors are defined as

$$d = \frac{R_0}{r} \quad (7)$$

$$\alpha = \frac{bQ}{2\pi m a R_0} \quad (8)$$

and

$$W(d, \alpha) = \frac{2\pi m}{aQ} s(r) \qquad (9)$$

respectively, and their substitution into Equation 6 leads to

$$W(d, \alpha) = 2.3 \log d + \alpha(d - 1) \qquad (10)$$

which for $\alpha = 0$ gives the Thiem solution. The graphical representation of Equation 10 is given in Figure 4 for a set of α values. It is important to notice that as the turbulence factor increases the drawdown also increases.

On a semilogarithmic paper, the late time-drawdown plots yield to a curve deviating from the Thiem straight line at small distances. Intitutively nonlinear flow is expected in fractured, coarse-grained, and karstic formations. Equation 7 brings out a set of conditions to have negligible nonlinear flow. These conditions are that either the pump discharge must be small or the aquifer thickness must be large, both of which imply small hydraulic gradients.

Example 1—A confined aquifer of 20-m saturation thickness is pumped at a constant rate of 0.5 m³/min and after 9 h of pumping the quasi-steady-state flow situation is observed at five observation wells. The layout of these observation wells are given in Figure 5. The distance of each observation well to the main well and the drawdowns are given in Table 2.

To understand whether the given data represent nonlinear flow it is useful to plot drawdown vs. inverse distance on a semilogarithmic paper as shown in Figure 6. A first glance indicates that the field points do not follow a straight line but a similar pattern to the type of curves in Figure 4 and therefore the flow is nonlinear.

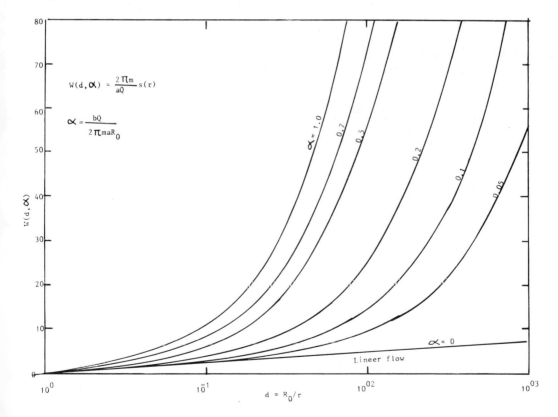

Figure 4 Dimensionless-distance-drawdown plot.

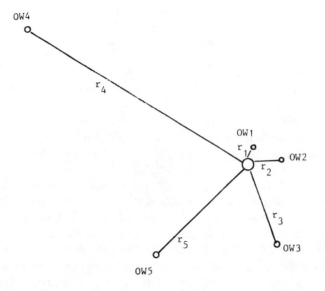

Figure 5 Well locations.

Numerical method—The unknowns a and b in Equation 6 can be determined from distance-drawdown readings at least from three observation wells. First, a relationship between a and b by using any two observation wells is obtained. In the example there are five observation wells and hence ten different combinations are possible for the application of Equation 4. These applications lead to the following simultaneous equation system: first for OW1 and OW2 pair

$$1.53 - 0.88 = \frac{2.3 \times 0.5 \times 60 \times 24a}{2 \times 3.14 \times 20} \log\frac{20}{12} + b\left(\frac{0.5 \times 60 \times 24}{2 \times 3.14 \times 20}\right)^2 \left(\frac{1}{12} - \frac{1}{20}\right)$$

or after simplification

$$0.65 = 2.82a + 1.09b$$

Likewise, calculations for other pairs lead to

OW1-OW3	$0.81 = 9.21a + 2.19b$
OW1-OW4	$1.52 = 13.17a + 2.46b$
OW1-OW5	$1.23 = 15.81a + 2.56b$
OW2-OW3	$0.61 = 6.29a + 1.09b$
OW2-OW4	$0.37 = 10.25a + 1.37b$
WO2-OW5	$10.58 = 12.88a + 1.47b$
OW3-OW4	$0.21 = 3.96a + 0.27b$
OW3-OW5	$0.42 = 6.60a + 0.37b$
OW4-OW5	$0.21 = 2.63a + 0.10b$

Table 2 Distance-Drawdown Data

Well no.	OW1	OW2	OW3	OW4	OW5
Distance (m)	12	20	60	120	190
Drawdown (m)	1.53	0.88	0.72	0.51	0.30

NONLINEAR FLOW AQUIFER TESTS

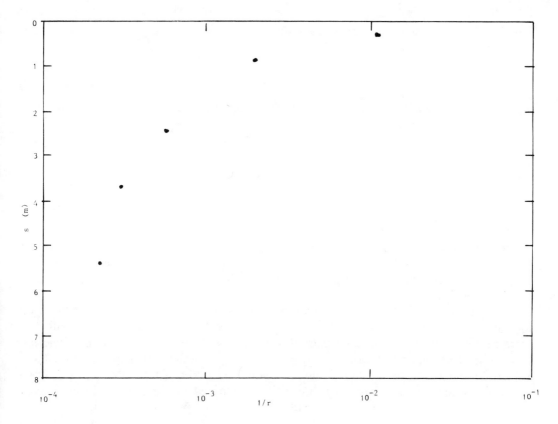

Figure 6 Distance-drawdown field data plot.

These 10 relationships between a and b can be solved pairwise and hence 45 different values for a and b are obtained. Their averages are adopted as the representative nonlinear flow parameters. The averages are $a = 0.20$ min/m and $b = 1.5$ (min)^2/m^2.

B. UNSTEADY-STATE FLOW

Unsteady-state nonlinear flow in confined aquifers during a pumping test was solved by Şen (1988a) with the Forchheimer flow equation. Physically, nonlinear flow causes more energy losses than the linear case and therefore the drawdowns are expected to have deviations from the Theis curve especially for large times or small distances. For small times aquifer response indicates that there are also deviations from the Theis curve with drawdowns smaller than the linear flow case.

All the assumptions that were set forward for the Theis solution in Chapter 9 are considered except that the flow within the aquifer complies with the Forchheimer law, Equation 2. Hence, the aquifer characteristics are the storage coefficient, S, hydraulic conductivity, K, and the Forchheimer flow coefficients a and b. Şen (1988a) obtained the type curve expression as

$$W(u,v) = \int_u^\infty F(x,v) \frac{e^{-x}}{x} dx + \frac{v}{\sqrt{\pi}} \int_u^\infty F^2(x,v) \frac{e^{-2x}}{x} dx \tag{11}$$

in which x is a dummy variable; u is the nonlinear flow dimensionless time factor which is defined explicitly

$$u = \frac{aSr^2}{4mt} \tag{12}$$

v is the dimensionless turbulence factor as

$$v = \frac{Qb}{2\sqrt{\pi m a r}} \tag{13}$$

and $W(u,v)$ is the nonlinear flow well function related to drawdown variable, $s(r,t)$, as

$$W(u,v) = \frac{4\pi m}{Q} s(r,t) \tag{14}$$

Finally, the nonlinear flow function $F(x,v)$ is given by

$$F(x,v) = [1 + v\sqrt{x}\Phi(\sqrt{x})]^{-1} \tag{15}$$

where $\Phi(x)$ indicates the area under the standard normal probability distribution function from $-\infty$ to x. It is obvious that for $b = 0$ and $a = 1/K$ all the above expressions reduce to the case of Theis solution. Numerical solution of Equation 11 yields a set of type curves for different v values as shown in Figure 7.

The general shapes of these curves are similar to the Theis curve in that for small u values the curves never become perfectly horizontal but although very small there is a continuous increase in the drawdown which indicates that the flow is always unsteady. For the same case increase in the turbulence factor gives rise to greater well function values than for small v values, implying that the drawdowns are comparatively larger than the Darcy case. On the contrary, for small times (u large) increase in the turbulence factor causes small well function values, i.e., smaller drawdowns than small v values, which proves that the drawdowns are relatively smaller than the Darcy flow case. The physical implication of this last statement is

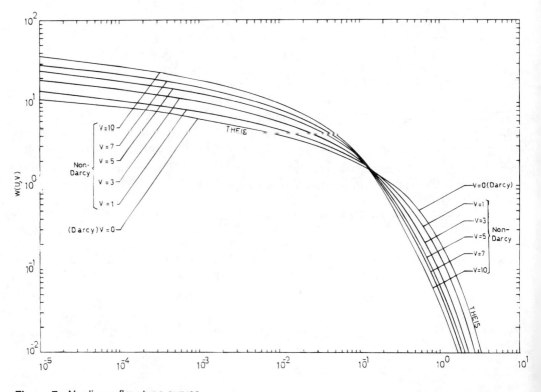

Figure 7 Nonlinear flow type curves.

that the response of a confined aquifer to pumping is delayed in the case of the nonlinear flow. As a result of the above-mentioned discussions there is a certain value of a for which the nonlinear and linear flow well functions are close to each other irrespective of v values. This critical region extends roughly between u values of 1×10^{-1} and 2×10^{-1} as in Figure 7. Aquifer parameters can be evaluated from a type curve matching to available field data.

During his studies Şen (1988a) presented a straight line method for the evaluation of aquifer data from a nonlinear flow at large times. As $x \to 0$ the difference in the error function in Equation 15 becomes practically equal to zero and hence $F(x,v) \cong 1$, the substitution of which into Equation 11 leads to

$$W(u,v) = \int_u^\infty \frac{e^{-x}}{x} dx + \frac{v}{\sqrt{\pi}} \int_u^\infty \frac{e^{-2x}}{x} dx \qquad (16)$$

The integration terms can be written succinctly as

$$W(u,v) = W(u) + \frac{v}{\sqrt{\pi}} W(2u) \qquad (17)$$

in which $W(.)$s are similar to the well functions for Darcy law as defined by Theis (see Equation 8 in Chapter 9). Approximately, for $u < 0.001$, Equation 17 can be written as

$$W(u,v) = -0.5772 - Lnu - \frac{v}{\sqrt{\pi}} [0.5772 + Ln2u] \qquad (18)$$

or in general

$$W(u,v) = A(v) + B(v)Lnu \qquad (19)$$

where

$$A(v) = -(0.5772 + 0.7167v) \qquad (20)$$

$$B(v) = -(1.0000 + 0.5642v) \qquad (21)$$

It is obvious from Equation 19 as shown in Figure 8 that the plot of $W(u,v)$ vs. u on a semi-logarithmic paper gives a set of straight lines.

The straight line slopes increase with increasing v values which implies that the drawdown rate is directly proportional with the turbulence factor for the same time period. Similar to the procedure as explained in Section V.B of Chapter 9 for the Jacob method, the intercept, t_0, on the time axis and the slope, Δs, of the straight line are equated to the theoretical intercept and slope values of the straight line in Equation 19 and hence,

$$S = \frac{4mt_0}{ar^2} e^{-\frac{A(v)}{B(v)}} \qquad (22)$$

and

$$\frac{m}{a} = -\frac{2.30Q}{4\pi\Delta s} B(v) \qquad (23)$$

The evaluation of the three aquifer parameters, namely, a, b, and S can be obtained from Equations 13, 22, and 23. First the ratio m/a is calculated from Equation 23 which is then

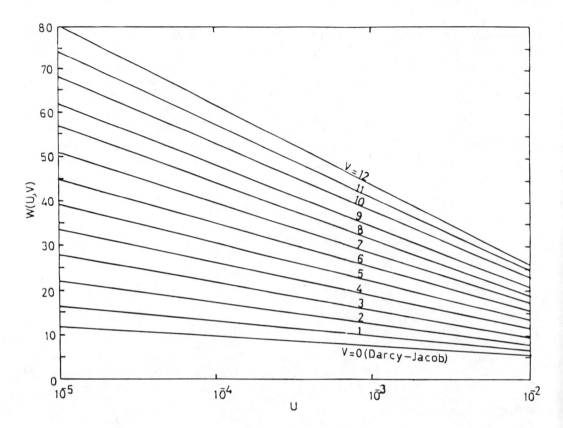

Figure 8 Forchheimer flow type straight lines.

substituted into Equation 22 so as to evaluate S. Of course, r in Equation 22 is the distance of the observation well to the main well.

It is interesting to notice at this stage that for the case of nonlinear flow condition m/a is equivalent to the transmissivity and $v = 0$ which implies that $A(0) = -0.5772$ and $B(0) = 1$, the substitutions of which into Equations 22 and 23 leads to the classical equations of Jacob for the linear flow as given in Equations 13 and 12 in Chapter 9, respectively.

On the other hand, if there were many observation wells around the pumping well, then Equation 19 can be employed to obtain valid expressions for the aquifer parameter determination in the nonlinear flow case for any constant large time instant from the distance-drawdown straight line. The two basic equations for the application of distance-drawdown line are

$$S = \frac{4mt}{ar_0^2} e^{-\frac{A(v)}{B(v)}} \qquad (24)$$

and

$$\frac{m}{a} = \frac{2.30Q}{2\pi \Delta s_r} \qquad (25)$$

For $v = 0$, these equations become identical with their counterparts for the linear flow as in Equations 16 and 15 in Chapter 9, respectively.

One of the most important distinctions between linear and nonlinear type curve matching and calculations thereof is that the nonlinear type curves do not yield the conventional transmissivity estimations as will be explained in Section VII.

Example 2—It is rather difficult to find readily available complete sets of aquifer test data in fractured or coarse-grained geological formations where the groundwater flow law might be nonlinear. Gale (1977) presented such aquifer test results from a single well in the field without subjecting them to matching for determining the aquifer parameters due to the fact that there was no type curve method available for nonlinear flow situation. The site of pumping well was in fractured and coarse-grained crystalline metamorphic rocks of Halifax City, Nova Scotia. This area is selected since it has a good combination of well-exposed bedrock, relatively simple fracture geometry, and no interference from pumping of existing domestic wells. One would expect that the nonlinear flow effects will be in the form of a skin around the pumping well. In addition for large times, in coarse-grained and/or fractured medium, the groundwater velocity becomes rather large to invalidate the application of any known type curve based on linear flow. The plot of Gale data for large times (l/t <0.01 1/min) on a semilogarithmic paper shows a straight line as in Figure 9. The well discharge was kept at a constant rate as 0.0272 m^3/min by using the discharge control tank. Comparison of the plotted data with the set of theoretically obtained straight lines in Figure 8 indicates that the best match is possible with $v = 9$. The substitution of this value into Equations 20 and 21 gives $A(v) = 7.03$ and $B(v) = -6.08$.

The slope of this straight line is $\Delta s = 3.96$ m. Hence the substitution of this value into Equation 23 yields $m/a = -2.3 \times 0.00272 \times (-6.08)/(4 \times 3.14 \times 3.96) = 7.65 \times 10^{-3}$ m^2/min. On the other hand, the intercept of the straight line on the time axis is about $1/t_0 = 3.35 \times 10^{-1}$ 1/min. The substitution of this value together with above calculated $A(v)$ and $B(v)$ values into Equation 24 yields $S = 2.9 \times 10^{-1}/r^2$. The value of r was not given by Gale. Therefore, provided that it is defined numerically, the storage coefficient can be obtained exactly. Because the aquifer thickness is about 50 m, one can find that $a = 65.3$ min/m and finally, the value of b can be estimated from Equation 13 as $b = 3.82 \times 10^6 r$ min^2/m^2.

III. VOLUMETRIC APPROACH

The basis of this method has been explained in Section VII.E in Chapter 9. By using the volumetric method Şen (1986) obtained type curves for nonlinear flow toward large-diameter wells in confined aquifers. The continuity equation for the well is

$$Q = Q_w(t) + Q_a(t) \tag{26}$$

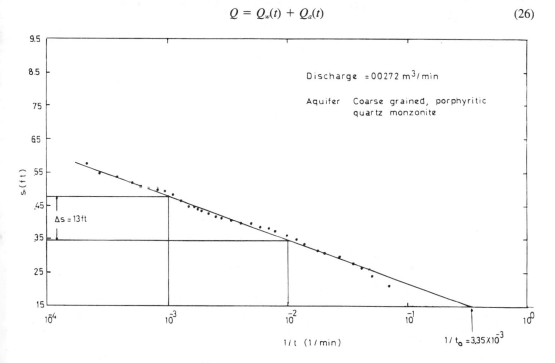

Figure 9 Time-drawdown field data plot.

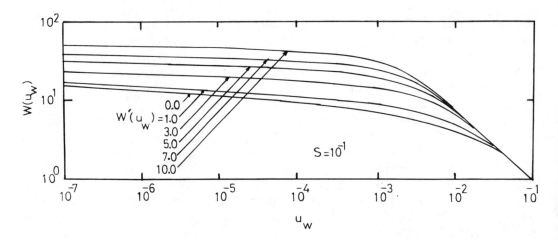

Figure 10 Nonlinear flow-type curves ($s = 10^{-1}$).

in which Q, $Q_w(t)$, and $Q_a(t)$ are the constant pumping discharge, time variant well storage, and aquifer storage discharges, respectively. On the other hand, the storage coefficient, S, can be expressed as the ratio of two volumes:

$$S = \frac{Qt - \pi r_w^2 s_w(t)}{V(t)} \qquad (27)$$

where r_w is the well radius, $s_w(t)$ is the drawdown in the well, and $V(t)$ is the volume of the depression cone. After some lengthy mathematical treatments the explicit form of the type curve equation is obtained as (Şen, 1986)

$$u_w = \frac{1}{\left(\alpha^2 + \dfrac{1}{S} - 1\right)W(u_w) - 4(\alpha - 1)^2 C - (\alpha^2 \ln\alpha^2 - \alpha^2 + 1)}$$

in which the dimensionless time factor, u_w, for large-diameter wells and the well function of the nonlinear flow are defined as

$$u_w = \frac{a r_w^2 S}{4mt} \qquad (28)$$

and

$$(u_w) = \frac{4\pi m}{aQ} s_w(t) \qquad (29)$$

respectively, and the dimensionless turbulence factor is

$$C = \frac{bQ}{4\pi a m r_w} \qquad (30)$$

Finally, radial distance ratio, α, is defined as

$$\alpha = \frac{R_0}{r_w} \qquad (31)$$

where R_0 is the radius of influence and r_w the well radius.

Curve types are presented in Figures 10 to 12 for different S values and a set of turbulence factors. The curves have three distinct portions. The initial straight line represents the abstrac-

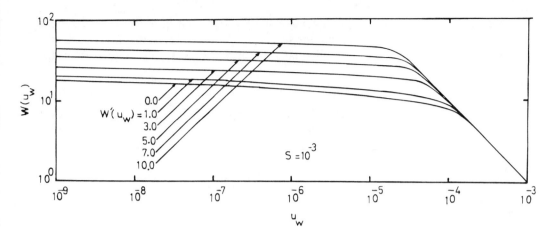

Figure 11 Nonlinear flow-type curves ($s = 10^{-3}$).

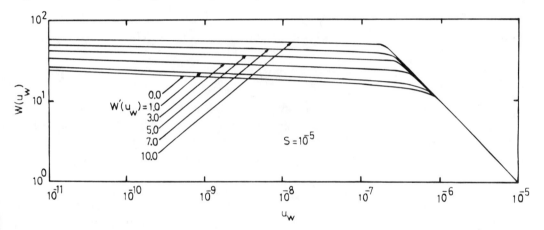

Figure 12 Nonlinear flow-type curves ($s = 10^{-5}$).

tion of pumping water from the well storage only. Therefore, similar to the Papadopulos and Cooper curves, this part does not yield any useful information about the aquifer parameters but can be employed in finding the pumping discharge estimation as was mentioned in Chapter 7, Section VI. As the hydraulic gradient increases the aquifer starts to respond which is reflected in the type curves as the curved intermediate portion. The effective duration of these portions depends mainly on the storage coefficient. The smaller the S value the longer is the duration. The last distinctive portions has very little curvature. The well storage effect is completely negligible in this portion.

IV. EXPONENTIAL LAW

After detailed laboratory studies Anandakrishnan and Varadarajulu (1963) concluded that an exponential flow law appears in sands after a threshold value and they have proposed the exponential law with two parameters as

$$[q(r,t)]^n = k' \frac{\partial s(r,t)}{\partial r} \tag{32}$$

where n is an exponent varying in the range $1 < n < 2$ and k' is the turbulence factor.

A fully penetrating well is considered in an aquifer for which all of the Theis assumptions listed in Chapter 9 are valid except that the Darcy law leaves its place to an exponential law. After the necessary mathematical calculations the well function $W(u,v)$ becomes (Şen, 1989):

$$W(u,v) = v \int_u^\infty \frac{1}{x} \left(-2 \frac{n-1}{n-3} xv + 1 \right)^{n/(1-n)} dx \qquad (33)$$

in which x is a dummy variable and the dimensionless time factor, u, is defined as

$$u = \frac{r^2 S}{4tm(k')^{1/n}} \qquad (34)$$

and the turbulence factor v is

$$v = \left[\frac{2\pi m r (k')^{1/n}}{Q} \right]^{1-n} \qquad (35)$$

and explicitly the well function in terms of exponential law parameters appears as

$$W(u,v) = \frac{4\pi m (k')^{1/n}}{Q} s(r,t) \qquad (36)$$

Equation 33 reduces to the classical Theis-type curve for the linear case when $n \to 1$.

Type curves from Equation 33 for nonlinear flow in confined aquifers are plotted on double-logarithmic paper in Figures 13 to 15 for some representative n values. For the sake of comparison all of the figures include the linear flow-type curve as it was proposed originally by Theis (1935). Similarly, type curves, for any desired set of v and n values, can be obtained by numerical integration of Equation 33. Although there exists a single-type curve for linear flow, an infinite number of type curves are available for exponentially nonlinear flow law. For any given turbulence exponent, n the form of type curves depends on the value of dimensionless turbulence factor, v. These curves are continuous with no inflection points. In

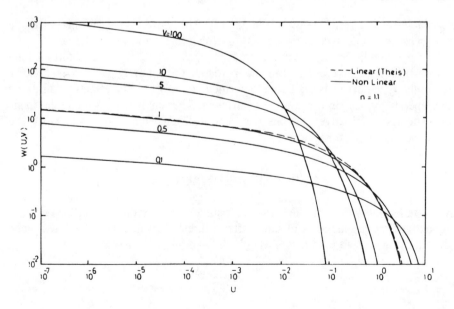

Figure 13 Exponential law-type curves, ($n = 1.1$).

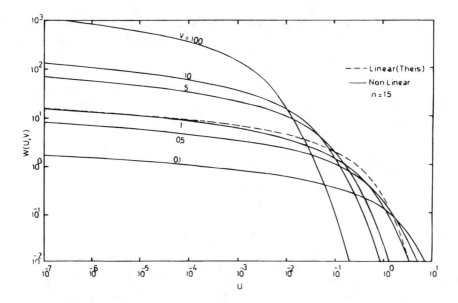

Figure 14 Exponential law-type curves, ($n = 1.5$).

general, the deviations from the Theis-type curve increase proportionally with the deviation of v from 1 for any n.

A common feature in all the type curves is that, for large times (u small) and $v > 1$ values, the curves are above the Theis-type curve which means that there are extra drawdowns due to the turbulence effects.

Another significant point that can be derived from Equation 35 is that, for the same discharge values, depending on the distances of the observation wells to the main well, the turbulence factor assumes different values. In fact, the closer the observation well is to the main well the more will be the intensity of turbulence, i.e., large v. It means that the intensity of nonlinear flow effects near the pumping well is more likely than at further distances. In

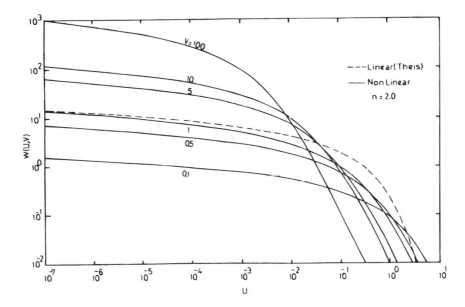

Figure 15 Exponential law-type curves, ($n = 2$).

the linear flow-type curve, the radial distance does not affect the shape of the type curve at all. On the contrary, in the nonlinear flow case, even though all of the other variables are kept constant, different distances require different type curves.

On the other hand, it is interesting to note that, for any fixed v value, the greater the turbulence exponent the smaller will be the drawdown for any u and Q. For small u values (large times) they all approach the same asymptotic value. The turbulence exponent is effective in the immediate portion of the type curves more than any other portion.

In practical applications decision on the best matching type curve for a given field data plays a dominant task in the aquifer parameter determinations. The late time portions of all the type curves in Figures 13 to 15 seem to match each other and hence it is not possible to make distinctions between them. However, when the type curves are all matched with each other, it becomes obvious that the initial and intermediate portions give to the individual curve its distinctive appearance. For example, this point has been shown in Figure 16 with $n = 1.01$. This indicates that for large times there is no distinctive difference between the linear and nonlinear flow-type curves in a matching procedure.

A significant result is that the nonlinear flow aquifer parameters, k' and n, cannot be determined from the late time portions of the type curves. Therefore, if there exist deviations from the Theis curve for the initial and intermediate time-drawdown data, then the most suitable, nonlinear flow-type curves must be matched to these portions for determining the nonlinear parameters.

Example 3—Application of the above-mentioned nonlinear flow-type curves is presented herein for the confined aquifer pumping test data by Kruseman and de Ridder (1979) in the "Oude Korendijk" polder south of Rotterdam, Holland. Although they have matched a linear flow-type curve to the available time-drawdown data, the following points make the applicability of the Theis-type curve suspect in this situation:

1. The major bulk of the confined aquifer in Oude Korendijk consists of rather coarse-grained sand with some gravel. Fine sand with clayey sediments are considered as impervious layers. The existence of coarse sand and gravel as the aquifer material implies large grain sizes and therefore, the possibility for the Reynolds number to be greater than one or perhaps ten may occur. This may give rise to the invalidity of Darcy's law and hence nonlinear flow occurrence

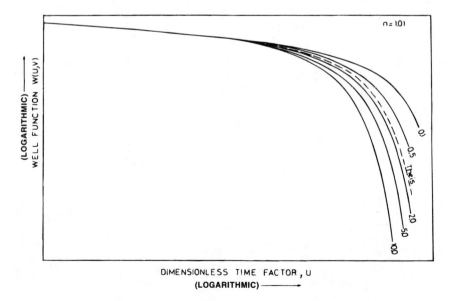

Figure 16 A set of exponential law-type curves.

in the vicinity of the well, considering the given discharge $Q = 788$ m³/d (9120 cm³/s), assuming a well radius of 3 in. ($\cong 7.62$ cm). The well is screened over the 7-m aquifer thickness and gives the water entrance area as $A = 2\pi rm = 33{,}514$ cm². Therefore, the specific discharge is $q = 9120/33{,}514 = 0.272$ cm/s. On the other hand, Todd (1980) estimated a medium gravel to be particles of average diameter 10 mm. The kinematic viscosity of water at 20°C is about 1.01×10^{-2} cm²/s. The substitution of all known quantities into Equation 1 yields a Reynolds number equal to 27. This implies that the nonlinear flow occurs around the vicinity of the well. However, the Reynolds number decreases with increase in the radial distance away from the well and consequently it may assume a value less than 1.
2. The deviation of time drawdown plot by Kruseman and de Ridder (1979) from the Theis-type curve especially for the small and moderate times are indications of nonlinear flow.
3. The comparison of these observation well data, namely, at distances 30, 90, and 215 m show that they have different curvatures and accordingly deviations from the classical Theis-type curve. It is, therefore, not possible to match these field data with a single-type curve. As mentioned earlier, only in the nonlinear flow case are there different type curves depending on the observation well's distance from the main well. Since the discharge is the same, the differences in curvature, according to Equation 35, might stem out from the distance only.

Theis-type curve application to Oude Korendijk data can be accepted as a first approximation only. The adaptation of these data for the application of the methodology developed herein is particularly beneficial because of the direct comparison possibility between the linear and nonlinear flow effects on the aquifer parameter estimations.

After many trials, the best type curve matching the given data is found to have the turbulence exponent as $n = 1.1$. Figures 17 and 18 show type curve matching to observation well time-drawdown data at 30- and 90-m distances, respectively. Due to the paucity of data in the observation well at distance 215 m, it has not been considered for type curve matching herein. The best match for the 30-m distance is achieved with the dimensionless turbulence factor 0.5, whereas for 90-m distance the similar value is 0.1. These results confirm with the previous statement that there exists an inverse relationship between the distance and turbulence factor. In Figure 17 the match point, M_1, is chosen with type curve sheet values $W(u,v) = 1$ and $u = 1$. The same match point has on the field data sheet the coordinates as $1/t = 7.7 \times 10^{-2}$ (1/min) and $s = 1.48$ m. On the other hand, in Figure 18 the match point, M_2, is chosen such that $W(u,v) = 10^2$ and $u = 10^{-2}$ with the corresponding coordinates on the field data sheet as $(1/t) = 2.6 \times 10^{-2}$ (1/min) and $s = 31$ m. Substitution of these values together with

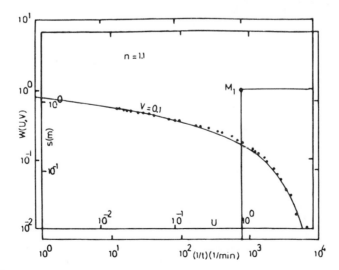

Figure 17 Type curve matching ($r = 30$ m).

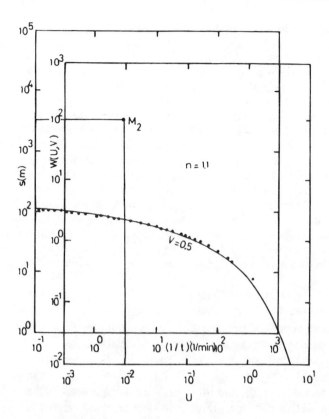

Figure 18 Type curve matching ($r = 90$ m).

the discharge $Q = 788$ m^3/day into Equations 34 and 36 yields the aquifer parameter estimations as shown in Table 3. The nonlinear flow transmissivity is defined as $T' = m(k')^{1/n}$.

It is interesting to note that the storage coefficient values are comparable with the estimations from the linear flow as were found by Kruseman and de Ridder (1979). The nonlinearity of flow does not make any significant change in the estimation of S. However, turbulence transmissivity coefficient values are smaller than transmissivities calculated with the linear flow.

V. TWO-REGIME FLOW

The groundwater flow that confirms to Darcy's law does not have significant inertial effects and consequently convergent, divergent, or curved flow paths do not cause any complication in the specific discharge-hydraulic gradient relationship. However, increase in the inertial effects and the possibility of turbulent flow occurrence in the vicinity of a well do modify the linearity property of such a relationship. In particular, the groundwater flow in the vicinity of small-diameter wells converges rather rapidly and, as a result of this, deviations from the conventional Darcy law appear. Due to this convergence, the flow cross-sectional area becomes smaller toward the well; consequently, the specific discharge increases. It may happen that the hydraulic

Table 3 Aquifer Characteristics

Flow type	Observation well	Match point	T' (m^2/d)	S	n
Nonlinear	30-m distance	M_2	202	2.3×10^{-4}	1.1
Linear	90-m distance	M_1	42	1.8×10^{-4}	1.1

gradient that is required to have a given specific discharge becomes greater than the one that the Darcy law predicts. Hence, it is expected that around the well there is a certain domain of the aquifer in which nonlinear flow occurs. The transition from the linear to a completely turbulent flow in this domain consists of the gradual spread of turbulence.

The critical well radius is defined as the radial distance between the center of main well and the critical points on a concentric circle in the aquifer where the first deviations from the Darcy law start. In other words, within the aquifer domain encircled by the critical radius, only the nonlinear flow is effective (see Figure 19). Such a situation gives rise to the occurrence of a two-regime flow in the aquifer.

For a given set of flow and aquifer material properties, a preliminary estimate of this critical radius may be obtained simply by equating the constant pump discharge, Q, to the aquifer discharge at the critical radius, r_c. Consideration of the steady-state flow yields

$$Q = 2\pi r_c m q_c \tag{37}$$

in which m is the aquifer thickness, and q_c is the critical specific discharge. Considering Equation 1, it is possible to obtain after some algebra

$$r_c = \frac{\left(\frac{Q}{m}\right)d}{2\pi \nu (Re)_c} \tag{38}$$

in which $(Re)_c$ is the critical Reynolds number, and the ratio (Q/m) will be referred to as the aquifer specific discharge. For a given aquifer, d and m are fixed values in addition to the kinematic viscosity, ν. The relationship between the critical well radius, aquifer specific discharge, and the Reynolds number are shown in Figures 20 and 21. In calculations, the kinematic viscosity of water is taken at 25°C as 9×10^{-7} m²/s (Albertson et al., 1960). Figure 20 is prepared for the critical Reynolds number equal to 10. Similar charts can be prepared on the basis of Equation 38 for any desired Reynolds number. However, they will all have the similar shapes in that on a double-logarithmic paper there is a linear relationship between the aquifer-specific discharge and the critical well radius. In practice, Figure 20 can be used

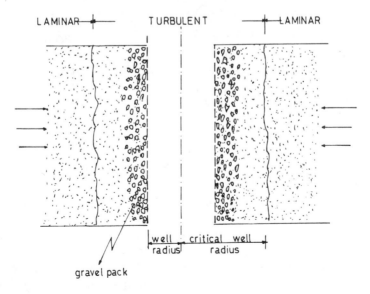

Figure 19 Turbulent flow domain near well.

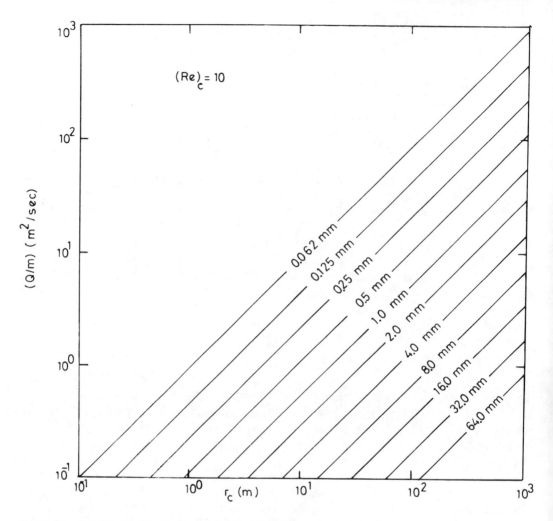

Figure 20 Specific aquifer discharge-critical well radius-grain size chart.

in determining an approximate estimate of minimum r_c, provided that the pumping discharge, aquifer thickness, and average grain size of the aquifer material are known. Generally, the flow in granular material aquifers is of linear regime, but it is possible that nonlinear flow may occur in the aquifer even beyond the gravel pack. However, the nonlinear flow does not extend very far because the Reynolds number is inversely related to the critical radius as in Equation 38.

The term "two-regime flow" has been coined by Huyokorn and Dudgeon (1962) where a turbulent flow may exist in the immediate vicinity of a well due to either screen resistance or gravel pack material or at high gradients in the aquifer material itself. Preliminary analytical studies have been presented by Kristianovich (1940) and Engelund (1953). On the basis of sample boundary conditions, Huyakorn and Dudgeon (1962) presented a mathematical model for steady-state two-regime well flow which led to a finite element solution. However, Basak (1977) presented an analytical solution for fully and nonpenetrating wells in two-regime flow domain.

The non-steady-state solutions have been presented by Şen (1988c) on the basis of Forchheimer-Darcy law combinations. The assumptions are the same as with the Theis approach

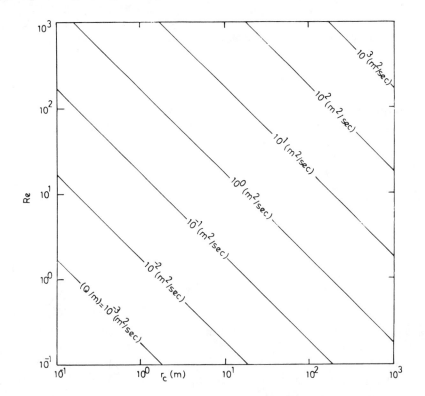

Figure 21 Critical Reynolds number-critical well radius-specific discharge chart.

except that nonlinear flow appears around the well. The type curve expressions for the linear-flow zone is given as

$$W(u_L) = \int_u^\infty F(\alpha\rho^2 x, \lambda) \frac{e^{-[1-\rho^2(1-\alpha)]x}}{x} dx \quad (0 < \rho < 1) \tag{39}$$

in which x is a dummy variable; $\rho = r_c/r$ is dimensionless distance factor; and λ is the turbulence factor defined as

$$\lambda = \frac{Qb}{\sqrt{\pi mar}} \tag{40}$$

is the dimensionless linear flow factor as

$$\alpha = aK \tag{41}$$

and $F(x,\lambda)$ is the dimensionless turbulence function which has an explicit form as

$$F(x,\lambda) = \left[1 + \frac{\lambda}{2}\sqrt{\alpha x}\Phi(\sqrt{\alpha x})\right]^{-1} \tag{42}$$

In terms of drawdown the well function is defined as

$$W(u_L) = \frac{4\pi T}{Q} s_L(r,t) \tag{43}$$

and finally, the linear flow domain dimensionless time factor is

$$u_L = \frac{r^2 S}{4tT} \tag{44}$$

However, for the linear flow domain the dimensionless distance factor varies as $1 < \rho < \infty$. Physically, it is expected that the total drawdown at a point in the nonlinear zone consists of the drawdown computed by assuming wholly linear flow and the additional drawdown due to the nonlinear flow. With these considerations the nonlinear zone well function is given as

$$W(u_N) = \int_{u_N}^{\infty} F(\rho^2 x, \lambda) \frac{e^{-\rho^2 x}}{x} dx + \alpha \int_{u_N}^{\infty} F(x,\lambda) \left[1 + \frac{\lambda F(x,\lambda)}{2\sqrt{\pi}} e^{-x}\right]^n \frac{e^{-x}}{x} dx \tag{45}$$

in which by definition

$$W(u_N) = \frac{4\pi T}{Q} s_N(r,t) \tag{46}$$

In Equation 45 the first term on the right-hand side signifies the linear flow contribution whereas the second term is basically due to nonlinear flow regime.

Identifications of aquifer parameters or drawdown predictions in terms of time-drawdown or distance-drawdown measurements during an aquifer test depend on the available or pertinent-type curves that reveal the functional relationship between the well function and the dimensionless time factor on a double-logarithmic paper.

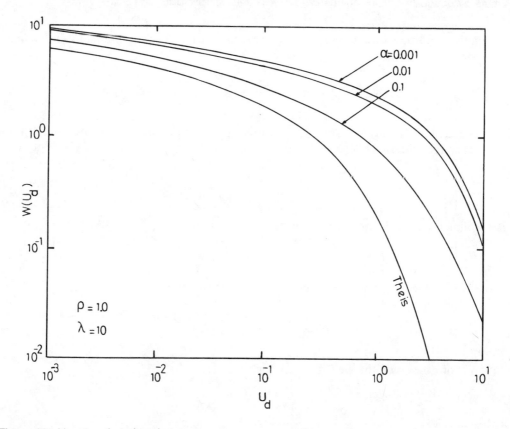

Figure 22 Linear regime domain-type curves.

The solution of Equations 39 and 45 have been achieved through the use of Simpson's numerical integration technique for a given set of α, ρ, and λ parameters. Some representative type curves are shown in Figures 22 and 23 for the linear and Figures 24 and 25 for the nonlinear flow regimes.

To provide a basis of comparison, the Theis curve is also drawn on the figures. It is possible to depict from these figures, in general, the following points for the two-regime flow in an aquifer.

1. The linear flow domain-type curves merge into the Theis-type curve for large times ($u_L \to 0$), whereas the nonlinear domain curves converges on the same curve but for very small times ($u_N \to \infty$), (see Figures 24 and 25).
2. For a given pair of ρ and λ values, all of the curves in the linear domain always remain above the Theis-type curve. However, decrease in λ value causes an increase in the magnitude of deviations from the Theis curve.
3. In the nonlinear domain, increase in ρ and λ gives rise greater (smaller) well functions for large (small) times.

Furthermore the following significant limiting cases are observable from these figures and precisely rederivable from Equations 39 and 45:

1. As $\alpha \to 0$, $(a \to 0)$, the turbulence function becomes equal to 1 irrespective of b or λ values, i.e., $F(o,\lambda) = 1$. Hence, Equations 39 and 45 reduce to

$$W(u_L) = \int_{u_L}^{\infty} \frac{e^{-(1-\rho^2)x}}{x} dx \quad (0 < \rho < 1) \quad (47)$$

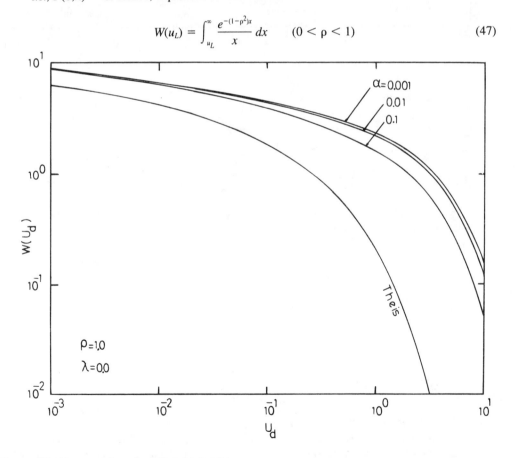

Figure 23 Linear regime domain-type curves.

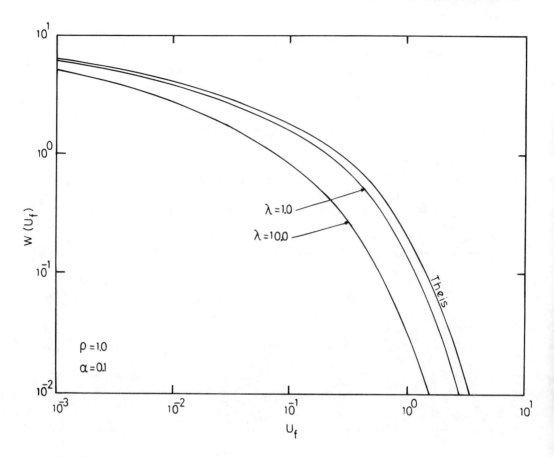

Figure 24 Linear regime domain-type curves.

and

$$W(u_N) = \int_{u_N}^{\infty} F(\rho^2 x, \lambda) \frac{e^{-\rho^2 x}}{x} dx \qquad (1 < \rho < \infty) \qquad (48)$$

respectively. Physically, this situation corresponds to an extreme two-regime flow, namely, linear regime and completely turbulent regime, since $a = 0$.

2. As $\alpha \to 1$, $(a \to 1/K)$, the linear flow component in the nonlinear flow regime becomes equal to the linear flow regime surrounding it. In such a case, Equations 39 and 45 yield

$$W(u_L) = \int_{u_L}^{\infty} F(\rho^2 x, \lambda) \frac{e^{-x}}{x} dx \qquad (0 < \rho < 1) \qquad (49)$$

and

$$W(u_N) = \int_{u_N}^{\infty} F(\rho^2 x, \lambda) \frac{e^{-\rho^2 x}}{x} dx + \int_{u_N}^{\infty} F(x, \lambda)\left[1 + \frac{\lambda F(x, \lambda)}{2\sqrt{\pi}}\right] \frac{e^{-x}}{x} dx \qquad (1 < \rho < \infty) \qquad (50)$$

3. As $\lambda \to 0$, $(b \to 0)$, the turbulence function again becomes equal to 1 (see Equation 42). This represents the physical case where there is a single regime flow that is linear but the aquifer consists of two concentric domains with different hydraulic conductivities, namely,

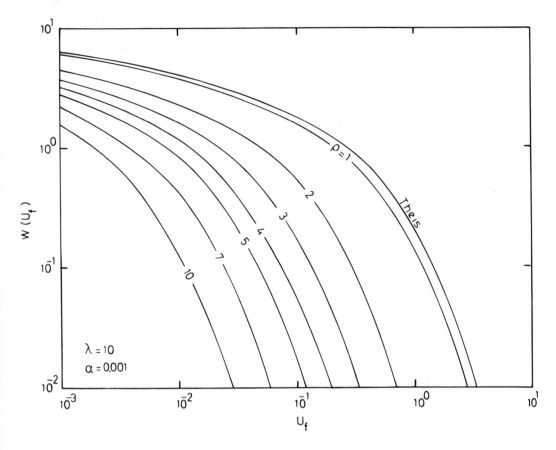

Figure 25 Nonlinear regime domain-type curves.

K and $1/a$. Hence, Equations 39 and 45 give

$$W(u_L) = \int_{u_L}^{\infty} \frac{e^{-[1-\rho^2(1-\alpha)]x}}{x} dx \qquad (0 < \rho < 1) \tag{51}$$

and

$$W(u_N) = \int_{u_N}^{\infty} \frac{e^{-\rho^2 x}}{x} dx + \alpha \int_{u_N}^{\infty} \frac{e^{-x}}{x} dx \qquad (1 < \rho < \infty) \tag{52}$$

4. Type curves depend also on the location of observation wells relative to critical well radius. Contrary to the single-regime flow (Theis curve), type curves in the two-flow regime case depend on the relative distance, ρ. The limiting cases within the linear flow domain are for $\rho \to 0$ ($r \to \infty$) and $\rho \to 1$ ($r \to r_c$) which render Equations 39 and 45 as

$$W(u_L) = \int_{u_L}^{\infty} \frac{e^{-x}}{x} dx \tag{53}$$

and

$$W(u_N) = \int_{u_N}^{\infty} F(\alpha x, \lambda) \frac{e^{-\alpha x}}{x} dx \tag{54}$$

respectively. It is obvious that, for large distances from the critical well radius in the linear flow domain, irrespective of aquifer parameter values, all of the curves approach the Theis-type curve. Equation 54 gives the maximum drawdown within the linear flow domain.

On the other hand, distance limits in the nonlinear flow domain are $\rho \to 1$, $(r \to r_c)$ and $\rho \to \infty$, $(r \to 0)$ which cause Equation 45 to be

$$W(u_N) = \int_{u_N}^{\infty} F(\alpha,\lambda) \frac{e^{-x}}{x} dx \tag{55}$$

and

$$W(u_N) = \alpha \int_{u_N}^{\infty} F(x,\lambda)\left[1 + \frac{\lambda F(x,\lambda)}{2\sqrt{\pi}} e^{-x}\right] \frac{e^{-x}}{x} dx \tag{56}$$

respectively. All of the aforementioned-type curves can be employed suitably for matching the field data through a similar procedure to the Theis-type curve matching and accordingly the aquifer parameter estimations can be achieved.

5. Another practically important limit is concerned with time durations of the aquifer test. For large times ($u_L \to 0$) Equation 39 reduces to

$$W(u_L) = \int_{u_L}^{\infty} \frac{e^{-[1-\rho^2(1-\alpha)]x}}{x} dx \tag{57}$$

However, as $u_N \to 0$ there exist two limiting cases in the nonlinear regime domain, namely, for large distances ($\rho \to 1$) as

$$W(u_N) = \int_{u_N}^{\infty} \frac{e^{-\rho^2 x}}{x} dx \tag{58}$$

and for small distances ($\rho \to \infty$) as

$$W(u_N) = \int_{u_N}^{\infty} \frac{e^{-\rho^2 x}}{x} dx + \alpha \int_{u_N}^{\infty} \left(1 + \frac{\lambda}{2\sqrt{\pi}} e^{-x}\right) \frac{e^{-x}}{x} dx \tag{59}$$

This final expression is valid in the near well zone. Neither Equation 57 nor Equation 46 is dependent on λ but Equation 59 is a function of λ which indicates that at any large distance there is only one value of the piezometric level. However, there exists a multitude of possible piezometric levels near the well depending on the λ value. This is unique indication of the nonlinear flow in the well vicinity.

Furthermore, similar to the Jacob straight line method, Equations 57 and 47 can be approximated for u_L, $u_N < 0.01$ as

$$W(u_L) = -0.5772 - Ln[1 - \rho^2(1 - \alpha)] - Lnu_L \tag{60}$$

and

$$W(u_N) = -0.5772 - Ln\rho^2 - Lnu_N \tag{61}$$

If the late time-drawdown data obtained from an observation well are plotted on a semilogarithmic paper as drawdown vs. logarithm of time, it appears as a straight line. If for the linear regime domain the intercept and the slope are, t_{0L} and Δs_L, respectively, then the substitution of Equations 43 and 44 into Equation 60 leads to aquifer parameters as

$$T = \frac{2.30Q}{4\pi \Delta s_L} \qquad (62)$$

and

$$S = \frac{2.25 T t_{0L}}{r^2 - r_c^2(1 - \alpha)} \qquad (63)$$

On the other hand, similar arguments are valid for the nonlinear flow domain and consequently Equation 61 yields

$$\frac{m}{a} = \frac{2.30Q}{4\pi \Delta s_N} \qquad (64)$$

and

$$S = \frac{2.25 t_{0N}}{r_c^2} \qquad (65)$$

in which t_{0N} and Δs_N are the time and slope of the straight line fitted to the late drawdown data in the nonlinear flow domain. The only limiting case for large times that helps to calculate λ is Equation 59, which requires data from the observation well in the near well zone.

Example 4—The field data with two-regime flow have been adopted from Dudgeon et al. (1973), who performed an aquifer test in Rosevale, Southeast Queensland, Australia. The aquifer material is composed mainly of gravel and coarse to medium sand. The sieve analysis has shown that the average grain size is 10 mm which corresponds to fine gravel domain. They have observed that, in addition to Darcy law, non-Darcy flow also occurs in close vicinity to the pumped well, resulting in significant well losses. The occurrence of non-Darcy flow was indicated by nonlinear drawdown-discharge relationships. Furthermore, the semilogarithmic plots of drawdown-distance data for late times also showed nonlinearity in the near well zone. The Forchheimer equation has been used to describe the nonlinear specific discharge-hydraulic gradient relationship.

The aquifer was pumped for 24 h with a constant discharge equal to 7050 gph (7.41 × 10^{-3} m³/s). The distance of observation well to the main well is 12 ft ($\cong$ 3.8 m) and the aquifer thickness has been found as 10 ft (3.2 m). The plot of time drawdown data is presented in Figure 26.

By adopting a minimum critical Reynolds number around 1 and the kinematic viscosity as 9 × 10^{-7} m³/s with the given data, Equation 38 yields approximately $r_c \cong 3.8$ m. Comparison of this value with the observation well distance indicates that its location is just in the linear flow regime domain. Furthermore, the dimensionless distance factor can be adopted approximately as $\rho \cong 1$. The best type curve matching has been achieved with $\alpha = 0.1$ and $\lambda = 10$ as shown in Figure 26. The arbitrarily chosen match point has coordinates on the type curve sheet as $W(u_L) = 10^0$ and $u_L = 10^{-1}$ which corresponds to coordinates on the field data sheet as $s_L = 4.5 \times 10^{-1}$ ft (1.44×10^{-1} m) and $l/t = 10^{-1}$ (1/min). Substitution of these numerical values into Equations 40, 41, 44, and 46 yields the aquifer parameter estimates as $S = 3 \times 10^{-2}$, $T = 4.1 \times 10^{-3}$ m²/s, $a = 1.3$ (min/m), and $b = 530$ (m/min)².

VI. DIMENSIONLESS STRAIGHT-TYPE LINES

The extension of the Theis (1935)-type curve for large dimensionless time factors leads to a straight line on a semilogarithmic paper as shown in Figure 16 of Chapter 9. There, this

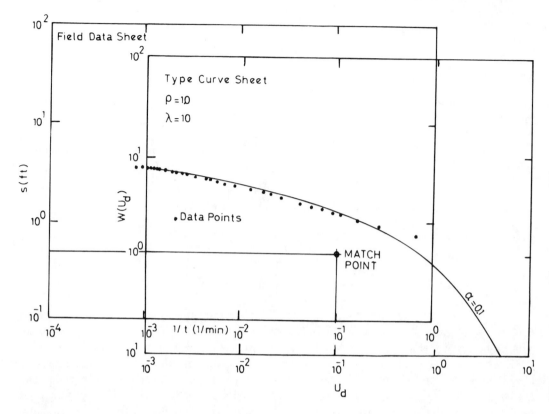

Figure 26 Field-data and type curve matching.

straight line is referred to as the Jacob line which is based on the Theis formula. However, the conditions for its application are more restricted than for the Theis method. For large values of time, t, or/and small values of radial distance, r, the classical Theis equation can be approximated with less than 5% error by the following expression:

$$W(u) = -Lnu - c \qquad (66)$$

or after simple algebraic manipulations one can find

$$W(u) = -Lnue^c \qquad (67)$$

or in terms of the ordinary logarithm

$$W(u) = 2.3 \log \frac{e^{-c}}{u} \qquad (68)$$

in which u and $W(u)$ are, respectively, the dimensionless time factor and the well function in the case of linear (Darcian) groundwater flow, and c is a constant equal to the Euler's factor, i.e., 0.5772. Equation 66 is valid for $u < 0.01$ and Jacob has shown that, if the drawdown, $s(r,t)$, in an observation well is plotted against the log of time since pumping started, then after some period the points will begin to describe a straight line as exemplified in Figure

27. The early points do not fall on the line because of the limitations of the Jacob modification. This is due to the fact that the implication of the Jacob method requires the aquifer test to extend over a minimum period of time (Walton, 1970). For confined aquifers, the time period is relatively few minutes, but in the case of unconfined aquifers it can be as much as several days. In most natural aquifer situations the minimum preferred time as determined by other considerations is more than sufficient to meet the requirements of the Jacob method. It is obvious from Equation 68 that the slope, $W(u)$, and the intercept, u_0, of the Jacob straight line is 2.3 and e^{-c}, respectively. The intercept for the linear flow case with $c = 0.5772$ is $u_0 = 0.5614$. In light of the Theis assumptions it is possible to derive a practical rule such that if the groundwater flow toward wells is linear, i.e., Darcian, then the dimensionless time drawdown plot of the late time-drawdown data on semilogarithmic paper must yield a straight line with slope equal to 2.3 (Şen, 1988d). Any deviation from this slope value warrants nonapplicability of the linear flow and therefore neither the Theis nor the Jacob methods are reliable in determining the aquifer constants. However, occurrence of a straight line on a semilogarithmic paper is taken for granted by many engineers as the prerequisite for the Jacob method application. In practice, due to its simplicity, engineers most often prefer the application of the Jacob method without giving any consideration to the slope value in a dimensionless plot. It is, therefore, not correct to make the statement that any straight line plot of late time-drawdown data on semilogarithmic paper justifies the use of the Jacob method. Instead, the appearance of a straight line on semilogarithmic paper implies that the flow is radial but not necessarily Darcian.

The expansion of the parenthesis on the right-hand side of Equation 33 into a convergent Binomial series leads

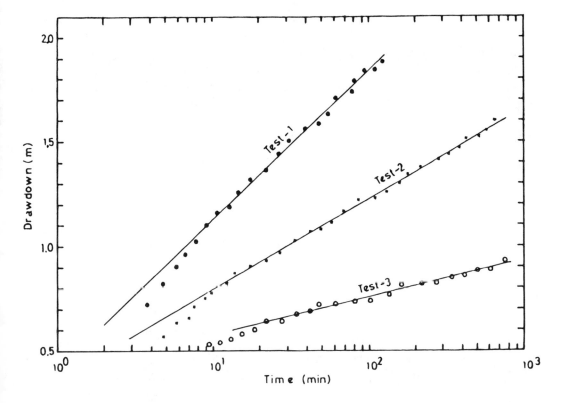

Figure 27 Late time-drawdown relationship.

$$W(u,v) = -c(v) - v\ln u + \frac{2n}{1!(n-3)} uv - \frac{2^2 n(2n-1)}{2!(n-3)^2} u^2 v^2 + \frac{2^3 n(2n-1)(3n-2)}{3.3!(n-3)^3} u^3 v^3$$
$$- \frac{2^4 n(2n-1)(3n-2)(4n-3)}{4.4!(n-3)^4} u^4 v^4 + \cdots - \quad (69)$$

in which $c(v)$ is a function depending on v only. For the Darcian flow case $n = 1$, which implies that $v = 1$ and hence this last expression becomes, keeping in mind that $c(1) = c$,

$$W(u) = -c - \ln u + u - \frac{u^2}{2.2!} + \frac{u^3}{3.3!} - \cdots + \cdots - \quad (70)$$

which was derived by Cooper and Jacob (1946). It is obvious from Equation 34 that u decreases as the time of pumping increases. Accordingly, for large values of t and/or small values of u the terms beyond $v\ln u$ in Equation 69 become negligible. Hence, for small values of u ($u < 10^{-9}$) the non-Darcian flow well function can be expressed asymptotically as

$$W(u,v) = -v L n u - c(v) \quad (71)$$

In fact, Equation 66 is a special form of this expression for the Darcian flow case with $n = 1$ which implies that $v = 1$ and correspondingly $c(1) = 0.5772$. Finally, in ordinary logarithm the straight line equation takes the form of

$$W(u,v) = 2.3v \log \frac{e^{-c(v)}}{u'^v} \quad (72)$$

This expression gives rise to a set of straight lines on semilogarithmic paper relating $W(u,v)$ to u for different v values. These lines will be referred to as type straight lines for nonlinear flow regime. It has been observed that type curve extensions from Equation 33 occur as straight lines on semilogarithmic paper for very small u' values, for $u < 10^{-9}$ (see Figure 28). Note that each straight line is labeled by two constant values, v and $c(v)$. On the other hand the slope and the intercept of the type straight lines can be found from Equation 71 as

$$\Delta W(u,v) = 2.3v \quad (73)$$

and

$$u_0 = [e^{-c(v)}]^{1/v} \quad (74)$$

respectively. Equation 73 indicates that the drawdown rate increases with the increase in the dimensionless turbulence factor, similarly with the turbulence exponent.

A. FLOW REGIME IDENTIFICATION

Aquifer parameter determination from large time-distance-drawdown data in non-Darcian flow can be achieved by straight line matching procedure on semilogarithmic paper. Such a procedure has two distinctive stages, namely, identification of the flow type, i.e., whether it is Darcian (linear) or non-Darcian (nonlinear) and if nonlinear then estimation of the aquifer parameters with the use of aforementioned type straight lines. There is no need to provide details for the Darcian flow because the classical Jacob method is described in standard hydrology textbooks.

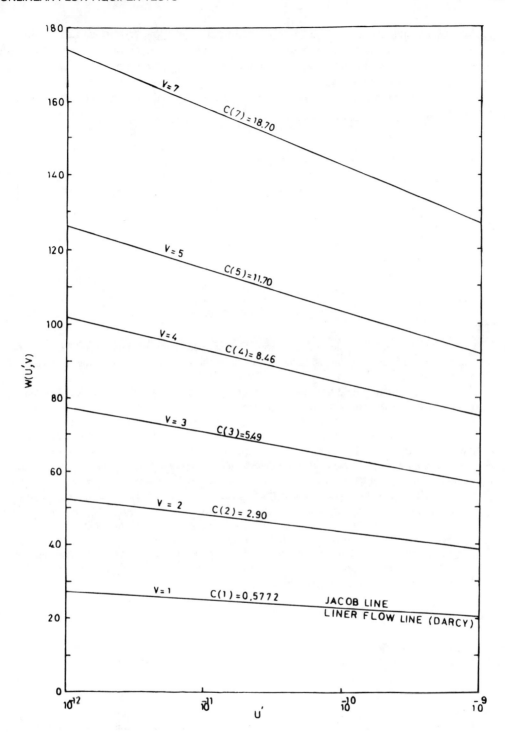

Figure 28 Type straight lines.

As mentioned above, any straight line on semilogarithmic paper implies radial flow and only the Jacob line indicates Darcian flow. Hence, in the first stage as a null hypothesis the groundwater flow is assumed to be Darcian with the implementation of the Jacob straight line method. The alternative hypothesis is that the groundwater flow is non-Darcian. The acceptance or rejection of this null hypothesis is achieved after the execution of the following steps in sequence.

1. Plot the large time drawdown data on a semilogarithmic paper.
2. Find the aquifer parameters, namely, S and T, by using the conventional Jacob method.
3. Calculate the dimensionless field values of time, u_f, and drawdown, w_f, values from the modified versions of the fundamental dimensionless time factor and the well function definitions as

$$u_f = \frac{r^2 S}{4 t_f T} \tag{75}$$

and

$$w_f = \frac{4\pi T}{Q} s_f \tag{76}$$

where t_f and s_f are the time and corresponding drawdown measurements in the field. Of course, the number of u_f and w_f values is equal to the number of measurements during an aquifer test.
4. Plot the dimensionless field time along the logarithmic axis vs. the dimensionless drawdown values on the same scale semilogarithmic paper as the type straight lines.
5. Draw the best straight line through the plotted dimensionless field time-drawdown points.
6. Compare this straight line with the Jacob straight line. Any difference in the slope suggests the rejection of the null hypothesis and as a conclusion the flow is non-Darcian. Consequently, the aquifer parameter estimates that are determined in step (2) are not representative, i.e., the Jacob method cannot be reliable for the set of data at hand. Otherwise, if the two slopes are practically the same then the null hypothesis is acceptable and the application of the Jacob method is justified. This procedure brings out the fact that any straight line match for the large time-drawdown data on a semilogarithmic paper does not guarantee the implication of the Jacob method. It is important to notice at this stage that the conclusion by Şen (1989) that the coincidence of late time nonlinear flow type curves with the Theis curve and the indication that the late portion of these type curves does not yield any clue about the flow regime is found to be rather misleading.

B. AQUIFER PARAMETER ESTIMATION

In a non-Darcian flow the three most important aquifer parameters are S, T', and n. In the case of null hypothesis rejection these parameters can be identified through the use of the type curves as presented by Şen (1988c, 1989). However, in practice, due to either the tediousness of type curve matching or availability of late time-drawdown data only or as a check of the type curve matching results or just for the quickness as well as convenience, the straight line method is preferred. In the nonlinear flow case the following steps are necessary for the parameter estimation.

1. Plot the dimensionless large time-drawdown values on a semilogarithmic paper.
2. Prepare straight line-type curves on the same scale logarithmic paper as in step (1) (see Figure 28).
3. Similar to any type curve matching, find the best matching straight line to the plotted data points. Read off the values of v and $c(v)$ from this type straight line.

4. Plot the late time field data on any semilogarithmic paper, fit a straight line, and extend it to reach the intercept on the time axis and finally read off the time intercept, t_0, which corresponds to zero dimensionless drawdown.
5. Find the slope, $(\Delta s)_t$, of the straight line for one full cycle of time on the logarithmic axis.
6. Calculate the nonlinear flow transmissivity value which is defined as $T' = m(K')^{1/n}$ from Equation 72 by its explicit expression through use of Equation 35 as

$$T' = \frac{2.3vQ}{4\pi(\Delta s)_t} \qquad (77)$$

7. With the values of v and T' it is possible to calculate the turbulence exponent, n, from Equation 35 and the storage coefficient from Equation 75 by considering Equation 36 as

$$n = 1 - \frac{\log v}{\log \dfrac{2\pi T' t}{Q}} \qquad (78)$$

and

$$S = \frac{4T't_0}{r^2}[e^{-c(v)}]^{1/v} \qquad (79)$$

respectively. Note that Equations 74 and 75 reduce to the classical Jacob equation for $v = 1$ for which $c(v)/v = 0.5772$.

Although all the aforementioned discussions were confined to the time-drawdown data, the type straight lines are usable equally for any other type of aquifer data such as the distance-drawdown or time-distance-drawdown composite data. The only difference occurs in the forms of analytical expressions. For instance, on the basis of the distance-drawdown data instead of Equations 77 and 79 the following expressions become valid:

$$T' = \frac{2.3vQ}{2\pi(\Delta s)_r} \qquad (80)$$

and

$$S = \frac{4T't}{r_0^2}[e^{-c(v)}]^{1/v} \qquad (81)$$

in which $(\Delta s)_r$ is the slope distance-drawdown data semilogarithmic paper plot with the distances on the logarithmic axis and r_0 is the intercept of the fitted straight line with distance axis. In fact, physically this distance corresponds to the radius of influence.

Finally, if the time-distance-drawdown data are used to implement the methodology developed in this paper, then the drawdown values are plotted against the composite variable in the form of ratio, t/r^2, on the semilogarithmic paper and the valid equations for the parameter estimations are

$$T' = \frac{2.3vQ}{4\pi(\Delta s)_{rt}} \qquad (82)$$

and

$$S' = 4T'\left(\frac{t}{r^2}\right)_0 [e^{-c(v)}]^{1/v} \quad (83)$$

where $(\Delta s)_n$ is the slope of large time-distance-drawdown straight line on a semilogarithmic paper and $(t/r^2)_0$ is the intercept on the (t/r^2) composite variable axis. However, the flow regime identification stage remains the same whatever the type of aquifer test data.

Example 5—Dudgeon et al. (1973) performed aquifer tests in southeast Queensland where the aquifer material is composed mainly of gravel and coarse-to-medium sand. In general, the overall material corresponds to the fine gravel domain. The non-Darcy flow has been observed in the vicinity of pumping well and therefore a nonlinear flow drawdown-discharge relation is expected to exist. They showed that the semilogarithmic plots of drawdown-distance data for large times yield a nonlinear flow pattern in the near well zone.

Table 4 Field and Dimensionless Data

Time (min) t_f (1)	Drawdown (ft) s_f (2)	Dimensionless Time u_f (3)	Dimensionless Drawdown w_f (4)
0	0	0	0
1	0.37		
2	0.63		
3	0.77		
4	0.88		
6	0.96		
8	1.05		
10	1.08		
15	1.29		
20	1.43		
25	1.51		
30	1.56		
40	1.62		
50	1.71	9.78×1^{-2}	2.30
60	1.79	7.32	2.41
80	1.93	5.49	2.60
106	2.10	4.14	2.83
122	2.14	3.60	2.88
150	2.28	2.92	3.08
180	2.37	2.44	3.20
215	2.47	2.04	3.34
241	2.52	1.82	3.40
303	2.64	1.45	3.57
360	2.79	1.22	3.77
415	2.88	1.05	3.89
480	3.13	9.15×10^{-3}	4.23
546	3.18	8.04	4.29
600	3.22	7.31	4.35
660	3.28	6.65	4.43
720	3.33	6.10	4.50
780	3.39	5.63	4.58
850	3.50	5.16	4.73
910	3.52	4.82	4.76
975	3.56	4.50	4.81
1050	3.63	4.28	4.90
1110	3.69	3.95	4.99
1180	3.74	3.72	5.05
1235	3.78	3.55	5.11
1293	3.81	3.40	5.15
1440	3.84	3.05	5.19

The aquifer was pumped continuously for 24 h with a constant discharge equal to 7050 gph (0.53 m³/min). The resulting time and corresponding drawdown values in an observation well are recorded as in the first two columns of Table 4. The distance of observation well to the main well is 12 ft (3.8 m) and the aquifer thickness has been found as 10 ft (3.2 m). The plot of time drawdown data on a semilogarithmic paper is presented in Figure 29. It is obvious that the large time-drawdown portion fits a straight line. The slope and the intercept of this line are $(\Delta s)_t = 1.7$ ft (0.52 m) and $t_0 = 7.8$ min, respectively. Hence, the transmissivity and storativity calculations from the Jacob equations as $T = 2.3Q/4\pi(\Delta s)_t$ and $S = 2.25t_0T/r^2$, respectively, yield $T = 0.186$ m²/min and $S = 0.227$. These values would be final results for an engineer who relies on the application of Jacob equations resulting in a straight line on semilogarithmic paper without thinking about the flow regime.

However, the methodology developed, herein, confirms from the straight line in Figure 29 that the flow is radial but the linearity of the flow is questionable. For this reason, the dimensionless time and drawdown values corresponding to the points around the straight line in Figure 29 are calculated from Equations 75 and 76. The numerical results are presented in columns 3 and 4 of Table 4.

The plot of dimensionless time on the logarithmic axis vs. the dimensionless drawdown on a semilogarithmic paper is given in Figure 30. In the same figure, the matching of type straight line is also shown. It is obvious that the slope of dimensionless field data straight line is greater than the Jacob straight line. This discrepancy in the slopes indicates that the flow is of nonlinear regime. The best matching type straight line has $v = 2.0$ and correspondingly $c(2) = 2.9$. With the substitution of these values at hand, into Equations 77 through 79, the necessary nonlinear flow aquifer parameters become $T' = 0.373$ m²/min, $n = 0.75$, and $S = 0.189$. Comparison of these results with the linear flow regime case indicates that the increase in transmissivity value is 50%, whereas in the storage coefficient about 83% decrease occurs. Hence, the differences in the parameter estimates based on linear and nonlinear flows are practically significant showing the necessity of checking the flow regime.

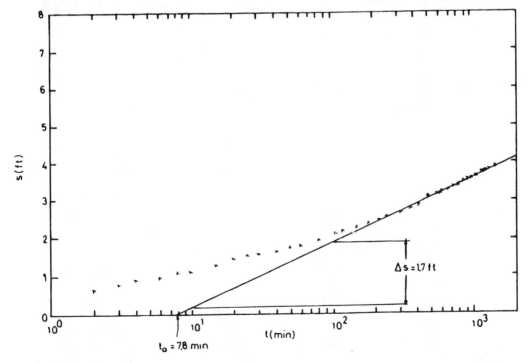

Figure 29 Field data plot.

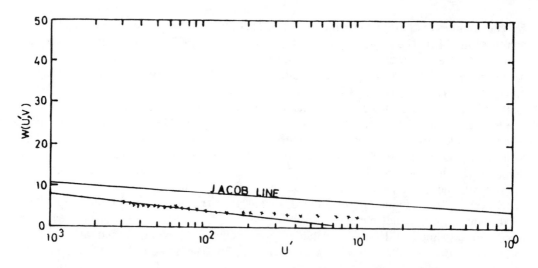

Figure 30 Dimensionless field data and type straight line matching.

VII. NONLINEAR FLOW TRANSMISSIVITY

Different definitions of transmissivity are presented in Chapter 5. The most widely used polynomial laws is due to Forchheimer (1901) as given by Equation 17 in Chapter 5 and if q is made subject the only physically plausible result emerges as

$$q = \frac{-a + \sqrt{a^2 + 4bi}}{2b} \tag{84}$$

the substitution of this into discharge equation in a cross section with width, w, and depth, m, gives

$$Q_F = wm \frac{-a + \sqrt{a^2 + 4bi}}{2b} \tag{85}$$

in which Q_F represents the Forchheimer discharge. This expression is completely different from the linear flow case given in Equation 18 in Chapter 4 and proves that in general the classical transmissivity definition is no longer valid for nonlinear flows. Only for very small b values does Equation 85 reduce

$$Q = wm \frac{i}{a} \tag{86}$$

in which case by considering that $K = 1/a$ this expression becomes equivalent to the linear flow transmissivity. Notice that for $b \to 0$, the flow regime approaches linear flow condition.

Similar deviations for any type of power law also indicate the invalidity of classical transmissivity in nonlinear flow regimes. For instance, the power law presented by Anandakrishnan and Varadarajulu (1963) leads to

$$Q_P = WD(k')^{1/n} i^{1/n} \tag{87}$$

After having proven that the transmissivity definition is not valid in nonlinear groundwater cases the question is how to decide that a nonlinear groundwater flow occurs in the flow domain and how to calculate the discharge that crosses through a given cross section.

The nonlinear transmissivity calculation plays an interim role in giving rise to some nonlinear flow parameters such as a and b in the Forchheimer and k' in the power law cases.

Example 6—Dudgeon et al. (1973) treated the aquifer data with numerical methods on the basis of the Forchheimer flow law. They calculated the relevant aquifer parameter estimations as $T_F = 2.85$ ft^2/min, $a = 0.285$ ft/min and $b = 6.5$ (min/ft)^2. The aquifer thickness is 10 ft. Let us find the actual discharge from a cross section, say, of width 5 ft under a hydraulic gradient 0.1. The traditional approach from Equation 17 in Chapter 5 leads to $Q = 1.425$ ft^3/min. However, by considering nonlinear flow Equation 85 gives after substitution of relevant quantities, $Q_F = 5.20$ ft^2/min. Hence, the linear flow transmissivity underestimates the actual discharge by almost 37% relative error.

Another example is presented for data from Section IV that were fitted by power law type curve resulting in aquifer parameters as $T_P = 200$ m^2/d with $n = 1.1$. If the conventional transmissivity definition is considered then the amount of groundwater flow through a cross section of 6-m width under hydraulic gradient 0.01 results from Equation 17 in Chapter 5 as $Q = 12$ m^3/d. However, the nonlinear flow consideration invalidates the conventional transmissivity usage in such conditions and rather Equation 87 must be adopted. Consequently, the substitution of relevant quantities into this equation gives $Q_P = 7.57$ m^3/d. In this case the linear flow over estimates the groundwater discharge by 59%.

REFERENCES

Albertson, M. L., Barton, J. R., and Simon, D. B., 1960. *Fluid Mechanics for Engineers*, Prentice-Hall, Englewood Cliffs, NJ.

Anandakrishman, M. and Varadarajulu, F. H., 1963. Laminar and turbulent flow of water through sand, *J. Soil Mech. Found. Eng.*, 89(5), 1.

Basak, P., 1977. Non-Darcy flow and its implications to seepage problem, *J. Irr. Drain. Div.*, ID-4, 459.

Bear, J., 1977. *Dynamics of Fluids on Porous Media*, Elsevier, New York.

Castany, G., 1967. *Traite Pratique des Eaux-Souterraines*, Dunell, Paris.

Cooper, H. H. and Jacob, C. E., 1946. A generalized graphical method for evaluating formation constants and summarizing well field history, *Trans. Am. Geophys. Union*, 27, 526.

Darcy, H., 1856. *Les Fountaines Publique de la Ville de Dijon*, Victor Dalmont, Paris.

Dugdeon, C. R., 1964. Flow of Water through Coarse Granular Materials. Report No. 76, University of New South Wales, New South Wales, Australia.

Dudgeon, C. R., Huyakorn, P. S., and Swan, W. N. C., 1973. Hydraulics of Flow Near Wells in Unconsolidated Sediments, University of New South Wales, New South Wales, Australia.

Engelund, F., 1953. On the Laminar and Turbulent Flows of Groundwater through Homogeneous Sand, Bull. No. 4, Technical University of Denmark, Copenhagen.

Fisher, V. C., 1935. Further tests on permeability with low hydraulic gradient, *Eos Trans.*, 16, 499.

Forchheimer, P. H., 1901. Wasserbewegung durch boden, *Zeitschrifft Ver. Dtsch. Ing.*, 49, 1736.

Gale, J. E., 1977. A Numerical Field and Laboratory Study of Flow in Rocks with Deformable Fractures, Scientific Series Environment Canada, Water Resources Branch, Montreal.

Huyokorn, P. and Dudgeon, C. R., 1962. Investigation of two-regime well flow, *J. Hydraul. Div.*, 100, 1119.

Kristianovich, S. A., 1940. Movement of groundwater violating Darcy's law, *J. App. Math. Mech.*, 4, 33.

Krusemann, G. P. and de Ridder, N. A., 1979. Analysis and Evaluation of Pumping Test Data. Int. Inst. Land Reclamation Bulletin, Wageningen, The Netherlands.

Meinzer, Q. E. and Fisher, V. C., 1934. Test of permeability with the hydraulic gradients, *Eos Trans.*, 15, 405.

Prinz, E., 1886. *Handbuch der Hydrologie*, Springer-Verlag, Berlin.

Sichard, W., 1929. Das Fassungsvermengen von bohrbrunnen, und seine Bedeutung fur grossere Abssenktiefen, Dissertation, Techniche Hochschule, Berlin.

Şen, Z., 1986. Volumetric approach to non-Darcy flow in confined aquifers, *J. Hydrol.*, 87, 337.

Şen, Z., 1988a. Analytical solution incorporating nonlinear radial flow in confined aquifer, *Water Resour. Res.*, 24(4), 601.

Şen, Z., 1988b. Type curves for two-regime well flow, *J. Hydraul. Eng.*, 114(12), 1461.

Şen, Z., 1988c. Dimensionless time-drawdown plots of late aquifer test data, *Ground Water*, 26(5), 615.

Şen, Z., 1988d. Dimensionless straight-type lines for aquifer tests, *Am. Soc. Civil Eng. J. Hydraul. Eng.*, 116(9), 1146.

Şen, Z., 1989. Nonlinear flow toward wells. *J. Hydraul. Div.*, 115(HY2), 193.

Şen, Z., 1990. Nonlinear radial flow in confined aquifer toward large-diameter wells, *Water Resour. Res.*, 26(5), 1103.

Theis, C. V., 1935. The relation between lowering of the piezometric surface and the rate and duration of discharge of a well using ground water storage, *Trans. Am. Geophys. Union*, Part 2, 519.

Todd, D. K., 1980. *Groundwater Hydrology*, John Wiley & Sons, New York.

Walton, W., 1970. *Groundwater Resources Evaluation*, McGraw-Hill, New York.

INDEX

Abstraction wells, 133
Acceleration gravity, 342
Adaptive models, 231–235, 288
Aeration (unsaturated) zone, 28
Aerial photographs, 115
Air lines, 142
Alluvial fans, 8–9
Alluvial fills, 9–11
Andesite, 20
Anhydride, 22
Anisotropic medium, 44
Anisotropic water-bearing layer, 82
Anisotropy, 87, 177, 303, 392
Aplites, 19
Approximate models, 260–264, see also specific types
Aquicludes, 29
Aquifers, 29, 69–88, 203, see also specific types
   assumptions about, 162
   barometric efficiency of, 156
   bounded, 88
   classifications of, 30–36
   composite hydraulic variables and, 82–86
   confined, see Confined aquifers
   confined-unconfined, 194–195
   continuous supply to, 62
   defined, 29
   depth variations in, 87
   discharge in, 37, 62, 174, 176
   extensiveness of, 87–88
   field measurements of, see Field measurements
   fractured, see Fractured aquifers
   fracture size distribution and, 77–82
   geometry of, 86–88, 161, 309
   grain size distribution in, 69–70
   heterogeneity of, 177, 288–299
   homogeneity of, 86–88, 136
   hydraulic conductivity and, 79–81, 83, 85
   hydraulic gradients and, 82, 87
   hydraulic properties of, 161, 203
   hydraulic resistance and, 85
   inflow through, 51
   input waters into, 63
   isotropy of, 86–88
   Karst, 111
   leakage factor and, 85–86
   leaky, see Leaky aquifers
   models of, 207–208
      for confined aquifers, see Confined aquifers, models of
      for leaky aquifers, see Leaky aquifers, models of
      for unconfined aquifers, 240–252
   multiple, 98, 135, 198–200
   outflow through, 51
   output waters into, 63
   parameters of, 91, 203
      dimensionless straight line method and, 426–429
      estimation of, 426–429
      relative values of, 205
   variables vs., 204–206
   patchy, 195–196
   perched, 33–35
   permeable water table, 135
   physical properties of, 205
   pitfalls in analysis of tests of, 299
   porosity and, 70–73, 79, 82
   potentiality of, 81
   recharge from, 37
   regional assessments of, 133
   retention and, 72–73
   specific capacity of, 377–378
   specific storage coefficients and, 73–77
   specific yield and, 72–73
   storage coefficients and, 73–77, 79, 80
   thickness of, 51, 52, 203, 209, 431
   transmissivity of, 83–84, 380
   types of, 30–36
   unconfined, see Unconfined aquifers
   uniformness of, 87–88
   unit-unconfined, 34
   vertical flow within, 135
Aquifuges, 29, 32–33
Aquitards, 29–30
Area of influence, 66
Askers, 12, 14
Atmospheric pressure, 154–156

Bailer test, 382–384
Barometric efficiency of aquifers, 156
Barometric pressure, 138
Barrenblatt et al. model, 310–313
Barrier effects, 270–279
Basalts, 20, 21, 71, 82
Bernouilli's theorem, 41
Bessel functions, 179, 268
Bierschenk method, 361–364
Blocks, 77, 78, 81, 82, 110, 303, 306
   discharge from to wells, 324
   drawdown in, 319
   finite, 319
   flow from to fractures, 111, 316
   fracture boundary with, 111
   fracture connection to, 303
   geometric idealizations of, 309
   geometry of, 304, 313
   groundwater flow within, 309–310
   hydraulic conductivity of, 110
   hydraulic heads of, 111
   recharge from, 307
   shape of, 303
   storage coefficient specific to, 111
   storativity of, 110, 307
   transmissivity of, 324
   type curves for, 319
   uniformly distributed, 311
   well connection to, 303
Bore hole logs, 6

Boulton delay index, 243
Boulton model, 242–246, 247
Boulton-Streltsova model, 317–321
Boulton-type curves, 243–245, 247
Boulton well function, 243
Boundary conditions, 91, 166–168
Bounded aquifers, 88
Breccia, 18
Buckling, 20

Capacity discharge, 358, 359
Capillary fringe, 28
Capillary-fringe drainage, 77
Cartesian coordinates, 137, 397
Cavity wells, 128
Cementation, 51, 71
Chalk, 26
Chemical solution, 25
Chow nomogram model, 225–226
Chow's function, 225–226
Circular cross sections of wells, 116–120
Clastic sedimentary rocks, 16–17, 71, see also specific types
Clays, 7, 80
Clouds, 40
Coastal plain deposits, 14–16
Collapse deposition, 14
Collector wells, 122–124
Compaction, 71, 74, 77
Composite variable model, 82–86, 220–222
Compression stress, 75
Conductivity
    defined, 79
    hydraulic, see Hydraulic conductivity
Confined aquifers, 32–34, 51
    adaptive models of, 231–235
    barometric pressure and, 138
    characteristics of, 75
    Chow nomogram model of, 225–226
    fully penetrating wells in, 168–174, 209
    horizontality of, 163
    Jacob model of, 213–225
    large-diameter wells in, 405, 406
    models of, 208–240
        adaptive, 231–235
        Chow nomogram, 225–226
        Jacob, 213–225
        Papadopulos and Cooper, 226–230
        slope matching method and, 235–240
        Theis, 209–213
    observation well distance and, 137
    overabstraction in, 194
    Papadopulos and Cooper model of, 226–230
    partially penetrating wells in, 190–194
    quasi-steady-state flow in, 171
    saturated portion of, 76
    slope matching method and, 235–240
    slug tests and, 385–386
    steady-state flow in, 168–174, 190–194, 377
    Theis model of, 209–213
    two-dimensional flow in, 61
    unsteady-state flow in, 63, 209
    water withdrawal from, 76
Confined-unconfined aquifers, 194–195
Conglomerates, 16
Connate water, 27
Conservation of energy, 37, 39
Conservation of mass, 37, 97
Consolidated sedimentary rocks, 21
Consolidation, 25
Constant energy, 50
Constant potential, 166
Container method, 144
Continuity equation, 209, 397, 405
Contractions, 91
Cooling joints, 20
Cracks, 20
Critical radius, 414
Crystalline rocks, 19, 303, 341, see also specific types
Cubic law, 343

Darcy flow, see Laminar flow
Darcy's law, 6, 94, 109, 164
    application of, 95–100
    defined, 403
    Forchheimer law combined with, 414
    fractured aquifers and, 310
    generalization of, 95
    in granular rocks, 111
    of groundwater flow, 93–100
    invalidity of, 410
    linear flow, 93–100
    in porous medium, 108
    Reynolds number and, 395
    for steady-state flow, 168
    two-regime flow and, 412, 413, 421
    validity of, 95, 102, 204, 410
Darcy velocity, 49
Darcy-Weisbach formula, 92
Darcy-Weisbach relation, 107
Decomposition, 51
Deep percolation, 137
DeGlee representation, 191
Deltas, 12–13, 15
Depression, 47
Depression cones, 66–67, 292, 293
Dewatering, 196
Diffusion equations, 310
Dikes, 37, 46
Dimensionless drawdown, 429
Dimensionless straight line method, 222–225, 421–429
    aquifer parameter estimation and, 426–429
    flow regime identification and, 424–426
Diorite, 19
Direct bore hole logs, 6
Discharge, 37, 47–50
    actual, 431
    aquifer, 176
    in aquifers, 37, 62, 174, 176
    block-to-fracture, 111
    block-to-well, 324
    calculations for, 152–154, 187

INDEX   **435**

capacity, 358, 359
defined, 144, 210
drawdown and, 355, 357, 358, 421, 428
equations for, 430
estimation of, 154
exponential law and, 409
field measurements and, 139
formula for, 196
fracture-to-well, 324
measurement of, 144–154
penetration ratio to, 193
predictions of, 203
pumping, 164, 168, 413, 414
at radial distance, 174
small, 216
specific, see Specific discharge
specific capacity and, 358
time-independent, 168
turbulent flow and, 395
well, 144–154
Discharge-time drawdown, 207
Distance-drawdown, 180–181, 206, 283, 400, 401, 404, 416, 421, 428
model of, 218–219
Distance-slope model, 261–263
Ditches, 161
Dolomite, 22, 82
Dolomitic rocks, 26
Double-porosity models, 79, 304, 305, 308, 310–324
Barrenblatt et al., 310–313
Boulton-Streltsova, 317–321
Kazemi et al., 315–317
SEN, 319–324
Warren-Root, 311–315, 317
Drainage, 77, 135, 242
Drawdown, 66, 76, 139, 144, 201
applicability of, 137
in blocks, 319
change in, 216
confined aquifers and, 169
corrected, 155, 187
defined, 137, 398
dimensionless, 429
discharge and, 355, 357, 358, 421, 428
discharge-time, 207
distance-, 180–181, 206, 218–219, 283, 400, 401, 404, 416, 421, 428
distribution of, 168
early time, 153
large, 359
maximum, 161, 378
measurement of, 137–143, 156, 174, 293, 376
in observation wells, 329
piezometers and, 322, 331
predictions of, 203
proportionality of, 205
during pumping periods, 142
in pumping wells, 330
quasi-steady-state, 377
at radial distance, 168
recovery, 140

residual, 140
specific capacity and, 357
square of, 175
steady-state, 182, 377
step, see Step drawdown
time-, see Time-drawdown
time-distance-, 215, 424, 427, 428
total, 194, 357, 416
two-regime flow and, 415, 416
zero, 308
Drilling logs, 115
Dunes, 11, 32
Dupuit-Forchheimer assumptions, 164–165, 174, 185
Dupuit-Forchheimer equation, 174, 177
Dupuit-Forchheimer expression, 175
Dynamic piezometric levels, 137–139

Early time drawdown, 153
Eden-Hazel method, 364–369
Elastic compaction, 77
Elastic storativity, 248
Electrical analog model, 192
Electrical resistivity, 6
Electric tape, 141–142
Elevation head, 42
Energy
conservation of, 37, 39
constant, 50
cycle of, 39–40
groundwater, 40–43
heat, 38, 41, 50
kinetic, 38–41, 91, 92, 101
loss of, 37, 39, 50, 92
increase in, 359
total, 355
turbulence and, 359
wells and, 115–117
well tests and, 359
potential, 38–41, 50, 91
specific, 41, 42
total, 39, 41
total loss of, 355
of water, 38–39
zero, 40
Energy transformation zone, 40
Equilibrium equations, 87
Equivalent porous medium, 110–111
Erosion, 17
Escande formulation, 105
Euclidean geometry, 304
Euler's factor, 422
Evaporite, 26
Evapotranspiration, 154, 155
Expansions, 91
Exponential law, 407–412
Extended wells, 121
Extensiveness condition, 168
Extensiveness of aquifers, 87–88
Extensive planar fractures, 121
Extrusive igneous rocks, 20

Fans, 8–9
Faults, 17–19, 37, 46, 47, 53, 78
Ferris method, 273–274
Field measurements, 133–159
  air lines in, 142
  container method in, 144
  discharge measurement and, 144–154
  drawdown measurements and, 137–143
  electric tape in, 141–142
  flat-bottomed weight in, 141
  graphical data and, 154–159
  jet stream method in, 146–147
  mercury meters in, 142–143
  observation wells and, 134–137
  orifice meters in, 145
  Parshall flume in, 147–148
  piezometers and, 134–137
  procedures in, 206–207
  time-drawdown, 203–205, 337
  water meters in, 144–145
  weirs in, 148–152
  wetted tape in, 140–141
Filling, 79
Fills, 9–11
Filter velocity, 49
Finite blocks, 319
Fissures, 17, 18, 78, 110, 319
Fixed hydraulic gradients, 61–62
Flat-bottomed weight, 141
Flow, see Groundwater flow
Flow exchange law, 309
Flow regime identification, 424–426
Fluid viscosity, 342
Force classification of flow, 64–65
Forchheimer equation, 421
Forchheimer flow, 401, 404
Forchheimer law, 401, 413
Fossil water, 27
Fractured aquifers, 303–351, see also Fractured medium
  anisotropy of, 303
  Barrenblatt et al. model of, 310–313
  Boulton-Streltsova model of, 317–321
  double-porosity models of, see Double-porosity models
  Gringarten-Witherspoon model of, 325–329
  idealization models of, 304–306
  Jenkins-Prentice model of, 329–331
  Kazemi et al. model of, 315–317
  natural, 315
  permeability of, 305
  pumping and, 306
  qualitative physical interpretation of, 306–309
  SEN model of, 319–324, 331–338
    application of, 337–338
  simplifying assumptions on, 309–310
  Warren-Root model of, 311–315, 317
Fractured medium, 17–21, see also Fractured aquifers; specific types
  idealization models of, 304–306
  laws of, 105–111
Fractures, 20, 46, 53, 54, 91
  apertures of, 79, 92, 101, 303, 343, 344
  block connection to, 303
  classification of, 79
  connectedness of, 303
  defined, 17
  degree of connectedness of, 303
  development of, 303
  direction of, 303
  discharge from to wells, 324
  enlarged, 80
  extension of, 303
  extensive planar, 121
  extent of, 71
  features of, 18
  geometric idealizations of, 309
  geometry of, 304
  in granite, 19
  Gringarten-Witherspoon model of, 325–329
  groundwater flow and, 105–111, 309–310
  horizontal, see Horizontal fractures
  hydraulic conductivity of, 18, 80, 81–82
  hydraulic properties of, 319, 339
  hyudraulic conductivity of, 110
  inclined, 325
  individual, 106–109
  Jenkins-Prentice model of, 329–331
  of limestone, 22
  long, 59
  of metamorphic rocks, 22
  models of, 324–338
    Gringarten-Witherspoon, 325–329
    Jenkins-Prentice, 329–331
  network of, 303
  number of, 303
  one-dimensional flow in, 106, 341
  open, 18–19
  patterns of, 21, 22
  porosity of, 18, 78–79, 303
  randum distribution of, 311
  Reynolds number in, 107
  roughness of, 107, 109, 341, 343, 344
  in sandstones, 54
  SEN model of, 331–338
  size of, 303
  size distribution of, 77–82
  solution, 26
  spacing of, 78
  steady-state flow toward or away from, 161
  storage capability and, 25
  storativity of, 110
  surface roughness of, 339
  transmissivity of, 324
  two-dimensional flow in, 58, 107, 339
  type curves for, 319
  types of, 78
  vertical, see Vertical fractures
  water flow in, 82
  water table and, 55, 56
  well connection to, 303
  well intersection with, 325
  wells and, 118, 303, 325

INDEX 437

Frequency distribution, 103
Friction, 39
Friction coefficients, 92, 107
Fully penetrating wells, 127, 168–184
    in confined aquifers, 168–174, 209
    in leaky aquifers, 178–184
    in unconfined aquifers, 174–178

Gabbro, 19
Gaussian distribution, 103, 233
Geometry
    of aquifers, 86–88, 161, 309
    of blocks, 304, 313
    Euclidean, 304
    of fractures, 304
    idealized flow, 335
    of partial penetration, 388
    of piezometric surfaces, 161
Glacial deposits, 11–12, 15
Gneiss, 21
Gradient flow, 310
Grain sizes, 7, 69–70
Granite, 19, 341
Granular porosity, 20
Granular rocks, 111, see also specific types
Graphical data treatment, 154–159
Gravels, 16–17, 70
Gravity, 73–75, 342
Gravity drainage, 135, 242
Green's function, 326
Gringarten-Witherspoon model, 325–329
Groundwater
    continuous monitoring of level of, 134
    flow of, see Groundwater flow
    geology and, 4–5
    hydrology and, 2–4
    monitoring of level of, 134, 135
    origin of, 27
    radial, 342
    regional trends in levels of, 154
    scientists and, 5–6
    seepage of, 59
    sources of, 27
    space classification of, 55–59
    velocity of, 47–50, 56, 65, 95
Groundwater basins, 37
Groundwater energy, 40–43
Groundwater flow, 3, 27, 31, 37–67, 69, 91–112
    assumptions about, 162–165, 309–310
    basis of laws of, 91–92
    block-to-fracture laws of, 111
    within blocks, 309–310
    classification of, 104
    constancy of, 61
    Darcy, see Laminar flow
    Darcy's law of, 93–100
    depression cones and, 66–67
    direction of, 55
    discharge and, 37, 47–50
    energy cycle and, 39–40
    force classification of, 64–65
    fractured medium laws and, 105–111
    within fractures, 309–310
    gradient, 310
    groundwater energy and, 40–43
    horizontal, 306
    hydraulic gradients and, 43–47
    interporosity, 313, 316
    karstic medium laws of, 111–112
    laminar, see Laminar flow
    linear, see Laminar flow
    non-Darcian, 102, 421, 424, 426, 428
    nonlinear, see Turbulent flow
    one-dimensional, 58–59, 106, 341
    piezometric surface and, 43
    polynomial laws of, 103–104
    power laws of, 104–105
    pseudo-steady-state, 64, 313
    quasi-steady-state, 63–64, 133, 140, 171, 176, 201
    regional, 61
    retardation of, 347
    Reynolds number and, 65–66, 92
    rough-laminar, 108
    rough-turbulent, 108
    smooth-laminar, 106
    smooth-turbulent, 108
    space classification of groundwater and, 55–59
    specific discharge and, 47–50
    spheric, 60
    steady-state, see Steady-state flow
    theory of, 303
    three-dimensional, 56–57
    time classification of, 59–64
    turbulent, see Turbulent flow
    two-dimensional, 57–58, 61, 107, 339
    two-regime, 412–421
    unsteady-state, see Unsteady-state flow
    velocity of, 91
    vertical, 135, 136
    water energy and, 38–39
Groundwater flow nets, 50–55
Groundwater potential, 69
Groundwater reservoirs, see Reservoirs
Groundwater velocity, 95
Gypsum, 22

Halite, 22
Hantush-Jacob model, 253–260
Hantush-Theis method, 263–267
Hantush-type curves, 282, 294, 295
Heat energy, 38, 41, 50
History, 1–2
Horizontal fractures, 324, 325, 339–351
    apertures of, 343, 344
    dimensionless time factor and, 344
    hydraulic conductivity of, 341
    hydraulic properties of, 339
    multi-, 341–351
    piezometric levels and, 341
    roughness of, 341, 343, 344
    two-dimensional flow in, 339
Hydraulic conductivity, 18, 51, 79–81, 83, 85, 105, 203

aquifer thickness and, 203
average, 97, 99
block, 110
confined aquifers and, 168
defined, 79–80
determination of, 203
dimensionless, 319
estimation of, 136
field measurements and, 136, 154
fracture, 110
of fractures, 18, 80–82
groundwater flow laws and, 92, 96, 97, 100, 101
high, 324–325
horizontal, 189
of horizontal fractures, 341
hydraulic gradients and, 397, 398
leaky aquifers and, 178, 181
Packer test and, 389, 391
patchy aquifers and, 195
porous medium, 93
slug tests and, 387
Theis model and, 209
turbulent flow and, 395, 397, 401
two-regime flow and, 418
unconfined aquifers and, 176, 177
unsteady-state flow and, 401
as velocity, 80
vertical, 100
as volume, 81
Hydraulic engineering, 5
Hydraulic gradients, 47, 51, 54, 66, 80, 82, 87, 102, 165
changes in, 109, 165
defined, 44, 46
field measurements and, 154
field setup for, 94
fixed, 61–62
groundwater flow and, 43–47
groundwater flow laws and, 91, 92, 94, 96, 98, 99, 101
horizontal, 99
hydraulic conductivity and, 397, 398
increase in, 407
in recharge areas, 45
specific discharge and, 92, 100, 397, 412, 421
timewise, 47
turbulent flow and, 395, 397, 430–431
two-regime flow and, 412–413
velocity and, 107
well tests and, 359
zero, 166
Hydraulic heads, 47, 75, 99, 111
block, 111
changes in, 44, 91, 140
corrections in, 156
differences in, 93
field measurements of, 50
horizontal fractures and, 342
losses of, 44, 99
measurements of, 134
relationship between, 43
rise and fall in, 155
total, 43

variation in, 95
Hydraulic resistance, 85, 86, 183, 359
Hydraulic variables, 82–86
Hydrodynamics, 37
Hydrogeophysical concepts, 282–288
Hydrographs, 155
Hydrological cycle, see Groundwater flow
Hydrostatic pressure, 42, 73–75, 341
Hydrostatics, 37
Hydrostratigraphy, 30

Ice, 12
Idealization models, 304–306
Idealized flow geometry, 335
Igneous rocks, 7, 18–20, 71, 303, see also specific types
Impermeable boundaries, 166–167
Impermeable reservoirs, 7
Impermeable rocks, 80
Inclined fractures, 325
Infiltration, 137
Inflection point model, 260–262, 282, 314, 408
Influence area, 66
Influence radius, see Radius of influence
Initial conditions, 166
Initial piezometric levels, 138
Interdisciplinary approach, 5
Interporosity flow, 313, 316
Intrusive igneous rocks, 19–20
Irregular cross sections of wells, 125–126
Isotropy, 86–88

Jacob model, 213–225, 282, 359, 379
  composite variable, 220–222
  dimensionless straight line method and, 222–225
  distance-drawdown and, 218–219
  in step drawdown test analysis, 360–361
  time-drawdown and, 216–218
Jenkins-Prentice model, 329–331
Jet stream method, 146–147
Joints, 19, 46, 71, 78, 82, 109
  cooling, 20
  defined, 17
  sheet, 21

Kalman algorithm, 234
Kalman filter technique, 231, 233, 234
Kalman gain elements, 233
Kames, 12, 15
Kanats, 124–126
Karst, 22, 23
Karst aquifers, 111
Karstic formations, 82, 127, 134, see also specific types
Karstic medium, 21–26, see also specific types
  flow laws for, 111–112
  piezometric levels in, 56
  piezometric surface in, 55
Karstic reservoirs, 21–26
Kazemi et al. model, 315–317
Kinematic viscosity, 19, 82, 107, 413
Kinetic energy, 38–41, 91, 92, 101
Kozeny's approach, 192

INDEX

Laminar flow, 64–65, 95, 108, 111, 209, 210, 223, 426
   coefficients of, 103
   Darcian, 95
   resistance and, 359
   rough-, 108
   turbulent flow vs., 102–103, 107
Laplace transformations, 310, 395
Large-diameter wells, 116–119, 136, 153, 226, 229
   in confined aquifers, 405, 406
   type curve matching for, 230–232
Leakage, 85–86, 135, 179, 203, 205, 207
   calculation of, 183
   classification of, 86
   coefficients of, 181
   potential, 184
   small, 205
   specific, 85
Leaky aquifers, 34–36, 62, 85, 223
   approximate models of, 260–264
   distance-drawdown plot for, 283
   field data on, 257
   fully penetrating wells in, 178–184
   Hantush-Jacob model of, 253–260
   Hantush-Theis model of, 263–267
   models of, 253–270
      approximate, 260–264
      Hantush-Jacob, 253–260
      Hantush-Theis, 263–267
      Neuman-Witherspoon, 264–267
      volumetric, 269–270
   Neuman-Witherspoon model of, 264–267
   steady-state flow in, 178–184, 282
   volumetric models of, 269–270
Least-squares method, 361
Leibnitz's rule, 293
Limestones, 7, 22, 23, 26, 82, 293, 303
Limiting conditions, 166–168
Linear flow, see Laminar flow
Longitudinal velocity, 58

Magmatized gneiss, 21
Main wells, 133–134, 137, 203
Manometers, 142
Marble, 21
Marly limestones, 26
Mass conservation, 37, 97
Mathematical models, 203, 208, see also specific types
Matrix, 78, 79, 82, 303, 330
Maximum drawdown, 161, 378
Mechanical analysis, 70
Mercury meters, 142–143
Metamorphic rocks, 10, 19, 21, 22, 71, see also specific types
Meteoric water, 27
Mica, 87
Miller-Weber model, 373–376
Minerals, 20, 51, 80, 82, 87
Models, see also specific types
   adaptive, 231–235, 288
   approximate, 260–264
   aquifer, see under specific types of aquifers

Barrenblatt et al., 310–313
Bierschenk, 361–364
Boulton, 242–247
Boulton-Streltsova, 317–321
Chow nomogram, 225–226
composite variable, 82–86, 220–222
of confined aquifers, see Confined aquifers
of distance-drawdown, 218–219
distance-slope, 261–263
double-porosity, see Double-porosity models
Eden-Hazel, 364–369
electrical analog, 192
of fractured medium, 304–306
Gringarten-Witherspoon, 325–329
Hantush-Jacob, 253–260
Hantush-Theis, 263–267
idealization, 304–306
inflection point, 260–262, 282, 314, 408
Jacob, see Jacob model
Jenkins-Prentice, 329–331
Kazemi et al., 315–317
of leaky aquifers, see Leaky aquifers
mathematical, 203, 208, see also specific types
Miller-Weber, 373–376
Neuman, 246–252
Neuman-Witherspoon, 264–267
Papadopulos and Cooper, 226–230, 356, 407
random double-porosity, 304
regular double-porosity, 305
Rorabough, 279, 360, 369–371
SEN, 319–324, 331–338
   application of, 337–338
Sheahan type-curve, 371–373
slope matching, 235–240, 268–269, 288
Theis, 209–213
of time-drawdown, 216–218
of unconfined aquifers, see Unconfined aquifers
volumetric, 269–270
Warren-Root, 311–315, 317
Multi-horizontal rough fractures, 341–351
Multiple aquifers, 98, 135, 198–200
Multiple wells, 196–198

Neuman model, 246–252
Neuman straight line method, 248–252
Neuman-type curves, 249, 250
Neuman-Witherspoon model, 264–267
No flow zone, 95
Non-Darcian flow, 102, 421, 424, 426, 428, see also specific types
Non-Darcian laminar zone, 95
Nonlinear flow, see Turbulent flow
Nonlinear pressure flow laws, 101
Nonpenetrating wells, 128–129, 135
Nonpumping aquifer tests, 380–384
Nonreservoirs, 7, see also specific types
Nuclear waste, 303

Oases, 23, 25
Observation wells, 129–132, 189
   distance-slope model and, 261

distances to, 137, 138
  drawdown in, 329
  large-diameter, 136
  location of, 133
  in multiple aquifers, 135
  piezometers and, 134–137
  single-aquifer, 130
  small-diameter, 136
  time-drawdown data for, 275
  time-drawdown field measurements in, 203
  transmissivity between, 181
  two-aquifer, 130
  type curves and, 419
One-dimensional flow, 58–59, 106, 341
Open fractures, 18–19
Orifice meters, 145
Outwash sediments, 12
Oxidation, 51

Packer test, 389–392
Papadopulos and Cooper model, 226–230, 356, 407
Parshall flume, 147–148
Partially penetrating wells, 127–128, 135, 184–194
  in confined aquifers, 190–194
  steady-state flow in, 184–194
  in unconfined aquifers, 186–190
Partial penetration, 194, 388
Particle size, 11, 16
Patchy aquifers, 195–196
Perched aquifers, 33–35
Percolation, 51, 137
Permeability, 51, 81, 87, 93, 303, 305, 307
Permeable reservoirs, 7
  fractured, 17–21
  karstic, 21–26
  porous, 7–17, 22
Permeable rocks, 80
Permeable water table aquifers, 135
Petrographical properties, 20
Photogeology, 4
Phyllite, 21
Piezometers, 131, 198, 331
  drawdown and, 322, 331
  observation wells and, 134–137
Piezometric levels, 37, 45, 46, 51, 55, 56, 66
  barometric pressure and, 138
  composite, 198
  discontinuities in, 355, 356
  dynamic, 137–139
  horizontal fractures and, 341
  initial, 138
  in karstic media, 56
  measurement of, 139
  reduction of, 196
  rises and falls in, 57
  static, 137–140, 209
  time intervals in measurement of, 139
Piezometric surface, 43, 50, 55, 66, 161
Polynomial flow laws, 103–104
Pore spaces, 7
Porosity, 18, 20, 50, 70–73, 79, 203
  defined, 70
  double-, see Double-porosity models
  effective, 395
  fracture, 18, 78–79, 303
  granular, 20
  of igneous rocks, 71
  of metamorphic rocks, 71
  primary, 71, 79, 82, 342
  secondary, 21, 71, 304
  single-, 306
  total, 71, 72
  vesicular, 20
Porous medium, 7–17, 22, 203–299, see also specific types
  alluvial fans as, 8–9
  alluvial fills as, 9–11
  aquifer paramters vs. variables and, 204–206
  barrier effects and, 270–279
  characteristics of, 93–100
  clastic sedimentary rocks as, 16–17
  coastal plain deposits as, 14–16
  Darcy's law and, 93–100, 108
  delta deposits as, 12–13
  equivalent, 110–111
  field test procedures and, 206–207
  general information on, 203–204
  glacial deposits as, 11–12, 15
  groundwater flow and, 53, 93–105
    Darcy's linear flow law, 93–100
    nonlinear flow laws, 100–105
  hydraulic conductivity of, 93
  hydraulic parameters of, 82
  hydrogeophysical concepts and, 282–288
  nonlinear flow laws, 100–105
  sand dunes as, 11
  storage coefficient determination and, 279–282
Postlinear non-Darcian laminar zone, 95
Potential energy, 38–41, 50, 91
Potential leakage, 184
Power laws, 104–105, 431
Precipitation, 71, 154
Prelinear non-Darcian laminar zone, 95
Primary porosity, 71, 79, 82, 342
Proportionality coefficient, 94
Pseudo-steady-state flow, 64, 313
Pumping discharge, 164, 168, 413, 414
Pumping wells, 129, 200, 240, 330

Quantitative analysis, 69, 78, see also specific types
Quartzite, 21
Quasi-steady-state drawdown, 377
Quasi-steady-state flow, 63–64, 133, 140, 171, 176, 201

Radial distance ratio, 406
Radial flow, 342
Radius of influence, 67, 161, 177, 178, 200–201
  storage coefficients and, 205
  turbulent flow and, 398
Random double-porosity model, 304
Recharge, 37, 45, 63, 87, 307
Recovery drawdown, 140

INDEX

Recovery period, 140
Recovery test analysis, 378–380
Recrystallization, 71
Rectangular cross sections of wells, 120–122
Rectangular weirs, 151–152
Reduction factor, 187
Reflection, 6
Refraction, 6
Regional groundwater flow, 61
Regular double-porosity model, 305
Reservoirs, 7–36, 75, 79
    aquifers and, 29–36
    defined, 7
    fractured, 17–21
    impermeable, 7
    karstic, 21–26
    permeable, see Permeable reservoirs
    porous, 7–17, 22
    setup of, 86
    storage capability and, 79
    subsurface water and, 26–27
    water-bearing formations and, 29–30
    water origins and, 27
    water released from, 76
    water storage capability of, 79
    water zones and, 28
Residual drawdown, 140
Resistance, 85, 86
    from flow medium, 37
    of grains, 79
    hydraulic, 85, 86, 183, 359
    laminar flow and, 359
    medium, 47
    from solids, 7, 8
Retention, 72–73
Reynolds number, 410
    critical, 103, 107, 413, 415, 421
    critical radius and, 414
    Darcy's law and, 395
    in fractures, 107
    groundwater flow and, 65–66, 92
    high, 395
    laminar vs. turbulent flow and, 102–103
    threshold, 92
    two-regime flow and, 414, 421
Rorabaugh method, 279, 360, 369–371
Rough-laminar flow, 108
Roughness, 92
    of fractures, 107, 109, 341, 343, 344
    of fracture surfaces, 339
    of horizontal fractures, 341
    patterns of, 109
    relative, 107
    surface, 79
Rough-turbulent flow, 108

Sand dunes, 11, 32
Sands, 95
Sandstones, 7, 16, 17, 54, 289, 303
Sandy loams, 80
Satellite images, 115
Saturated zone, 40
Saturation thickness, 83, 95, 165, 174, 187, 192, 240, 241, 399
Saturation zone, 28, 31, 62, 115
Secondary porosity, 21, 71
Sedimentary rocks, 16–17, 21, 71, 81, 87, 303, see also specific types
Sedimentation, 25
Sedimentology, 4
Sediments, 12, 22
Seepage, 59
Seismic reflection, 6
Semi-logarithmic paper, 159
Semipermeable rocks, 80
SEN model, 319–324, 331–338
    application of, 337–338
Shales, 17
Sheahan type-curve method, 371–373
Shear zones, 18
Sheet joints, 21
Shist, 21
Shrinkage cracks, 20
Siltstones, 16, 17
Simpson's numerical integration method, 417
Single-aquifer observation wells, 130
Single-porosity, 306
Sinkholes, 23, 24
Slate, 21
Slope matching method, 235–240, 268–269, 288
Slug tests, 384–388
Small-diameter wells, 118–120, 136
Smooth-laminar flow, 106
Smooth-turbulent flow, 108
Soil moisture zone, 28
Solar radiation, 40
Solid boundary conditions, 91
Solids, 7, 8, 69, 75, see also specific types
Soluble rocks, 22, see also specific types
Solution cavities, 25, 46, 91
Solution fractures, 26
Spacing, 79
Specific capacity, 357, 358, 377–378
Specific discharge, 47–50, 55–56, 58, 65, 79–80, 82, 210
    critical, 413
    estimation of, 154
    groundwater flow laws and, 91
    hydraulic gradients and, 92, 397, 412, 421
    increase in, 412
    small, 216
    time-independent, 168
    turbulent flow and, 395
    two-regime flow and, 413
    well tests and, 359
Specific discharge hydraulic grandients, 100
Specific energy, 41, 42
Specific leakage, 85
Specific retention, 73
Specific storage coefficients, 73–77
Specific yield, 72–73, 75, 79, 203
Spheric groundwater flow, 60
Springs, 47

Stagnant water zone, 53
Stallmann method, 274–278
Stallman-type curves, 275, 278
Static piezometric levels, 137–140, 209
Steady-state drawdown, 182, 377
Steady-state equations, 198
Steady-state flow, 61–64, 94, 109, 110, 136, 161–201
   aquifer assumptions and, 162
   assumptions about, 161–166
   boundary conditions and, 166–168
   in confined aquifers, 168–174, 190–194, 377
   in confined-unconfined aquifers, 194–195
   Darcy's law for, 168
   away from and toward ditches, 161
   Dupuit-Forchheimer assumptions and, 164–165, 174, 185
   field measurements and, 140
   away from and toward fractures, 161
   initial conditions and, 166
   in leaky aquifers, 178–184, 282
   limiting conditions and, 166–168
   multiple aquifers and, 198–200
   multiple wells and, 196–198
   in partially penetrating wells, 184–194
   in patchy aquifers, 195–196
   practical significance of, 161
   significance of, 161
   specific capacity and, 377
   turbulent, 397–401
   in unconfined aquifers, 174–178, 186–190
   well assumptions and, 165–166
   away from and toward wells, 161
   well tests and, 355
   yields of, 413
Step drawdown test analysis, 360–376
   Bierschenk method in, 361–364
   Eden-Hazel method in, 364–369
   Jacob method in, 360–361
   Miller-Weber model in, 373–376
   Rorabaugh method in, 369–371
   Sheahan type-curve method in, 371–373
Storage capability, 25, 79
Storage capacity ratio, 316
Storage coefficients, 73–77, 79, 80, 203, 204, 209, 234
   block-specific, 111
   calculation of, 268
   defined, 76
   determination of, 223, 279–283, 338
   dimensionless straight lines and, 427
   estimation of, 240
   fractured aquifers and, 330
   radius of influence and, 205
   during recovery, 379
   unsteady-state flow and, 401
   variation in, 290
   well tests and, 357
Storativity, 76, 77, 81, 204, 207, 380
   block, 110, 307
   calculation of, 252
   changes in, 293
   elastic, 248

   error variance for, 234
   fracture, 110
   high, 304
   increase in, 295, 299
Streamlines, 50, 53, 54, 56, 60, 185
Streamtubes, 53
Stress, 75, see also specific types
Structural geology, 4
Subsurface water, 26–27, see also Groundwater
Surface roughness, 79
Surface tension, 73
Swamps, 126
Swelling, 20
Syenite, 19
Syncline, 34

Tectonic forces, 17, 19, 22, 82
Terraces, 12
Theis assumptions, 408
Theis equation, 273, 293
Theis model
   for confined aquifers, 209–213
   recovery and, 379–380
Theis-type curves, 211, 213, 260, 277, 329, 408–411, 417
   fractured aquifers and, 306
   modified, 324
Thiem equation, 175, 186, 386
Thiem method, 171
Thiem solution, 399
Three-dimensional flow, 56–57
Tides, 154
Tills, 12
Time classification of flow, 59–64
Time-distance-drawdown, 215, 424, 427, 428
Time-drawdown, 207, 326, 410, 416, 426, 427
   curves for, 212
   dimensionless data on, 288
   distance behavior of, 330
   field measurements of, 203–205, 337
   model of, 216–218
   in observation wells, 275
   patterns of, 307
Tortuosity, 48
Total energy, 39, 41
Total porosity, 71, 72
Transmissivity, 20, 83–84, 87, 97, 102, 136, 199, 203, 207
   aquifer, 83–84, 380
   block, 324
   calculation of, 252, 431
   changes in, 293
   coefficients of, 279
   confined aquifers and, 169, 210, 223
   conventional usage of, 431
   defined, 84, 96, 430
   determination of, 223
   error variance for, 234, 236
   estimation of, 172
   fracture, 324
   in leaky aquifers, 181, 184

low, 333
of matrix, 330
nonlinear, 431
turbulence, 412
turbulent flow, 412, 430–431
unconfined aquifers and, 187
variation in, 289
Tunnel deposition, 14
Turbulence, 223
calculation of, 427
defined, 398–399
dimensionless, 402, 408, 411
energy loss and, 359
intensity of, 409
Turbulence transmissivity coefficient, 412
Turbulent flow, 65, 101, 102, 111, 395–405
exponential law and, 407–412
laminar flow vs., 102–103, 107
laws of, 100–105
rough-, 108
smooth-, 108
steady-state, 397–401
transmissivity of, 412, 430–431
two-regime, 412–421
unsteady-state, 401–405
volumetric approach to, 405–407
Turbulent zone, 95
Two-aquifer observation wells, 130
Two-dimensional flow, 57–58, 61, 107, 339
Two-regime flow, 412–421
defined, 414
Type curve-matching procedure, 212–213, 230–232, 258, 426, see also specific types
exponential law and, 411
field data and, 290
for horizontal fractures, 339, 345–348
linear, 404
nonlinear, 404
observation wells and, 419

Unconfined aquifers, 31–32, 34, 51, 77
Boulton model of, 242–247
fully penetrating wells in, 174–178
horizontality of, 164
models of, 240–252
Boulton, 242–247
Neuman, 246–252
restrictive, 241
Neuman model of, 246–252
observation well distance and, 137
partially penetrating wells in, 186–190
quasi-steady-state flow in, 176
restrictive models of, 241
saturated portion of, 76
slug tests and, 386–388
steady-state flow in, 174–178, 186–190
water table in, 156
Unequilibrium equations, 87
Unified field plotting, 207
Uniformness of aquifers, 87–88
Unit-unconfined aquifers, 34

Unsaturated (aeration) zone, 28
Unsteady-state flow, 63, 64, 94, 95, 110, 203
assumptions about, 161
in confined aquifers, 209
turbulent, 401–405
toward wells, 208
well tests and, 355–356

Vadose zone, 28
Velocity
Darcy, 49
filter, 49
groundwater, 47–50, 56, 65, 91, 95
hydraulic conductivity as, 80
hydraulic gradients and, 107
longitudinal, 58
real, 50
Vertical flow, 135, 136
Vertical fractures, 324–331
hydraulic properties of, 339
Vesicular porosity, 20
Viscosity, 19, 82, 107, 342, 413
V-notch weirs, 150–151
Void spaces, 7, 69, 71, 75, 91
Volcanic activity, 21
Volcanic rocks, 71
Volumetric models, 269–270

Walton method, 274, 278
Warren-Root model, 311–315, 317
Water, see also Groundwater
connate, 27
demand for, 118
energy of, 38–39
flow of in fractures, 82
fossil, 27
kinematic viscosity of, 19
meteoric, 27
origin of, 27
subsurface, 26–27, see also Groundwater
viscosity of, 19
zones of, 28, see also specific zones
Water-bearing capacity, 21, 71, 72
Water-bearing formations, 29–30
Water engineering, 5, 6
Water meters, 144–145
Water-resisting rocks, 80
Water storage capability of reservoirs, 79
Water-storage capacity, 71–72, 303
Water table, 54–56, 156
Water-transmitting properties, 21
Water wells, see Wells
Weathering, 20, 21, 51, 71, 82
Weirs, 148–152
Well-aquifer continuity, 167–168
Well loss, 355, 360, 361, 376–377
Wells, 115–132, see also specific types
abstraction, 133
assumptions about, 165–166
block connection to, 303
cavity, 128

circular cross sections of, 116–120
collector, 122–124
configuration of, 309
cross sections of, 115–121
   circular, 116–120
   irregular, 125–126
   rectangular, 120–122
diameter of, 213
discharge in, 144–152
distances between, 203
efficiency of, 357
excavation of, 115
extended, 121
fracture connection to, 303
fracture intersection with, 325
fractures and, 118, 303, 325
fully penetrating, see Fully penetrating wells
irregular cross sections of, 125–126
in karstic formations, 127
large-diameter, see Large-diameter wells
location of, 133–134, 171, 176, 400
main, 133–134, 137, 203
multiple, 196–198
nonpenetrating, 128–129, 135
observation, see Observation wells
partially penetrating, see Partially penetrating wells
penetration of, 127–129
piezometers and, 134–137
pumping, 129, 200, 240, 330
purpose of, 129–132
rectangular cross sections of, 120–122
single-aquifer observation, 130
small-diameter, 118–120
steady-state flow toward or away from, 161
tests of, see Well tests
two-aquifer observation, 130
Well tests. 355–392
   Bailer, 382–384
   Bierschenk method in, 361–364
   defined, 355
   Eden-Hazel method in, 364–369
   Jacob method in, 360–361
   Miller-Weber model in, 373–376
   nonpumping, 380–384
   Packer, 389–392
   recovery, 378–380
   Rorabaugh method in, 369–371
   Sheahan type-curve method in, 371–373
   slug, 384–388
   step drawdown test analysis and, 360–376
      Bierschenk method in, 361–364
      Eden-Hazel method in, 364–369
      Jacob method in, 360–361
      Miller-Weber model in, 373–376
      Rorabaugh method in, 369–371
      Sheahan type-curve method in, 371–373
Wetted tape, 140–141
Wind, 11

Yields, 72–73, 75, 79, 203, 413

Zero energy, 40
Zones of water, 28, see also specific zones

GB 1003.2 .S45 1995
Sen, Zek^ai.
Applied hydrogeology for scientists and engineers

DATE DUE